T0325789

Single-Molecule Cellular Biophysics

Recent advances in single-molecule science have led to a new branch of science, single-molecule cellular biophysics, combining classical cell biology with cutting-edge single-molecule biophysics. This textbook explains the essential elements of this new discipline, from the state-of-the-art single-molecule techniques to real-world applications in unravelling the inner workings of the cell.

Every effort has been made to ensure the text can be easily understood by students from both the physical and life sciences. Mathematical derivations are kept to a minimum, whilst unnecessary biological terminology is avoided, and text boxes provide readers from either background with additional information. 100 end-of-chapter exercises are divided into those aimed at physical science students, those aimed at life science students, and those that can be tackled by students from both disciplines. The use of case studies and real research examples makes this textbook indispensable for undergraduate and graduate students entering this exciting field.

MARK C. LEAKE is a biophysics group-leader at the University of Oxford, heading an interdisciplinary team in live cell single-molecule research using cutting-edge biophotonics and state-of-the-art genetics. His work is highly cited and he has won many fellowships and prizes.

Single-Molecule Cellular Biophysics

MARK C. LEAKE
Clarendon Laboratory, University of Oxford

Shaftesbury Road, Cambridge CB2 8EA, United Kingdom

One Liberty Plaza, 20th Floor, New York, NY 10006, USA

477 Williamstown Road, Port Melbourne, VIC 3207, Australia

314–321, 3rd Floor, Plot 3, Splendor Forum, Jasola District Centre, New Delhi – 110025, India

103 Penang Road, #05–06/07, Visioncrest Commercial, Singapore 238467

Cambridge University Press is part of Cambridge University Press & Assessment, a department of the University of Cambridge.

We share the University's mission to contribute to society through the pursuit of education, learning and research at the highest international levels of excellence.

www.cambridge.org
Information on this title: www.cambridge.org/9781107005839

First published 2013

A catalogue record for this publication is available from the British Library

Library of Congress Cataloging-in-Publication data
Leake, Mark C.
Single-molecule cellular biophysics / Mark C. Leake.
p. cm.
ISBN 978-1-107-00583-9 (Hardback)
1. Biophysics–Textbooks. 2. Cytology–Textbooks. I. Title.
QH505.L36 2012
571.4–dc23

2012017427

ISBN 978-1-107-00583-9 Hardback

To Pumpkin, Big Boy, Ganny, Mr Blue Sky and the Vampire Squid (wherever he may be)

Contents

Preface *page* xi

1 Once upon a (length and) time (scale)... 1

1.1 Introduction 1
1.2 There are already many informative 'multi-molecule' methods 2
1.3 American versus European coffee 15
1.4 Scales of length, force, energy, time and concentration 17
1.5 Some basic thermodynamics of life 22
1.6 The concept of 'functionality' 23
1.7 Test tube or cell? 24

The gist 25
References 26
Questions 27

2 The molecules of life – an idiot's guide 29

2.1 Introduction 29
2.2 The atomic components of single biological molecules 30
2.3 Cell structure and sub-cellular architecture 31
2.4 Amino acids, peptides and proteins 34
2.5 Sugars 38
2.6 Nucleic acids 40
2.7 Lipids 44
2.8 Miscellaneous 'small' molecules 49
2.9 The 'central dogma' of molecular biology 50
2.10 Molecular simulations 52
2.11 Importance of non-covalent forces 54

The gist 55
References 56
Questions 57

3 **Making the invisible visible:** part 1 – methods that use visible light 60
3.1 Introduction 60
3.2 Magnifying images 62
3.3 Generating optical contrast using scattered light or fluorescence 64
3.4 Organic dyes, FlAsH/ReAsH, fluorescent amino acids and quantum dots 66
3.5 Fluorescent proteins, SNAP/CLIP-Tags and HaloTags 71
3.6 Illuminating and detecting fluorescent tags 74
3.7 Fluorescence correlation spectroscopy (FCS) 79
3.8 Fluorescence lifetime imaging (FLIM) 81
3.9 'Super-resolution' techniques 81
3.10 'Multi-dimensional' imaging 95
The gist 96
References 96
Questions 99

4 **Making the invisible visible:** part 2 – without visible light 102
4.1 Introduction 102
4.2 Scanning probe microscopy 102
4.3 Electron microscopy 106
4.4 Ionic currents through nanopores 109
4.5 Raman spectroscopy 114
4.6 Interference-based detection 115
The gist 116
References 117
Questions 118

5 **Measuring forces and manipulating single molecules** 121
5.1 Introduction 121
5.2 Optical tweezers 121
5.3 Magnetic tweezers 134
5.4 Atomic force spectroscopy 136
5.5 Using force spectroscopy to explore non-equilibrium processes 139
5.6 Electric dipole induction in polarizable particles for torque and trapping 140
The gist 143
References 143
Questions 145

6 **Single-molecule biophysics: the case studies that piece together the hidden machinery of the cell** 149
6.1 Introduction 149
6.2 What makes a 'seminal' single-molecule biophysics study? 149
6.3 Carving up the cell into a few sensible themes 154
The gist 155
References 156
Questions 156

7 **Molecules from beyond the cell** 159
7.1 Introduction 159
7.2 Receptor molecules and ligands in the cell membrane 159
7.3 Endocytosis and exocytosis 168
7.4 Viral invasion 173
The gist 178
References 178
Questions 180

8 **Into the membrane** 183
8.1 Introduction 183
8.2 Molecular transport via pores, pumps and carriers in membranes 183
8.3 Rotary motors – the rise of the (bionano) machines 192
8.4 Energizing the cell 202
8.5 Architecture and shape of the cell surface 207
The gist 212
References 212
Questions 215

9 **Inside cells** 220
9.1 Introduction 220
9.2 Free, hindered and driven molecular diffusion in the cytoplasm 220
9.3 Chromosomes and DNA: their architecture and replication 232
9.4 Translating, transcribing and splicing the genetic code 238
The gist 247
References 247
Questions 249

10 Single-molecule biophysics beyond single cells and beyond the single molecule 253

10.1 Introduction 253

10.2 Single-molecule biophysics in complex organisms 253

10.3 Bionanotechnology and 'synthetic' biology 258

10.4 The outlook for single-molecule cellular biophysics 261

The gist 262

References 263

Questions 263

Index 265

Preface

Life, from the bottom up

A biological system can be exceedingly small. Many of the cells are very tiny, but they are very active; they manufacture various substances; they walk around; they wiggle; and they do all kinds of marvelous things – all on a very small scale. (FEYNMAN, 1959)

Richard Feynman, celebrated physicist, science communicator and bongo-drum enthusiast, gave a lecture in Caltech, USA, a few days after Christmas 1959, that would come to be seen by future nanotechnologists as essentially prophetic. His talk was entitled 'There's plenty of room at the bottom', and was concerned primarily with discussing the feasibility of a future ability to store information and to control and manipulate machines on a length scale which was tens of thousands of times smaller than that of the macroscopic world of things like typical books and electric motors of that day. It was essentially a clarion call to scientists and engineers to develop a new field, which would later be termed *nanotechnology* (see Taniguchi, 1974). But in one aside, Feynman alluded to the very small scale of biological systems, and how cells used these to do 'all kinds of marvelous things', which in its own small way has been wisely prescient for the subsequent seismic shifts in our understanding of how biological systems really work. We now know that the fundamental minimal functional unit which can adequately describe the properties of these systems is the single biological molecule. That is not to say that the constituent atoms at smaller length scales do not matter, nor the sub-atomic particles that make up the individual atoms, nor smaller still the quarks that make up the sub-atomic particles. Rather that, in general, we do not need to refer to a length scale smaller than the single molecule to understand biological processes.

Developments in experimental physics have primarily been the driving force in establishing the field of single-molecule biology, and even though the science of single-molecule biophysics is barely more than a generation old it is now at the traditionally incongruous interfaces between the physical and the life sciences, and at the scale of the single molecule, that many of the most fundamental questions concerned with living systems are being addressed. There is essentially a new emerging scientific discipline of *single-molecule cellular biophysics*. Although there are many existing reviews and research compilations for the expert reader concerned with the biophysics of single-molecule biology, which appear to multiply yearly with an efficiency of which the most potent virus might be proud, there is a paucity of written material to guide the informed novice into this exciting area of science.

For example, undergraduate students in both the physical and life sciences taking courses relating to this area, as well as new graduate students and post-docs moving or merely considering a move from another field, lack suitable introductory material. It is with this in mind that motivation for this handbook primarily evolved. Some of the material is based upon a lecture course in single-molecule methods which began in 2010, which myself and colleagues gave in the Clarendon Laboratory in the Physics Department of Oxford University. It has also been stimulated by several discussions with both my group leader biophysicist peers in the Oxford Physics and Biochemistry Departments, and a number of apparently willing students and post-docs who were eager to offer ideas and feedback as to what should be included and how the material might best be structured for clarity of reading.

The main difficulty with such a text is keeping both physicists and biologists suitably happy, but not at the expense of dumbing-down either of these areas of science. Maths is a particular sticking point – some of my physicist colleagues argued that biologists simply 'don't have the math' to understand many of the fundamental concepts of physics underlying biophysics. However, I took the view that a good experimentalist communicator does not need to hide behind reams of complex equations in order to get the basic physics of an experimental technique across. Similarly, some of my biologist colleagues believed that the depth of language and terminology of the life sciences was simply too substantial to be penetrated by the physicist.

Again, I took a different view in that, although clearly some essentials of the language of biology are needed to make any sensible discussion about molecules in the context of cells and living organisms, this should not prevent an enthusiastic reader native to the physical sciences from understanding the essential, core features of a given biological system. That being said, what I was at pains to prevent was a disruption of the flow of the text by explaining things from either physics or biology in greater detail for those wishing to become more expert, or similarly by explaining relatively basic concepts which some might find too elementary. My solution throughout this book has been to create aside sections which readers may decide to explore or avoid as they wish, without significantly impairing the prime material in the rest of the book.

This book is arranged broadly into three parts of introduction, techniques and applications. It is intended to be read primarily as an entry-level text by students and post-docs new to the field of single-molecule biophysics, as well as acting as a good source of reference for those more expert in the field, especially course lecturers teaching aspects of modern biophysics. It can also be used as an excellent companion text for students embarking on related courses in biophysics/biological physics, bio-engineering, bionanotechnology, cell and molecular biology, nanoscience/nanophysics and even experimental systems and synthetic biology.

The book has been structured to accompany a ca. 10 lecture course in the subject, and over one hundred questions are included, in a section at the end of each chapter, divided as a very crude guide into those that might appeal more to life scientists or physical scientists as well as a more general set of questions that could be attempted by members from either camp, but the enthusiasts are strongly encouraged to attempt them all. Model answers to some of the problems are available upon request in a separate downloadable file to course tutors/lecturers, and if there is a demand and significant bribery then more may feature in subsequent editions. These form a series of both quantitative and discussion questions which lecturers, tutors and course-coordinators are welcome to use and modify as they wish.

Similarly, throughout each chapter there are multiple 'aside' comments included so as not to disrupt the flow of the text, divided into categories of BIO-EXTRA and PHYSICS-EXTRA which give more advanced information for those seeking it, as well as KEY POINT sections which are the essential details which need to be understood. An extensive and internally complete reference list is included at the end of each chapter in case readers wish to explore further, which is heartily recommended of course. For those students in particular who wish to be reminded of the key messages of each chapter then a brief section called 'The gist' is included at the end of each chapter to give a bullet-point list of the vital concepts covered to aid revision, along with the prime message of that chapter of the 'take-home message'.

There was a need to include both an introduction to the conceptual foundations of single-molecule cellular biophysics and a concise description of the complementary *multi-molecule* standard biophysics methods currently in use in research labs around the world to address biological questions, which appear in Chapter 1. There is a *biology orientation* in Chapter 2, which explains some of the relatively basic types of molecules and biological concepts and terms required to understand the subsequent technique chapters properly; this may be omitted by those with formal life sciences training without serious impairment, though I have discarded a classical biology hierarchical emphasis in favour of a more holistic approach from systems biology so it may be worth a quick read all the same. Chapters 3–5 include an outline of the principal biological single-molecule experimental methods in use to date, whilst the remaining chapters primarily include discussions of real case studies in which biological single-molecule methods have been applied to address a variety of different biological questions. As opposed to being technique driven, the emphasis here is on the utility of the technique to help us understand some very fundamental features of living things. We end with some speculation in Chapter 10 as to where this new generation of physiologically relevant single-molecule biophysics may lead in the near future, and the impact it is likely to have on a multitude of different fields.

I convey my thanks to several of my students, staff and colleagues for their ideas for this book, most especially the members of my own lab group. In addition, there are several of my peers who offered useful feedback and/or original data, to whom I pay a special tribute, including: Clive Bagshaw, Christoph Baumann, Richard Berry, Kirstine Berg-Sørensen, Michael Börsch, Ashley Cadby, Sheng-Wen Chiu, Pietro Cicuta, Charlotte Fournier, Sarah Harris, Oliver Harriman, Adam Hendricks, Jamie Hobbs, Erika Holzbaur, Akihiro Ishijima, Isabel Llorente Garcia, Achillefs Kapanidis, David Klenerman, Stuart Lindsay, Cait MacPhee, Conrad Mullineaux, Teuta Pilizota, Lijun Shang, Paul O'Shea, Gerhard Schütz, John Sleep, Yoshiyuki Sowa, Stephen Tucker, Andrew Turberfield, Tom Waigh, Denis Wirtz, Gijs Wuite, Toshio Yanagida and Zhaokun Zhou. Finally, I extend my thanks for the patience and diligence of the production and editorial staff and associates of Cambridge University Press, both past and present, including Simon Capelin, Katrina Halliday, Christopher Miller, Lindsay Barnes, Antoaneta Ouzounova, Siriol Jones and John Fowler.

I hope you enjoy reading the book as much as I gained pleasure from writing it, and any reader feedback will be received with the utmost grace!

References

- Feynman, R. P. (1959). *'There's Plenty of Room at the Bottom' lecture, transcript deposited at Caltech Engineering and Science*, Volume 23:5, February 1960, pp. 22–36 (quote used with kind permission), available at www.its.caltech.edu/~feynman/plenty.html
- Taniguchi, N. (1974). On the basic concept of 'nano-technology', *Proc. Intl. Conf. Prod. Eng. Tokyo*, Part II, Japan Society of Precision Engineering.

1 Once upon a (length and) time (scale)...

Truly it has been said, that to a clear eye the smallest fact is a window through which the Infinite may be seen. (T. H. HUXLEY, *THE STUDY OF ZOOLOGY*, 1861)

GENERAL IDEA

Here we discuss the conceptual foundations of single-molecule biophysics in the context of cellular biology. We provide an overview of existing ensemble average techniques used to study biological processes and consider the importance of single-molecule biology experiments.

1.1 Introduction

Some of the most talented physicists in modern science history have been led ultimately to address challenging questions of biology. This is exemplified in Erwin Schrödinger's essay 'What is Life?' (see Schrödinger, 1944). It starts with the question 'How can the events in space and time which take place within the spatial boundary of a living organism be accounted for by physics and chemistry?' In other words, can we address the big questions of the life sciences from the standpoint of the physical sciences. The ~60 years following the publication of this seminal work has seen a vast increase in our understanding of biology at the molecular scale, and the physical sciences have played a key role in resolving many central problems. The biological problems have not been made easier by the absence of a compelling and consistent definition of 'life' – writing from the context for a meaning of 'artificial life', the American journalist Steven Levy noted 48 examples of definitions of life from eminent scientists, no two of which were the same (Levy, 1993). From a physical perspective, life is a means of trapping free energy (ultimately from the sun or, more rarely, geological thermal vents) into units of increased local order, which in effect locally decrease entropy, while the units maintain their status in situations which are generally far from thermal equilibrium.

This notion of a *life unit* is robust since it links a physical concept to one of the greatest discoveries in biology, made in the nineteenth century, that living organisms were composed of cells. Modern biochemistry as we know it today is largely about trying to understand the molecular mechanisms that make cells do what they do. The very first cellular structures were observed as far back as the seventeenth century when Robert Hooke published the first images from a functional microscope in *Micrographia.* He observed the porous microstructure of cork and first coined the term *cell* as the basic unit of life. Thus, even from the very conception of cell biology, our understanding of the nature of living organisms has progressed hand in hand with many groundbreaking technological developments of the day in physics – one only has to look at the pioneering studies of nerve and muscle tissue, as well as the earliest developments at the molecular level in our knowledge of the structures that drive biology, such as proteins and DNA.

Biological physics, or *biophysics*, is still a comparatively young but fearlessly ambitious discipline which has grown from its childhood, essentially involving biological investigations using physical instruments, to being now a handsome and youthful adult with much freedom of thought as to its purpose and future. It is a subject which can apply concepts from all the physical and mathematical sciences towards understanding how living matter really functions. Physicists often struggle with the language of biology, and biologists often toil with the mathematics that physicists use to describe the natural world. The meeting of these two worlds can often be uncomfortable at first, and sometimes stormy, but the key to understanding and enjoying biophysics is for physicists to make the effort to learn at least the basics of core biological concepts, while biologists try to come to grips with the intellectual concepts of physical models even if some of the mathematics may seem intractable. There are now several university level courses emerging across the globe which attempt to integrate life and physical sciences.

Modern experimental single-molecule biology research essentially combines many examples of cutting-edge biological approaches with state-of-the-art physics technology and physical insight, and is arguably the most active area of biophysics research at present. The purpose of this handbook is to guide both biologists and physicists, with relatively little or no experience of single-molecule approaches, through the real, practical methods in use today for investigating molecular level biological questions, and to do so in a manner accessible to people from both physical and life sciences backgrounds. In this first chapter you will find an overview of existing single-molecule approaches as well as relevant ensemble (i.e. many molecules) level techniques. There is also an 'orientation' chapter to follow which is especially important for readers from a background more rooted in the physical sciences, which describes some core biological concepts. The subsequent three chapters are devoted to a description of the biological single-molecule techniques in current use. In its most reduced form, a single-molecule method will tell us something about either *position* or *force*.

The ways in which positional information is obtained from single biological molecules can be sensibly divided into methods which primarily utilize *visible light*, and those which do not. A separate chapter is devoted to each, as is the discussion of techniques which allow us either to *measure* molecular forces or to use force to *manipulate* single molecules controllably. The rest of the book is arranged broadly by themes along the lines of different biological processes in the living cell, with descriptions of real experimental single-molecule approaches in use from a variety of both recent and seminal investigations.

KEY POINT

A single-molecule method will either (1) give us information on the position of a molecule in space at a given time, in other words it has the ability to detect the presence of a single molecule at a specific location, or (2) allow the measurement and/or control of forces exerted by/on the molecule – or sometimes both.

1.2 There are already many informative 'multi-molecule' methods

There exist many reliable, standard biophysical experimental approaches for investigating biological processes at the level of typically many thousands of molecules, so-called *bulk ensemble* techniques. Although a full discussion of these approaches is beyond the scope of this handbook (for a more comprehensive description see Van Holde *et al.*, 2005 and Nölting, 2009), it is first important to understand the basics of what these approaches are, and why they are useful in complementing single-molecule investigations.

1.2.1 Calorimetry

Changes in a variety of *thermodynamic potentials*, for example in the *Gibbs free energy G*, *enthalpy H* and *entropy S*, can be measured directly and/or deduced by carefully measuring temperature changes using a specifically calibrated sample chamber in which the sample is undergoing some chemical or physical transition of interest. One of the most useful forms of this technique is *isothermal titration calorimetry* (ITC) which is often used to study binding affinity and stoichiometry, as well as changes in thermodynamic potentials, of small ligand molecules binding to larger biopolymers such as proteins and DNA (see Chapter 2 for a full explanation of the various different types of biological molecules).

Being able to measure thermodynamic changes is enormously important; they allow us to predict the likelihood of processes occurring under given environmental conditions, and also may indirectly give us insight into the underlying mechanisms of these processes.

PHYSICS-EXTRA

Classical thermodynamics uses ensemble parameters, assuming a system with many, many particles (for example, a single 'microlitre' of water contains $\sim 10^{19}$ molecules). To apply the same concepts to a single molecule requires the **egodic hypothesis** – that all accessible microstates are equally probable over a long time, i.e. that the average properties determined over many molecules ('classical thermodynamics') are equal to the properties of any given single molecule averaged over a long time ('single-molecule thermodynamics'). In other words, the ensemble state equals the time-averaged state, i.e. if one allows the system to evolve in time indefinitely, the system will eventually pass through all possible states.

1.2.2 Chromatography and dialysis

A general chromatography device is used to separate different molecular components present in an heterogeneous sample. There is a wide range of different techniques such as *gas chromatography*, *high-performance liquid chromatography* (HPLC) and standard solid–liquid phase methods such as *gel filtration*, *thin-layer chromatography* (TLC) and even simple paper chromatography. Each molecular component in the sample will bind with some characteristic affinity to an immobile substrate in the given chromatography device, to form a *stationary phase*. The binding lasts for a characteristic *dwell time*, dependent upon factors such as the physical and chemical nature of the immobile substrate and the surrounding molecules.

Once unbound, a particular molecular component will enter the *mobile phase*, and thus can move through the chromatography device either by diffusion or by being driven in a given direction via, typically, a pressure gradient. This process of binding and unbinding can occur through the physical extent of the chromatography device, until the given molecular component emerges at the exit of the device, where it can be detected, typically using some form of optical absorption technique, and collected. The end result is the separation of different molecular components on the basis of their relative binding strengths to the immobile substrate and of their mean speeds of translocation through the chromatography device.

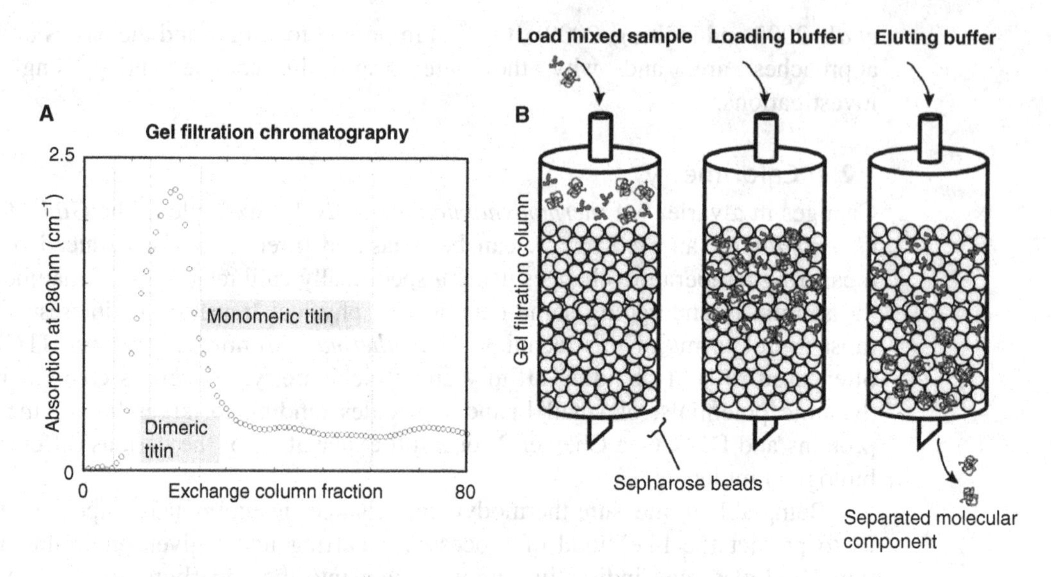

FIGURE 1.1 **(A)** Column-based solid–liquid phase gel filtration chromatography, shown here for a heterogeneous 'oligomer' mix (i.e. different numbers of monomer subunit molecules stuck together) of the muscle protein titin in which the heavier fractions (i.e. dimers and heavier multimers) run off in the first column 'fractions' (a small ~1 ml extracted volume which is then tested for its absorbance at wavelength 280 nm). **(B)** A mixed sample can be separated into different molecular components by eluting the column using different chemical solvent buffers.

The immobile substrate itself often takes the physical form of small tightly packed beads, having a typical diameter in the range 40–400 μm, which is designed to produce optical separation of the given molecular components in the sample. A common substance used for these beads involves a large sugar molecule extracted from seaweed called *sepharose*. These beads are typically packed into a glass column (for *gel filtration* chromatography), or there may be chemical binding, often using specific *antibodies*, to the bead surface of the sample (so-called *affinity* chromatography). In the mobile phase, typically, the sample is solvated and swept through the device in controlled, directed flow, which might at its simplest be gravity driven, or it may involve a pump-controlled pressure gradient. The likelihood for a given molecular component in the sample being in the stationary phase is dictated by issues such as molecular mass, ionic charge, electrical surface features and the presence of exposed chemical groups – for example, *ion-exchange* chromatography is a form of liquid chromatography in which the stationary phase consists of charged beads (Figure 1.1A).

These factors determine the typical dwell time of the substance bound in the immobile stationary phase before unbinding into the mobile phase and being swept along in the flow. Thus the mean *drift speed* through the device of a substance is determined by these factors. Hence this can be used to separate out different substances, either for purification or to measure these factors directly using suitable calibration. Often, since the binding strength to the immobile substrate is dependent on the conditions of the sample solvent, such as ionic strength and pH, different molecular components can be *eluted* selectively through a chromatography column by running through different solvents which are known to reduce selectively the binding between the immobile substrate and a particular molecular component (Figure 1.1B).

Dialysis, or *ultrafiltration*, has similar principles in that the sample mobility is characterized by the same factors as for chromatography, except here the solvated sample sits on one side of a dialysis membrane with a pre-defined pore size. This sets an upper limit on what molecules can pass through, in terms of molecular weight. If the other side of the membrane contains only solvent then, given sufficient time, high mass molecules can be separated from small mass molecules. This process can be speeded up by applying a forced pressure gradient, and by repeated use of different pore sizes a sample can be purified from a mix of different molecules.

1.2.3 Circular dichroism and optical rotation

Optically active samples will rotate the polarization plane of linearly polarized light, and the degree to which this occurs can provide mean average structural information about the sample. A wave of light is really a propagating oscillation of electric and magnetic intensity. These force-fields have directionality, or *polarization*. If this directionality stays constant with respect to the wave itself, polarization is *linear*. If this directionality rotates as the wave propagates, the polarization is *elliptical*; a special case of this is *circular polarization* when the relative components at right angles to each other are equal.

If circularly polarized light is shone through certain optically transparent samples, as is the case for circular dichroism, the *ellipticity* of the polarization may change, typically due to the presence of repeating molecular structures of a given shape. In this way, changes in ellipticity may be indicative of certain structural motifs in the sample. The extent to which fine structural detail can be explored is generally limited, but the technique can often give an indication of the relative proportions of different *secondary structure*.

As we will see later in Chapter 2, the primary structure of a biological molecule describes how the subunit chemicals forming the molecule are linked directly via strong *covalent* chemical bonds. Secondary structure describes how the primary structure forms a three-dimesnsional shape using weaker non-covalent bonds (typically *hydrogen bonds*). *Beta sheets* and *alpha helices* are the main secondary structures of a protein (see Chapter 2), and circular dichroism can discriminate reasonably well between regions of β sheet, α helix and *random coil* secondary structures in proteins, and also of different helical forms of DNA.

1.2.4 Electron microscopy

Structural features in fixed biological samples may be visualized by introducing some form of heavy-metal contrast agent into the sample, shining a beam of electrons through it (Figure 1.2A) and imaging either the scattered beam (as in *scanning electron microscopy*, or SEM) or the transmitted beam (*transmission electron microscopy*, or TEM). The principle here is that different structures within the sample take up the contrast agent to different extents and so will transmit or scatter different proportions of electrons, thereby allowing different structural features to be seen. *Negative staining* may be used in which the contrast agent (typically a heavy-metal chemical salt, for example uranyl acetate) stains everything apart from the structures of interest (Figure 1.2B), or *positive staining* (typically evaporating a heavy metal such as platinum onto the sample to create 'shadow' features from the local topography) may be employed. Fixation and staining often involve multiple steps and an issue is the prevalence of not uncommon staining/fixing artifacts, in addition to radiation damage to the sample from the electron beam.

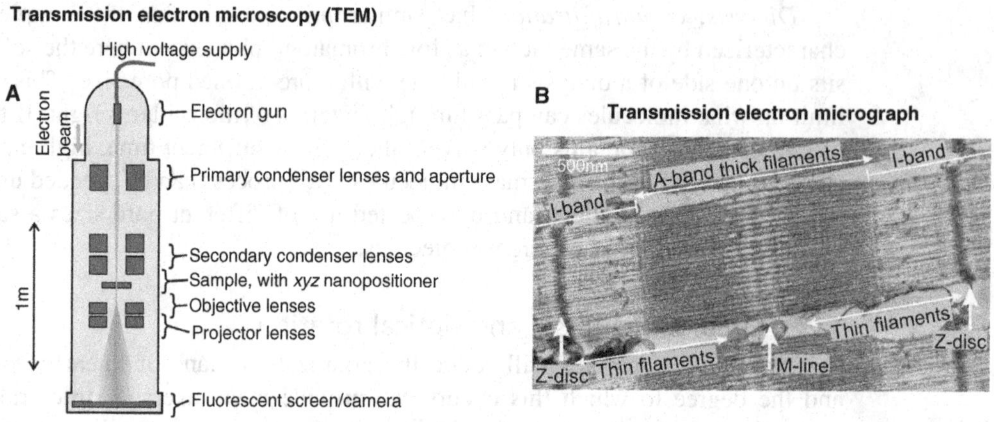

Transmission electron microscopy (TEM)

FIGURE 1.2 (A) Schematic cross-section through a transmission electron microscope. (B) Negatively stained transmission electron micrograph of a thin section of a 'myofibril' found in muscle tissue (see Chapter 4).

Similarly, a significant disadvantage is that the sample is dead, so this approach cannot be used to monitor physiological processes directly (though by careful *snap-freezing* at different time points these can sometimes be inferred indirectly). The technique began as a bulk ensemble method (for example, discriminating different relatively macroscopic length scale tissue features, such as occur in muscle) but with finesse can now provide information at the level of single molecules.

1.2.5 Electrophysiology

Many processes in biology involve the transmission of an electrical current, for example, conduction through nerves, and the flow of ions through channels into and out of cells. These can be monitored using sensitive electrical recording devices either by measuring the changes in voltage inside the cell in question relative to the outside as a function of time, or by measuring the flow of ion current directly. For example, *patch clamping* can be used to extract a patch of several ion channels on the surface of a cell onto the end of a micropipette, and the ion flux through these channels can be monitored directly. Traditionally this has involved monitoring the flow through multiple ion channels simultaneously, but recent improvements now allow the ultrasensitive measurement of single ion-channel currents, and so at the cutting-edge this can be used as a single-molecule technique (see Chapter 4).

1.2.6 Fluorimetry

Here a cuvette of a sample is excited into fluorescence using typically a broadband light source such as a mercury or xenon arc lamp, with individual wavelengths of the fluorescence often measured at 90° to the light source to minimize unwanted detection of non-fluorescence light. The emission spectrum as a function of excitation wavelength is an indication of electronic energy level transitions within the sample, and thus is a characteristic of the underlying molecular structure as well as the physical and chemical environment. This has particular use for natural or engineered fluorescent dye molecules. *Tryptophan fluorescence* involves measuring the native fluorescence of the aromatic amino acid tryptophan.

Tryptophan is a very *hydrophobic* molecule – this means that it has no significant electrical polarity and so will not attract polar solvent water molecules. As a result, tryptophan is generally found buried at the centre of folded proteins so it will not be exposed to surrounding water. Its fluorescence changes upon exposure to water, and so this may be used to indicate whether a protein is in a folded or unfolded state.

A further extension to fluorescence involves *Förster energy resonance transfer* (FRET). This is a non-radiative energy transfer between a donor and an acceptor molecule, which often but not always are fluorescent molecules, whose electronic energy levels overlap significantly. This process only occurs efficiently over length scales less than typically a few nanometres, which is of the order of the length scale of small molecular machines and protein complexes, and so FRET measurements may be used to indicate co-localization/molecular interaction between regions of one or more such molecules tagged with a suitable donor and acceptor molecule.

Also, in *fluorescence anisotropy*, the polarization of the emission signal from a fluorescently labelled sample changes with time if the molecules are free to rotate, and so this technique has been used to estimate the rough shape of large molecules as well as to determine binding constants in the case of interacting molecules.

1.2.7 Gel eletrophoresis

Here, a sample, often containing a mixture of different biological molecules, is injected into a semi-porous gel and exposed to an electric field gradient. The gel is typically either *polyacrylamide* (for protein samples) or *agarose* (for samples containing nucleic acids such as DNA or RNA). The sample can be denatured first by heating and combined with a charged surfactant such as SDS, or it can be run in a non-denatured native state. The mobility of the sample is a function of its molecular weight as well as its shape and (if run in the native state) the presence of charged residues on the surface. Thus, a mixture of different biological molecules can be separated by running the gel for a sufficient time. The positions of the different molecules appear as bands at different distances from the start position (Figure 1.3A), which can be visualized using appropriate staining, either directly in visible light (for example, *Coomassie Blue* is a standard stain of choice for proteins, though *silver staining* may also be applied if greater sensitivity is required) or using fluorescence via excitation of a stain from ultraviolet light (for example, *ethydium bromide stain*, used for nucleic acids). Each band may also be carefully extracted to reconstitute the original, now purified, sample. Thus this technique may be used both for purification and for characterization (for example, to estimate the molecular weight of a sample by interpolation of the band positions from a reference calibration sample).

1.2.8 Mass spectrometry

For the analysis of bulk ensemble average biophysical properties, mass spectrometry is one of the most quantitatively analytical techniques. Here, a small quantity of sample (typically $\sim 10^{-15}$ kg) is injected into an ionizer, and ionized (note though that although this quantity appears meagre this could still equate to typically millions of molecules, so this is far from a single-molecule method). This ionization process generates fragments of whole molecules with different masses and charges. The simplest machine is the *sector* mass spectrometer which accelerates these ion fragments in a vacuum using an electric field sector, and deflects them using a magnetic

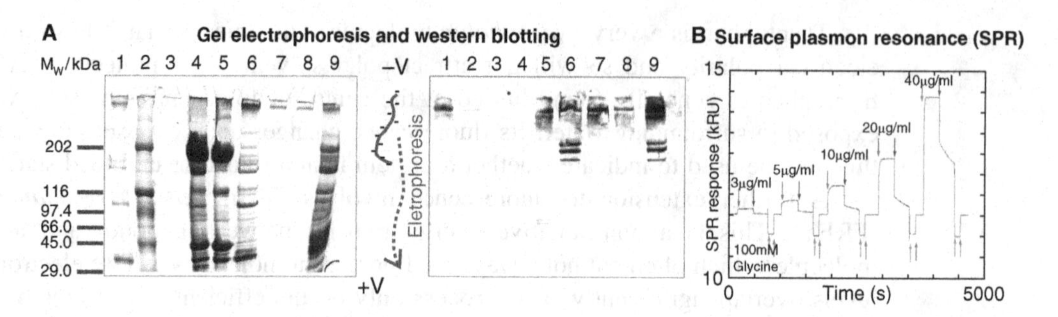

FIGURE 1.3 **(A)** Polyacrylamide SDS gel electrophoresis of a mix of different muscle proteins stained with Coomassie Blue (left panel) with molecular weight markers indicated (lane 1) and lanes 2–9 corresponding to sample extract taken at different stages of a titin preparation from raw muscle tissue; the right panel shows the corresponding western blot probed with an anti-titin antibody. The protein molecules run in a denatured state from cathode to anode, covered with negative surface SDS charges (inset), so lighter proteins will be found further down the gel in general, but there is also a mobility effect related to molecular shape. **(B)** Surface plasmon resonance showing the effect of adding increasing quantities of an anti-titin antibody ligand which binds to titin molecules that have been immobilized on the surface of the sample flow-cell; the binding can be disrupted via addition of the amino acid glycine.

field sector at right angles to this so that the ions follow a roughly circular path. The radius of this circle is a function of the *mass-to-charge ratio* of the particles in the beam. Different ionized molecular fragments can be collected and analysed depending upon the location of the detector in relation to the circular ion path, generating a *mass spectrum* which yields detailed information concerning the relative proportions of different ionic species in the sample.

Other similar types of machine include the *Fourier transform* mass spectrometer (ions are injected into a cyclotron cell and resonated into orbit using an oscillating electric field; this generates a radio-frequency signal from the ions which is detected and subsequently Fourier transformed to yield the mass spectrum), *ion trap* mass spectrometer (injected ions are trapped in a cavity using electric fields and are ejected and subsequently analysed on the basis of their mass-to-charge ratio), *time-of-flight* mass spectrometer (an ion vapour pulse is created, typically using a high-energy laser, and the ions are accelerated using an electric field with the time taken to travel a given distance for this cloud of ions measured; this can then be related back to the mass-to-charge ratio) and *quadrupole* mass spectrometer (the accelerated ion beam is passed between four metal rods to which DC and AC potentials are applied; this causes resonance to the ion beam such that only ions with a narrow range of mass-to-charge ratio will pass through the rod cavity into the detector unit, so by methodical variation of the electric potentials applied to the rods a mass spectrum can be generated).

The biophysical applications of these instruments are significant and include highly sensitive detection/biosensing of biological particles (for example, although the sensitivity is not strictly speaking at the single-molecule level, the minimum concentration level is only around one particle per litre, which is very sensitive relative to other bulk ensemble average techniques, which can be detected in around a few minutes), quality control of sample purity, detecting relatively subtle mutations in nucleic acids such as RNA and DNA, protein conformation and folding studies and proteomics experiments investigating protein–protein interactions. In fact, when NASA's robotic space rover *Curiosity* touched down in Gale Crater on the surface of Mars on 6 August, 2012, one of its key onboard devices was a portable mass spectrometer for the detection of biological-related material.

1.2.9 NMR and ESR spectroscopy

Nuclear magnetic resonance (NMR) is a powerful technique utilizing the principle that magnetic atomic nuclei will undergo resonance by absorbing and emitting electromagnetic radiation in the presence of a strong external magnetic field. A magnetic nucleus implies a non-zero spin angular momentum, which is the case for all stable *isotopes* which have an odd number of protons and/or neutrons. The most common isotopes used for biological samples are ^{1}H (the most sensitive stable isotope) and ^{13}C (relatively low natural abundance compared to the non-magnetic ^{12}C and also a low sensitivity but used since carbon is the key component of all organic compounds). Other lesser used isotopes include ^{15}N (low sensitivity but used since nitrogen is a key component in proteins and nucleic acids), ^{19}F (rarely present in most natural organic compounds so in general needs to be chemically bound into the sample first, but has a high sensitivity) and ^{31}P (moderate sensitivity and phosphorus is a key element of many biological chemicals).

In the presence of an external magnetic field, the different spin states of the nuclei have a different energy; by absorbing or emitting electromagnetic radiation of the right frequency the nucleus can resonate between the different spin energy states. However, not all atomic nuclei in a sample will have exactly the same differences in spin energy states because there is a small shielding effect from the surrounding electrons which causes subtle differences in the absolute level of the external magnetic field sensed in the nucleus. These differences are related to the physical distribution of the local electron cloud, which in turn is a manifestation of the local chemical environment. In other words, this shift in the resonant frequency, sometimes referred to as the *chemical shift*, can be used to deduce the chemical structure of the sample.

Typically, the resonance will be in the radio-frequency range (of the order of 100 MHz) for the high fields of ~10 Tesla used (almost a million times the Earth's magnetic field) with the detected chemical shift being typically in range of ~10 parts per million (ppm) of the non-shifted value. This technique has been used to great effect in obtaining atomic level structures of several important biological molecules, especially of membrane proteins. In general these proteins are very difficult to crystallize which is a requirement of the competing atomic level structural determination technique of *x-ray crystallography*. The concentration of the sample for NMR needs to be relatively high to obtain good signals for structural studies, and necessarily therefore this is an in vitro technique. However, at a lower spatial resolution it can be used in vivo in, for example, *magnetic resonance imaging* (MRI), which is widely used in medical physics but also utilized in biophysical applications to follow in vivo biochemistry. The spatial resolution is typically of the order of millimetres at best.

Electron spin resonance (ESR), also known as *electron paramagnetic resonance* (EPR), relies on similar principles to NMR but here the resonance is from the absorption and emission of electromagnetic radiation due to transitions in the spin states of the electrons. This only occurs for unpaired electrons, since paired electrons have a net spin (and hence a magnetic strength) of zero. Unpaired electrons are unstable and associated with highly reactive species, such as free-radicals. These reactive species are short lived, which limits the application of ESR, though this can be used to advantage in that standard solvents do not produce a measurable ESR signal and so the relative strength of the signal from the actual sample above this background *solvent noise* can be very high.

1.2.10 Optical interferometry

There are two principal multi-molecule optical interferometry techniques: *dual polarization interferometry* (DPI) and *surface plasmon resonance* (SPR). In DPI, a reference laser beam is guided through an optically transparent sample support, while a sensing laser beam is directed similarly through the support but at a very oblique angle to the surface. At this steep angle the laser beam is totally internally reflected away from the surface, but with the by-product of generating an *evanescent field* into the sample, generally solvated by water for the case of biophysical investigations, with a characteristic depth of penetration of ~100 nm. Small quantities of material from the sample which bind to the surface have subtle but measureable effects upon the polarization in this evanescent field. These can be detected very sensitively by measuring the interference pattern produced between the sensing and reference beams. DPI gives information on the thickness of the surfaced-adsorbed material, and its refractive index.

SPR works in a similar manner in that an evanescent field is generated, but here a thin layer of metal, ~10 nm thick, is first deposited onto the outside surface. At a certain angle of incidence to the surface the sensing beam reflects slightly less back into the sample. This is due to a resonance effect via the generation of resonant oscillations in the electrons at the metal surface interface, so-called *surface plasmons*. This angle is a function of the absolute amount of the adsorbed material on the metal surface from the sample, and so DPI and SPR are essentially complementary techniques. Both ultimately can yield information on the stoichiometry and kinetics of binding of biological materials. This can be used to understand how cell receptors work, for example. If the surface is first coated with purified receptor proteins and the sample chamber contains a ligand thought to bind to the receptor, then both DPI and SPR can be used to measure the strength of this binding, and subsequently unbinding, and to estimate the relative numbers of ligands that bind for every receptor protein (Figure 1.3B).

1.2.11 Optical microscopy

Here, visible light is either transmitted from, or scattered by, a sample. Different features within a biological sample, for example in a cell, will absorb different relative proportions of light and so the intensity of the image is some indication of different underlying structural features (so-called *brightfield* microscopy). However, the inside of most biological samples is generally mostly composed of water, which is the same as the surrounding solution, and so the imaging contrast on the basis of transmitted intensity alone is often poor.

To improve the contrast several methods have been employed. Optical interference enhancement methods include *phase contrast* (increases the sensitivity to changes in refractive index in the sample) and *differential interference contrast* or DIC (the image is enhanced for areas of high spatial gradients in the refractive index, for example at the perimeters of cells or organelles inside the cells). *Darkfield* microscopy provides enhancement to image contrast by imaging the scattered features from the sample (so that the non-sample regions appear dark). This is useful for investigating surface features of cells and tissues, and also finds application in single-molecule methods.

Staining methods are also employed to success. These include the relatively crude methods involving coloured organic stains which bind differentially to different biological material (so they can be used for example to differentiate cell and/or tissue types, but with a disadvantage that the sample is killed in the process, and non-specific binding can reduce the ultimate contrast), and also include those methods which use fluorescence

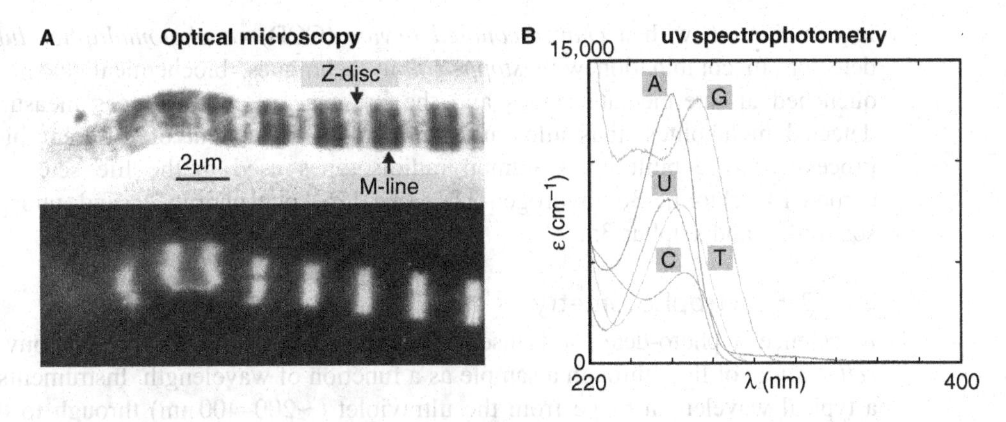

FIGURE 1.4 **(A)** Brightfield (upper panel) and fluorescence microscopy (lower panel) images of a single muscle myofibril which has been incubated with a fluorescent antibody which binds to titin. **(B)** Ultraviolet absorption spectrum for the different nucleotide bases used in nucleic acids such as RNA and DNA (see Chapter 2). (Original data, author's own work, for full details see Leake, 2001.)

to provide far better contrast. In fluorescence microscopy, as with fluorimetry mentioned above, fluorescent components are excited in the sample either using a broadband source which has been blocked at all wavelengths apart from a narrow band over which a fluorescent tag is known to be efficiently excited, or more commonly now using a narrow-band laser as an excitation source. Back reflections from the laser are blocked so that only fluorescence emissions from the sample are imaged.

An organic fluorescent dye can be introduced using crude chemical incubation of the sample (which generally has a detrimental physiological effect because of the high concentrations required combined with non-specific binding) or better still can be used in conjugation with a specific antibody which has been labelled with fluorescent dye molecules (Figure 1.4A) or even more exotic tags such as *quantum dots* (see Chapter 3). The challenge here is how to introduce the tag into the sample without damaging it, especially important if the sample is a living cell which you want to keep alive. Strategies here have involved forms of chemical and physical permeation (basically creating small holes in the outside of the sample to allow the tagged molecules to enter) and more subtle approaches which utilize the natural ability of certain cells to engulf particles.

A more refined approach still is to use so-called *fluorescent proteins* as tagging molecules; this will be discussed in later chapters when describing their significant use with single-molecule experiments. In essence, here an 'alien' fluorescent protein not normally made by a cell is genetically engineered so that the living cell will produce it from its DNA. This fluorescent tag is fused with 100% specificity to a native non-fluorescent protein of interest. This means that features inside the cell made up of this protein can be imaged to very high contrast.

1.2.12 Radioactivity

Here, a radioactive isotope is introduced in place of a normal relatively non-radioactive isotope to visualize components or metabolites in a biological system. Radioisotopes have less stable atomic nuclei and their presence can be detected from their emission of radiation during radioactive decay. The type of radiation produced depends on the isotope but can typically be detected by a *Geiger counter* or *scintillation phosphor screen*, often in combination with a *charge-coupled device* (CCD) or *photomultiplier tube* (PMT) detector. In combination with *stopped-flow* techniques, biochemical reactions can be quenched at intermediate stages and the presence of radioisotopes measured in the detected metabolites, thus allowing a picture of the extent of different biochemical processes to be built up. Common radioisotopes used in the life sciences include carbon-14, chlorine-36, hydrogen-3, iodine-125, phosphorus-32 and phosphorus-33, sodium-22 and sulphur-35.

1.2.13 Spectrophotometry

In essence a photo-detector is used to monitor the *transmittance* (or conversely the *reflectance*) of light through a sample as a function of wavelength. Instruments can have a typical wavelength range from the ultraviolet (~200–400 nm) through to the visible (~400–700 nm) up into the infrared (~700–2000 nm) generated from one or more broadband light sources in combination with a monochromator. Light is then directed through a solvated sample held in a sample cuvette generally made from glass or plastic, for visible and infrared, or quartz, for ultraviolet, since glass and plastic have a low transmittance to ultraviolet light (Figure 1.4B).

The transmitted light intensity from the sample is then amplified and measured by a photo-detector, typically a *photodiode*. More expensive machines will include a second reference beam using an identical reference cuvette with the same solvent (generally water, with some chemicals to stabilize the pH) but no sample, which can be used as a 'blank' baseline against which to reference the sample readings. This method is used for measuring the density of samples containing relatively large biological particulates (for example cells in suspension, to determine the so-called 'growth stage') and much smaller ones, such as molecules in solution. This method can also be extended into *colorimetry* in which an indicator dye is present in the sample, which changes colour upon binding with a given chemical. This can then be used to report on whether a given chemical reaction has occurred or not, and so monitoring the colour change with time will indicate details of the kinetics of that chemical reaction.

Scanning infrared spectrophotometers work on similar principles to those centred around visible light wavelengths, but scan only, and deeper, into the infrared. A variant of this is the *Fourier transform infrared* (FTIR) spectrometer which, instead of selecting one probe wavelength at any one time as with the scanning spectrometer, utilizes several in one go to generate a *polychromatic interference pattern* from the sample which has some advantage in terms of signal-to-noise ratio and spectral resolution. This can then be inverse Fourier transformed to yield the infrared absorption spectrum. Such spectra can be especially useful for identifying different organic chemical motifs in samples, since the vibrational energies of the covalent bonds used correspond to infrared wavelengths and will be indicated by measurable absorption peaks in the spectrum. Although broad-band infrared sources are still sometimes used in older machines, it is more common now to use infrared laser sources, with samples being supported typically by sodium chloride or silicon plates to minimize chemical bond vibrational interference from standard glass/ plastic or quartz cuvettes.

1.2.14 Sedimentation methods

The speed at which a sample will form a sediment or pellet when it is spun in a centrifuge relates to both the frictional viscous drag of the sample and its mass. Quantitative measurements may be made using so-called *analytical ultracentrifuges* which generate centripetal forces ~300,000 times that of gravity and also have controlled cooling to avoid localized heating in the sample which may be damaging in the case of biological material. By estimating the sedimentation speed we can deduce details of the size and shape of biological molecules and large complexes of molecules, as well as their molecular mass.

A heterogeneous mix of different biological molecules (for example, several different *enzymes*) may sometimes be separated on the basis of sedimentation rates, and typically a *density gradient* of suitable material (sucrose and caesium chloride are two commonly used agents) is created, such that there is a higher concentration/density of substance towards the bottom of a centrifuge tube. By centrifuging the mix into such a gradient the different chemicals may separate out as bands at different heights in the tube, and subsequently be extracted as appropriate.

BIO-EXTRA

The simplest conventional definition of an **enzyme** is as a biological catalyst, but this does little due justice to these remarkable molecular complexes. Enzymes are better viewed as **molecular machines** composed primarily of protein subunits which facilitate biochemical reactions via intermediate sub-steps involving transient binding to various parts of the enzyme, often with molecular conformational changes, such that the summed free energy changes from all such sub-steps (sometimes referred to as the **activation energy**) is less than that required for the single spontaneous reaction in the absence of any enzyme. In general, an enzyme reverts back to its initial state following the last reaction sub-step, and so in this regard can have substantial catalysing effects at relatively meagre concentrations. Most complex chemical reactions in the living cell occur far too slowly spontaneously to be viable, and so these enzymatic molecular machines are at the heart of facilitating biological processes. In the first half of 2011 ca. 4500 different enzymes had been classified.

1.2.15 X-ray, neutron and electron diffraction

X-rays are high energy electromagnetic waves which have a typical range of wavelength of 0.1–10 nm. This is very similar to the length scale for the separation of individual atoms in a biological molecule, and also for the size of certain larger scale periodic features at the level of molecular complexes and higher ordered molecular structures, which makes x-rays ideal probes of bio-molecular structure. Monochromatic x-rays can be generated locally, from a relatively small beam generator which can fit into a typical small research lab, which accelerates high energy electrons onto a metal plate, or they can be generated from a larger scale facility synchrotron source; the latter type are generally brighter and allow for better resolution.

When a beam of x-rays is passed through a crystal composed of a regular, periodic arrangement of several thousand individual biological molecules, the beam is diffracted due to the interference of scattered x-rays from the different crystal layers. The scattering effect is due primarily to the interaction of an x-ray with a free outer-shell *valence* electron (so-called *Thompson elastic scattering*) and so is influenced mainly by electron density as opposed to the atomic nuclei. The angle of a given emergent beam due to diffraction is inversely related to the length separation within the periodic structure involved in the scattering of that particular electromagnetic ray; the pattern of all of the scattered rays appears as periodic spots of varying intensity and can be recorded behind the crystal using either photographic film or a charge-coupled device (CCD) camera. Typically, the crystal will be rotated on a stable mount so that diffraction patterns can be collated from all possible orientations.

The intensity and spacing of the spots in the diffraction patterns can then be used in intensive computational analysis to deduce the underlying positions of atoms in the biological molecule, and hence to solve the molecular structure. The typical resolution is quoted as a few Ångströms (an Ångström, Å, is a non-SI unit used frequently by crystallographers which equals 0.1 nm, and is useful since it is of a similar length scale as typical chemical bonds). An essential additional requirement in this analysis is information concerning the phase of scattered rays. The intensity of the spots in the diffraction pattern alone does not provide this since there is no x-ray 'lens' as such to form a direct image as can be done using visible light wavelengths for example (crystallographers refer to this as the *phase problem*), but it can be obtained using a variety of additional methods, such as doping the crystals with heavy metals at specific sites which have known phase relationships.

X-ray crystallography has been at the heart of the development of modern biophysics. For example the first biological molecular structure to be solved was that of *cholesterol* as early as 1937 by Dorothy Hodgkin, and the first protein structures solved were *myoglobin* in 1958 (John Kendrew and others) followed soon after by *haemoglobin* in 1959 (Max Perutz and others). The prime disadvantage of the technique, as with the other techniques of diffraction described below of neutron and electron crystallography that require an artificially tightly packed spatial ordering of molecules, is that it is clearly intrinsically non-physiological and it is sometimes far from clear how the crystal structure relates to the actual functional structure in the living organism. In addition, the approach is reliant upon being able to manufacture very pure crystals which are relatively large (typically of the order of a few tenths of a millimetre long, containing $\sim 10^{15}$ molecules). This crystal manufacture process is technically non-trivial, and many important biological molecules which are integrated into cell membranes are very difficult if not impossible to crystallize because the requirement of adding solvating detergents affects the process. In addition, since the scattering is due to interaction with regions of high electron density, the positions of hydrogen atoms in a structure cannot be observed by this method directly since the electron density is too low, but rather need to be deduced from knowledge of typical bond lengths.

Variants of the technique include *powder* and *fibre diffraction*. The latter has repeating subunits arranged periodically in just one dimension as opposed to three dimensions for crystals, and was used to great effect in solving the double-helical structure of DNA from the work of Crick, Franklin, Wilkins and Watson ca. 1953. Another variant is *small-angle x-ray scattering* (SAXS). In SAXS a crystalline sample is not needed, and the range of scattered angles explored is small (typically less than 10°) and is used to determine the spacing of relatively large scale structures up to ~ 150 nm (for example, to study periodic features in muscle fibres).

With *inelastic scattering* the wavelength of the emergent x-rays is greater than that of the incident beam (i.e. the scattered beam has a lower energy) whereas in elastic scattering there is no change, since some energy has been transferred from the beam to energize a process in the sample, for example to excite an inner-shell electron to a higher energy level. This is not directly useful in determining atomic structures but has been applied in the form of *resonant inelastic soft x-ray scattering* (RIXS) to various biological questions including probing solvation effects in chemo-receptors, investigating the electronic structure of transition metals involved in photosynthesis and studying the dynamics of phospholipid bilayers (see Chapter 2).

Neutron diffraction works on similar principles to that of x-ray diffraction but here the incident beam is composed typically of *thermal neutrons*, generally from a nuclear reactor, with an effective wavelength of ~0.1 nm. Here, the scattering is due to interaction between the atomic nuclei as opposed to the electron cloud. It has a significant advantage over x-ray diffraction in that hydrogen nuclei (i.e. a single proton) will measurably scatter a neutron beam, and this scatter signal can be further enhanced by chemically replacing any solvent-accessible labile hydrogen atoms with deuterium, D (i.e. a proton and a neutron in each atomic nucleus) typically by solvating the target molecule in heavy-water (D_2O) rather than normal water (H_2O) prior to crystallization. This allows the position of the hydrogen atoms to be measured directly, resulting in more accurate bond length predictions, but has the disadvantages of requiring larger crystals (length ~1 mm) and a nearby nuclear reactor.

Electron diffraction again works on similar scattering principles, but here the incident beam of accelerated electrons interacts more strongly with matter than x-rays do. This makes three-dimensional crystals opaque to an electron beam, however two-dimensional crystal arrays can generate a strong emergent scatter pattern which can be utilized to solve atomic level crystal structures. Electron beams can be focused using electromagnetic lenses, which means that the phase information of the beam is retained directly instead of having to be deduced using other techniques as in the case of scattered x-rays. The main application advantage with this approach over x-ray or neutron crystallography is that structures can be solved for membrane proteins for which three-dimensional crystals are difficult to manufacture, but for which the manufacture of close-packed two-dimensional lipid–protein arrays is much more feasible. The first atomic level structure of a membrane protein to be solved using electron diffraction was that of bacteriorhodopsin, by Henderson and others in 1990, which was only the second ever atomic level structure of a membrane protein to be determined. (The first was the protein subunits of the photosynthetic reaction centre, determined using x-ray diffraction by Diesenhofer and others just a few years earlier in 1985.) However, the strong interaction with matter of the electron beam has the disadvantage of resulting radiation damage of the sample, and consequently samples need to be cooled dramatically, using liquid nitrogen or sometimes even liquid helium.

1.3 American versus European coffee

So, why should we wish to perform single-molecule biological experiments which are, as a general rule, technically highly demanding requiring the measurements of often tiny signals in a background environment of significant noise, in all but rare cases suffering from poor experimental yields and, apart from a few exceptions, being not remotely ‘high-throughput’? As we have seen from the previous section, there already exists a substantial selection of bulk ensemble average biophysical methods which can illuminate many aspects of the structure and function of cellular systems and do so using well-characterized experimental apparatus, with an effect of averaging over many molecular events, typically resulting in low measurement noise.

Comparing single-molecule experiments with those using bulk ensemble averaging has been likened to drinking American versus European coffee. European coffee is typically strong, containing many thousands of active caffeine molecules. American coffee in comparison contains far fewer active molecules, and it is often challenging to find them, at least for those who like strong coffee. (Post-doc chit-chat over coffee from American Gregory Harms, ca. 1999 in the lab of the European Thomas Schmidt, University of Linz, Austria – that being said, more recent post-doc chit-chat in ca. 2012 in Princeton University involving Teuta Pilizota in the lab of Joshua Shaevitz has concluded that some of the coffee available in Philadelphia is actually stronger than that found in Greece or Italy, and so my coffee analogy may soon become redundant!) There have been many reasons discussed in previous texts for the importance of performing single-molecule experiments, but this has often led to confusion in the student's mind. This should be clarified, as the key point below indicates.

KEY POINT

The prime reason for studying biology at the level of single molecules is the prevalence of **molecular heterogeneity**.

Intuitively, one might think that a mean average property of many thousands upon thousands of molecules should be an adequate description for any given single molecule. In some very simple, or exceptional, molecular systems this is in fact the case. However, in general this is not strictly true. The reason is that single biological molecules often exist in multiple states, which is in general intrinsically related to their biological function. A state here is essentially a measure of the energy locked in to the molecule, which is a combination of mainly chemical binding energy, so-called *enthalpy*, and energy associated with how disordered the molecule is, or *entropy*. There are many molecules which exist in several different spatial conformations; a good example is *motor proteins* which will be explored later in Chapters 8 and 9.

With these tiny molecular machines, although there may be one single conformation which is more stable than the others, several other shorter-lived conformations still exist, and these are utilized in different stages of force and motion generation in the molecule. The mean ensemble average conformation would look something close to the most stable of these many different conformations, but this single average parameter tells us very little of the behaviour of the other shorter-lived, but functionally essential, conformational states. What cannot be done with bulk ensemble average analysis, irrespective of what experimental property is being measured, is to probe such multi-state molecular systems. The strength of single-molecule experiments is that these sub-populations of molecular states can be explored. Such sub-populations of states are, as will be seen from subsequent case studies in this book, in general a vital characteristic feature of the proper functioning of natural molecular machines; there is essentially a fundamental instability in molecular machines which allows them to switch between multiple states as part of their underlying physiological function.

BIO-EXTRA

There are several experimental methods which can be employed in bulk ensemble investigations to **synchronize** a molecular population: these include thermal and chemical jumps such as stopped-flow reactions, electric and light field methods to align molecules, as well as freezing of a population. A danger with such approaches is that the normal physiological functioning may be different. Some biological tissues, for example muscles and cell membranes, are naturally ordered on a bulk scale and so these have historically generated the most physiologically relevant ensemble data.

Then there is also a potential issue of synchronicity in ensemble experiments, or rather the lack of it. The problem here is that different molecules within a large population may be doing different things at different times, for example molecules may be in different conformational states at any given time, so the mean ensemble average snapshot from the large population encapsulates all such *temporal fluctuations*, resulting in a broadening of the distribution of whatever parameter is being measured. The root of the problem of molecular asynchrony is that in a typical ensemble experiment the population is in chemical equilibrium, or steady-state; the rate of change between forward and reverse molecular states is the same. If the system is momentarily taken out of equilibrium then transient molecular synchrony can be obtained, for example by forcing all molecules into just one state. However, by definition this is a short-lived effect so practical measurements are likely to be very transient.

Clearly this is related to the issue of molecular inhomogeneity above, since in a completely homogeneous one-state molecular system there is no issue with a lack of synchronicity. Some ensemble techniques, for example crystallography, in effect overcome this problem by forcing the majority of the molecules in a system into a single state. But in general this widening of the measurement distribution is difficult to deal with since there is no clear way to discriminate between expected widening of a measurement distribution due to, for example, experimental error, and more physiologically relevant widening of the distribution due to *molecular asynchrony*.

Thermal fluctuations are in general the driving force for molecules switching between different states. This is because the typical energy difference between different molecular states is very similar to the heat energy associated with any molecule at a given temperature. However, it is not so much the heat energy of the biological molecule itself which drives change into a different state, but rather the heat energy associated with each surrounding water molecule. The local density of water molecules is much higher in general than the density of the biological molecules themselves, and so each biological molecule is bombarded by frequent collisions with water molecules (of the order of $\sim 10^9$ per second), and this heat energy from the momentum of bombarding water molecules can be transformed into mechanical energy of the biological molecule, which may be sufficient to drive a change of molecular state. Biological molecules are often described as existing in a *thermal bath*.

A final point which should be mentioned here is that although, as we will see from the next section, there is a broad range in concentration of biological molecules inside living cells, the actual number of molecules that are directly involved in any given biological process at any one time is generally low. At this level, biology occurs under minimal stoichiometry conditions in which individual, stochastic molecular events become important. Paradoxically, it can often be these rarer, single-molecule events that are the most significant to the functioning of cellular processes, and so it becomes all the more important to strive to monitor biological systems at the level of single molecules.

KEY POINT

Temporal fluctuations in the molecules from a large population often result in a broadening of the distribution of a parameter measured from a bulk ensemble experiment, which can be difficult to interpret physiologically. Thermal fluctuations are mainly driven by **collisions** from surrounding water molecules which can drive biological molecules into different states. In an ensemble experiment this can broaden the measured value which makes reliable inference difficult; in single-molecule measurements molecules can be measured individually in just one, or sometimes a few, of these states.

1.4 Scales of length, force, energy, time and concentration

The physical parameters of biological molecules in living cells can be characterized using, in effect, suitable scale bars.

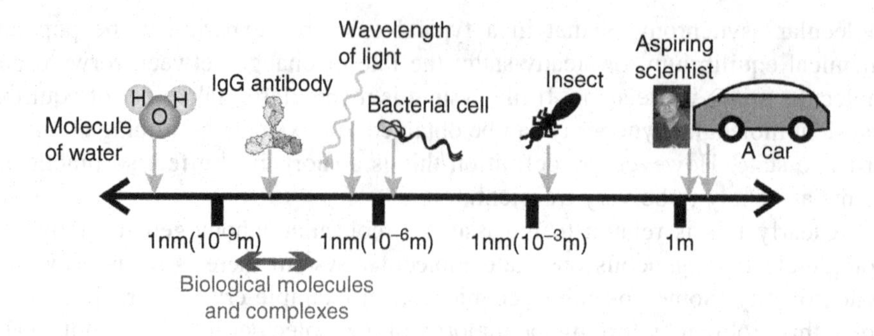

FIGURE 1.5 The length scale of biological molecules. Single biological molecules and 'molecular machines' made up of single-molecule components are typically 10–100 times smaller than the wavelength of light.

1.4.1 Length

Most single biological molecules, or biological complexes made from single-molecule components, are characterized by a length scale of typically a few *nanometres*, unit 'nm' (that is 1 metre divided by one thousand million). The term *nanoscale* is often used to refer to this nanometre length scale. That being said, there are exceptions to this, seen principally in a type of biopolymer called a nucleic acid – as we will see later in Chapter 2, these polymers can have up to tens of millions of molecular subunit components. For example, the genetic code in a human cell is made from a type of nucleic acid called DNA, with the largest single DNA molecule in principle having a contour length when fully extended of up to a few millimetres. In practice, DNA in the cell is not extended in this way but is tightly packed, but even so such condensed *chromosome* structures made from DNA can still have a length scale of a few hundred nanometres. Figure 1.5 gives an idea of the typical scale of single biological molecules and complexes, compared with objects which are better known on an everyday basis.

1.4.2 Force

Similarly, the forces exerted by molecular components as part of their natural function in these nanometre length scale biological machines are typically around the level of a few *piconewtons*, unit 'pN'. For example, the force exerted by a single molecule called myosin against filaments made up of molecules of actin in muscle tissue as part of the contraction of muscle is around 5 pN. The piconewton is equivalent to 1 Newton unit of force divided by 10^{12}, or one million million, which is clearly very small compared with macro scale objects – for example a healthy adult human weighs the equivalent of $\sim$700 N. This piconewton level of force should be compared with the typical force required to break a covalent chemical bond such as a carbon–carbon bond of $\sim$1000 pN, or a weaker non-covalent biological bond such as those of an *antibody* bound to its *antigen* of $\sim$100 pN, and is more comparable to the force required to break single very weak hydrogen bonds, such as those that glue together the separate helical strands of DNA (see Chapter 2 for a general discussion on all such bonding).

There are also *electrostatic forces* relevant to biological molecules, however these tend to be very short range. The length scale of electrostatic forces can be characterized by the *Bjerrum length*, $\lambda_B = e^2/4\pi\varepsilon_0\varepsilon_r k_B T$, where e is the unitary charge on a single electron, and ε_0 and ε_r are the vacuum and relative electrical permittivity respectively. This is equivalent to the separation distance for which the electrostatic interaction potential energy between two elementary charges is comparable to the thermal energy scale of k_BT. For physiological solutions λ_B is ~0.5 nm, smaller than the length scale of typical molecules, so electrostatics for single-molecule biology is really only relevant for close surface interactions.

The very weakest forces in biology arise from the random *Brownian* type fluctuations of the surrounding water molecules hitting biological structures. This is called the *Langevin force* and ultimately depends upon the size of the structure and the time over which you observe it, but to give an idea for a small size of effective diameter ~1 μm (i.e. 10^{-6} m) observed over a single second this random Langevin force accounts for a small force discrepancy of ~0.01 pN (see section 5.2 for a practical analysis of this force).

Most revealing is if we compare the piconewton level of force relevant to molecular machines with the weight (i.e. gravitational force) of a typical biological molecule itself. This weight is around 10^{-9} pN. In other words, although the level of *functional* molecular force at the piconewton level seems small by *macro* scale standards, such as the weight of a human body, it is significantly larger than the molecule's own weight, and so gravitational force is not a significant factor for single biological molecules.

PHYSICS-EXTRA

The Bjerrum length is closely related to the Debye length, also known as the **Debye–Hückel length** λ_D, which is a measure of the length scale of any electrostatic screening effect (i.e. for a charged ion at distances much above λ_D from a molecule's surface the screening of electrostatic charges due to the binding of counter-ions on the surface results in negligible electrostatic force interaction). For an electrolyte, as in the cell, $\lambda_D = 1/\sqrt{(8\pi\lambda_B N_A I)}$, where N_A is Avogadro's number and I is the ionic strength. In other words, as the concentration of ions increases the Debye length goes down since there is an increase in electrical screening.

1.4.3 Energy

These piconewton functional forces generally lead to spatial displacements of the molecular components over the nanometre length scale. The product of the force and the length displacement is a measure of the energy scale of single biological molecules, which is therefore equivalent to a few *piconewton nanometres*. The piconewton nanometre unit, or 'pN nm', is equivalent to 10^{-21} joule, J. As a reminder, the small energy dE required to move an object a small distance dx against a force F is given simply by $dE = F dx$. The total energy E required to move through a large distance is then the sum of all of these small dE values as the object moves through the large distance, which equates to the integral of F with distance, or $E = \int F dx$. The SI unit is the joule, J, which is equivalent to the energy required to move an object 1 m against a force of 1 N.

As we have seen, biological molecules are often described as existing in a thermal bath of surrounding water molecules. This means that they are in effect in constant thermal contact with a large *thermal reservoir* of water molecules which constantly bombard the biological molecules, typically with around a thousand million collisions occurring every second. Thermal physics predicts that the mean energy of one of these water molecules has a magnitude of roughly k_BT where k_B is the Boltzmann constant at T the absolute temperature in units of kelvin, K (i.e. the normal Celsius scale minus 273.15 °C). This means that the energy state of any given single biological molecule is likely to fluctuate randomly with time by typically a few units of k_BT.

To a physicist, organisms have a *high* temperature in that quantum energy transitions are small relative to classical levels, and so the classical *equipartition theorem* is a good

TABLE 1.1 Typical energy values for common transitions encountered in cellular processes

Energy transition	Energy (k_BT)	Energy (pN nm)
Hydrogen bond breakage	7	28
Hydrolysis of one ATP molecule	20	80
Folding of typical denatured protein	25	100
Moving a single electron through 1 V (1 eV)	40	160
Absorption of a photon of sunlight (wavelength ~500 nm)	100	400
Carbon–carbon single covalent bond	150	600

approximation: each independent quadratic term (*degree of freedom*) in the energy equation for a molecule in an ensemble of molecules at absolute temperature T is associated with an average energy of $(1/2)k_BT$. A water molecule has three translational, three rotational and three vibrational modes, and since each vibrational mode is associated with two degrees of freedom (one for potential, one for kinetic energy), the mean energy is $\sim 6k_BT$, though the presence of hydrogen bonds in liquid water raises this to $\sim 9k_BT$. During a collision with a biological molecule some of the kinetic energy is transferred, both to and from the molecule, resulting in $\sim k_BT$ energy fluctuations.

Because of this, some biological physicists find it useful to refer to k_BT directly as the actual unit of energy, instead of either the joule or the piconewton nanometre referred to previously. One k_BT unit at a normal 'room' temperature at which living organisms exist, is equivalent to roughly 4 pN nm, which is therefore about the same energy scale as a biological molecule undergoing a displacement transition of a few nanometres through a force of a few piconewtons as part of their normal function. This is no coincidence; many biological molecules that undergo such transitions as part of their function have evolved in effect to *feed off* the thermal fluctuations of the surrounding water to energize their activity. In understanding the energy transformation processes of life it really helps to have an appreciation of the characteristic energy values embodied in typical transition processes that may be encountered in the cell; some common values are given in Table 1.1.

1.4.4 Time

The characteristic time scale over which single biological molecules function is, as we shall see from the experimental case studies described later in this book, very broad. At the lower limit, the fastest events are characterized by the average separation of collisions from the surrounding water molecules, or roughly the nanosecond, 10^{-9} s. However, biological processes which ultimately arise from single-molecule activity can happen much more slowly than this. For example, some processes require the transfer of one or more electrons from one molecule to another, and the characteristic time scale of this is more like 10^{-6} s. Similarly, other processes require conformational changes to components in molecular machines, and these are slower still with a time scale of more like 10^{-3} s.

Some molecular complexes contain components which in effect turn over, i.e. older subunits are replaced by newer ones, and the time scale for this can be of the order of a few seconds. Ultimately, the upper limit of time scale is set by the lifetime of a molecule. Some cells break down older molecules controllably, and the time scale for this again varies considerably depending on the type of cell and the molecule in question, but several minutes are not atypical. However, some molecules in certain types of cell may have a longer lifetime, and in some cases may be limited by the average lifetime of the cell itself.

Potentially the time scale for the activity of single biological molecules, in the context of participation in biological processes, can span something like *18 orders of magnitude*! In exceptional cases some biological molecules go into quiescent states and may remain dormant for a considerable time (for example, bacterial spores have been discovered which are likely to be over 250 million years old but which can still grow functional bacteria). At a push some imaginative scientists consider the absolute lower limit of the time scale relating to biological function to be the characteristic oscillation time of an individual covalent bond within a single molecule, which is of the order of 10^{-15} s, and the absolute maximum to be the time over which life has known to exist, ca. 4 billion years, or 10^{18} s. Either way, the time scale is clearly enormous.

BIO-EXTRA

In the human body, **neutrophil** white blood cells survive a few days, red blood cells live for on average 120 days, cells associated with structural tissues such as the skeleton can live up to ca. 10 years, and some cells (those in the **cerebral cortex** of the brain, the **heart** and the **lens** of the eye) can in principle last until the body itself dies.

1.4.5 Concentration

The scale of molecular concentration in a cell obviously relates to the size of a cell and the number of molecules that exist in that cell. This range was assessed before the days of single-molecule experimentation using bulk ensemble average biochemical measurements on cellular extracts. Some small cells such as bacteria have a characteristic effective diameter of a micrometre or 10^{-6} m, whereas others are ca. 10 times larger. Some types of molecule may be present at certain points in a cell's lifecycle in exceptionally low numbers (~1 per cell), whereas other types of molecule may be very tightly packed and present in numbers of more like 10^3–10^4 per cell. The most concentrated molecules in any cell are two proteins called *HU* and *Fis* which are known to bind non-specifically to DNA in bacterial cells and are present at numbers of ~30,000 molecules per cell. The range of molecular concentration is therefore 1–30,000 molecules per 10^{-18} m^{-3}.

Biochemists often think in terms of *molarity* in regard to concentrations, or the number of moles of that substance present in 1 litre of water solvent, denoted by the unit 'M', where 1 *mole* is equal to *Avogadro's number* of particles ($\sim 6\times 10^{23}$, formally defined as the number of atoms of the carbon-12 isotope present in 12 g of pure carbon-12). The *atomic mass unit* (amu) is defined as 1/12 the mass of 1 mole of carbon-12, and so is equal to precisely 1 g, but for historical reasons is referred to as 1 dalton (Da). The *molecular weight* of any general molecule, usually denoted M_w, is equal to the mass, expressed in units of Da, of 1 mole of that molecule. Many biomolecules have M_w values of several 1000 Da, so it is common to refer to these in units of *kilodalton*, kDa. Molecular concentrations in cells are often quoted as 10^{-9} molar or *nanomolar*, nM (see Q1.3).

Another less commonly encountered form of concentration is *molality*. This is defined as the number of molecules measured in moles in a given volume of solution divided by the mass in kilograms of solvent used in the solution volume. The units are therefore mol kg^{-1}. In general, however, for biological applications molality and molarity can be used synonymously; for example, a typical physiological solution might have a molarity of 150 mM ('150 *millimolar*') of sodium chloride, which is equivalent to 150 millimoles of sodium chloride in a volume of 1 litre of water, which at room temperature also has a mass of 1 kg, so we could also say that this is a '150 *millimolal*' solution.

In addition, some biochemists refer to a *mass concentration*, that is the total mass of molecules present in a given volume of water solvent. The normal units of this are milligrams per millilitre, or mg/ml (note, however, if you wish to demonstrate how clever you are to professional biochemists it is important to pronounce the unit correctly as 'miggs per mill'), and can be useful because the usual concentration scale when purifying molecular components from cells, such as a specific types of protein, can be of the order of a few mg/ml.

1.5 Some basic thermodynamics of life

As we saw from the previous section, the energy scale of molecular-level processes that occur in living cells is that of $\sim k_B T$. But what sort of energy is this? The most useful concept here is that of *free energy*. This has a strict definition from thermal physics which means the energy in a *system* which can be converted to do some form of *thermodynamic work*. Such energy therefore can then be used to fuel a variety of cellular processes.

For many applications in physics the so-called *Helmholtz free energy*, often referred to by the letter F, is most useful. This is defined from thermodynamic theory as $U - TS$ where U is the *internal energy* and S the *entropy* at absolute temperature T (as a reminder to physical scientists the reason why F is so useful is because it has the advantage of varying in a very simple way with the logarithm of the *partition function*, from which many other thermodynamic potentials may be derived).

However, for applications involving biochemical and biophysical processes, the so-called *Gibbs free energy*, often referred to by the letter G, is in general more appropriate. This is defined as $H - TS$, where H is the total heat energy or enthalpy. Whether we use F or G as the relevant free energy really depends upon what parameters in the thermal system are being held constant. If the volume is being held constant then F actually gives the best estimate for the available free energy. However, if the pressure in the system is held constant then G is the best estimate for the available free energy. For biochemical reactions in a test tube it may seem sensible that the reactions occur under constant pressure conditions. However, observant readers may want to consider that certain cells in the natural world in effect contain a constant volume environment in which the pressure may vary significantly, and in such cases it may be more appropriate not to use G (see Q1.7).

Living entities, as with all physical systems, obey the *Second law of thermodynamics*. For biological systems, this means that any molecular process occurring in the cell cannot result in a net decrease in the entropy of the *whole universe* (though fluctuation-type decrements in entropy can occur locally, as predicted by the *fluctuation theorem*, see section 8.3). When entropy is combined with the effects of thermal energy in the form of enthalpy then the free energy change before and after a given molecular process in the cell can be used to predict the likelihood for that process to occur or not, as embodied in the concept of *detailed balance* or *microscopic reversibility* (discussed later in section 5.5).

A word of caution concerns some of the basic theory in thermodynamics applied to the cell. Much of this is applied loosely to understanding biochemical reactions and implicitly assumes that the system is in *thermal equilibrium*, which the Second law of thermodynamics interprets as the available free energy being a *minimum*. However, for a cell, this situation corresponds to *death* – most biochemical reactions in the cell do not occur spontaneously but rather require some *external energy input* if there is minimal free energy available to drive them then the reactions might not occur over a sufficiently fast time scale. Rather, life, as we know it, requires a state relatively *divorced from thermal equilibrium* (this is a by-product of molecular heterogeneity, since such a collection of distinct energy *microstates* implies non-zero free energy transitions between them which, by definition, is a system not in thermal equilibrium). Later we will see, with examples from single-molecule experiments, how this distance from equilibrium is maintained, which fundamentally involves establishing and sustaining non-equilibrium concentration gradients of ions across *phospholipid dielectric membranes* (see Chapter 8). This maintenance of a transmembrane electrochemical potential energy must be coupled to a constant input flux of energy, which in turn results in the generation of entropy, and hence all is well with the Second law!

KEY POINT

Thermal equilibrium equates to **death** for a cell, and cellular processes in general are far from thermal equilibrium. This should not be confused with chemical equilibrium which states that the forward and backward rates for a reversible chemical reaction are equal, so that reactants and products appear to be in **steady-state**; many biological systems exist in effect in such steady-state conditions but behind this there can be significant **turnover** of molecular components.

1.6 The concept of 'functionality'

One significant difference between the intellectual approach of life and physical scientists is the concept of *biological function*. Physical scientists in general have a relatively modern tradition of referring to *immutable laws*, whereas life scientists often refer to this concept of function, which some physical scientists find confusing. Historically, the concept that some general entity may have a function or *purpose* arguably precedes the more modern notion of scientific laws. Classical Greek philosophers such as Aristotle espoused the idea over 2000 years ago in so-called *teleological* arguments.

In its modern biological form, the function can span different length scales from that of the single molecule (for example, what is the 'function' of a protein molecule called haemoglobin) through to complexes and sub-cellular structures (for example, what are the 'functions' of the cell membrane), and on to the cell, populations of cells in tissues and organs, and ultimately the whole organism. Arguably one might even go beyond this and consider the function of collections of organisms in their environment, and even then of the collection of environments up to the level of an entire planet, as in the well-known *Gaia hypothesis* developed primarily by the physical scientist James Lovelock and the life scientist Lynn Margulis in the 1970s (there are several good general references which the reader can easily find, but for a seminal discussion see Lovelock and Margulis, 1974).

The most accepted view of the meaning of biological function for a given entity is really a consensus notion of what the entity appears to do in order to optimize the viability of the living organism in which the entity normally exists. For example, haemoglobin functions as an oxygen transport molecule (which ultimately is manifest in establishing a more viable human body). A subtle but important point to note is that this view of function is largely imposed by the observer(s), and has two obvious and related weaknesses: one is that the view of function is clearly *data limited* (i.e. new as yet unseen future data may change our concept of the function, so in this sense function cannot be regarded as being *absolute*), and the other is that the organism itself does not necessarily have the same view of function as we observers do.

Scientists and philosophers have argued about the length scale for the ultimate *grand purpose* relating to biological function, whether the level of optimal viability of the organism is the best one to use, or whether it is more sensible to go down the length scales either to the level of the viability of the individual cell in which a specific entity exists, or even to the level of the *genes*, or even individual segments of DNA making up a gene, ultimately responsible for making the entity in the cell (this idea is best embodied in the *selfish gene hypothesis*, see Dawkins, 2006).

The most robust definition for biological function is that it is the activity of a given biological entity which appears best to achieve its own future reproduction into subsequent generations of cells. As such it goes hand-in-hand with theories of evolution, since this then becomes the ultimate *driving force* behind evolutionary change of an organism in the light of a background of random *genetic mutation* that can generate *biological variability*.

KEY POINT

Biological function is a difficult concept even for those who refer to it on a daily basis, but it is such a fundamental one that it is important for those fresh to this notion to give it due consideration. Without the concept of function, one could argue, there is no special distinction between biology and the other sciences.

1.7 Test tube or cell?

Single-molecule experimentation is still a relatively young discipline. Single particle detection began in non-biological samples, first involving physically trapping single elementary particles in a gaseous phase in the form of a single electron (1973), and subsequently (1980) single atomic ions (interested physical scientists should look at the work of Dehmelt, who was the pioneer in this field, see Wineland *et al.*, 1973, and Nagourney *et al.*, 1986). However, the first definitive single-molecule biological investigations actually were earlier still and used pioneering electron microscopy techniques (see Chapter 3) to produce metallic shadow replicas of relatively large, filamentous molecules including DNA and a variety of proteins (Hall, 1956). These investigations used dried samples in a vacuum.

The first single-molecule biological investigation in which the surrounding medium included that one compound which is essential to all known forms of life, water, came later in 1961 with the fluorescence detection of the activity of single protein molecules of an enzyme called β-*galactosidase* by chemically modifying one of its reaction *products* to make it fluorescent, and observing the emergence of these fluorescent molecules during the enzyme-catalysed reaction inside microscopic water droplets (Rotman, 1961). Fluorescence observations were made over a decade later in aqueous solution without the need for microdroplets using the organic dye *fluorescein*, similar in structure to the fluorogenic component in the 1961 study, bound via antibodies to label single *globulin* protein molecules each with 80–100 individual fluorescein molecules attached (Hirschfield, 1976). The decade that followed involved significant improvement in measurement sensitivity, including fluorescence detection of single molecules of a liquid-phase solution of a protein called *phycoerythrin* labelled with the equivalent of

~25 molecules of the orange organic dye *rhodamine* (Nguyen *et al.*, 1987), as well as the detection of single molecules in solids using *optical absorption* of a non-biological sample (Moerner and Kador, 1989).

The pioneering biological single-molecule work that came in the subsequent 10 years all involved in vitro studies, essentially experiments done, in effect, in the test tube, and it was only as recently as the year 2000 that the first definitive biological single-molecule investigation involving a living sample was performed (Sako *et al.*, 2000 were the first to perform single-molecule live-cell imaging on the cell membrane; Byassee *et al.*, 2000 were the first to perform single-molecule live-cell imaging inside the centre of a cell). Again, fluorescence microscopy was used to monitor single proteins, but now with a much enhanced sensitivity compared to the earlier studies, enabling the detection of single fluorescent molecules attached to single proteins in single live cells.

But why should we care about studying molecular details in living cells when we can do so in the test tube? The test tube environment is significantly more controlled, potentially less contaminated and generally comes with less measurement noise. Well, the best answer is the most obvious: cells are not test tubes. A test tube experiment is a much reduced version of the native biology with only the components which we think, or hope, are important included. But we now know definitively that even the simplest cells are not just bags of chemicals, but rather have localized processes in both space and time. A related issue is that of localized effects of the physical and chemical environment in the cell, and these localized features are essentially impossible to replicate faithfully in the test tube.

Also, the effective numbers of molecules involved in many cellular processes are often low, sometimes just a few per cell, and these minimal stoichiometry conditions are not easy to reproduce in the test tube without incurring a significant reduction in physiological efficiency. However, this is not to say that test tube experiments are intrinsically bad and in vivo experiments, that is experiments either on the living organism or involving a subset of living cells from an organism, are implicitly good. Rather, they provide *complementary* information.

Test tube experiments are obviously divorced from a true physiological setting, but the level of environmental control is relatively high. In vivo experiments on the other hand are generally more demanding technically and are subject to both greater experimental noise and *intrinsic biological variation*. The fact of being in a native physiological environment is appealing at one level but can cause difficulty in interpretation since there is a potential lack of control over other biological processes which are not directly under study but which may influence the experimental outcome. The best approach therefore is to devise genuinely complementary in vitro and in vivo single-molecule biology experiments to navigate a path of best understanding of the molecular functions of the cell.

This handbook should be thought of as both an introduction to the field of single-molecule biology, whether utilized as a companion to a formal lecture course or for new researchers to the discipline, and also as a stimulus for more experienced researchers in formulating new, pioneering single-molecule approaches which complement live-cell and test tube information.

THE GIST

- There exist several standard 'ensemble' biophysical and biochemical approaches to the investigation of biological processes, which can generate very useful information.
- Ensemble experiments typically output a mean value from many thousands of individual molecules and may mask underlying molecular heterogeneity.

- Many different types of biological molecule have heterogeneous states as part of their normal physiological function.
- Single-molecule biology experiments are technically very challenging and generally low throughput, but permit heterogeneous molecular distributions to be measured.
- The physical scales for length, force and energy of single biological molecules are the nm, pN and k_BT, and the time scale has a very broad range of 10^{-9}–10^9 s.
- Single biological molecules in functional molecular machines typically feed off 'thermal noise' energy of the surrounding water.
- Cellular processes in general are far from thermal equilibrium, but may often be in chemical equilibrium (i.e. steady-state).
- Test tube and living cell single-molecule biology experiments both have advantages and disadvantages, and the ideal approach is to combine the two in complementary studies to maximize physiological relevance and clarity of measurement signal.

TAKE-HOME MESSAGE Single-molecule experimental biology using biophysical methods is technically challenging but, when it works, allows us to explore directly the underlying molecular mechanisms of many processes which are fundamental to the living cell.

References

GENERAL

- Dawkins, R. (2006). *The Selfish Gene*, 3rd edition. Oxford University Press.
- Hall, C. E. (1956). Method for the observation of macromolecules with the electron microscope illustrated with micrographs of DNA. *Biophys. Biochem. Cytol.* **2**: 625–628.
- Hirschfeld, T. (1976). Optical microscopic observation of single small molecules. *Appl. Opt.* **15**: 2965–2966.
- Levy, S. (1993). *Artificial Life: A Report from the Frontier Where Computers Meet Biology*. Vintage.
- Lovelock, J. E. and Margulis, L. (1974). Atmospheric homeostasis by and for the biosphere – the Gaia hypothesis. *Tellus* **26**: 2–10.
- Nölting, B. (2009). *Methods in Modern Biophysics*. Springer.
- Rotman, B. (1961). Measurement of activity of single molecules of beta-D-galactosidase. *Proc. Natl. Acad. Sci. USA* **47**: 1981–1991.
- Schrödinger, E. (1944). *What is Life? The physical aspect of the living cell*. Folio Society, First Thus edition: 2000.
- Van Holde, K. E. *et al.* (2005). *Principles of Physical Biochemistry*. Pearson Education.

ADVANCED

- Byassee, T. A., Chan, W. C. and Nie, S. (2000). Probing single molecules in single living cells. *Anal. Chem.* **72**: 5606–5611.
- Leake, M. C. (2001). *Investigation of the extensile properties of the giant sarcomeric protein titin by single-molecule manipulation using a laser-tweezers technique*. PhD dissertation, London University, UK.
- Moerner, W. E. and Kador, L. (1989). Optical detection and spectroscopy in a solid. *Phys. Rev. Lett.* **62**: 2535–2538.
- Nagourney, W., Sandberg, J. and Dehmelt, H. (1986). Shelved optical electron amplifier: observation of quantum lumps. *Phys. Rev. Lett.* **56**: 2797.

- Nguyen, D. C., Keller, R. A., Jett, J. H. and Martin, J. C. (1987). Detection of single molecules of phycoerythrin in hydrodynamically focused flows by laser-induced fluorescence. *Anal. Chem.* **59**: 2158–2161.
- Sako, Y., Minoguchi, S. and Yanagida, T. (2000). Single-molecule imaging of EGFR signalling on the surface of living cells. *Nature Cell Biol.* **2**: 168–172.
- Wineland, D., Ekstrom, P. and Dehmelt, H. (1973). Monoelectron oscillator. *Phys. Rev. Lett.* **31**: 1279.

Questions

FOR THE LIFE SCIENTISTS

Q1.1. 'A typical tissue in the human body may contain many thousands of molecules, so there is not a need to perform single-molecule biology experiments to understand tissue behaviour.' Discuss.

Q1.2. What are the standard bulk ensemble average methods available to investigate the structure of single biological molecules and molecular complexes? Would it be better to measure their structure using single-molecule techniques directly?

Q1.3. Calculate what the actual range of typical molarity will be in a cell in units of nM. How does this compare with the molarity of pure water? (Hint: Think about the formal definition of molarity.)

FOR THE PHYSICAL SCIENTISTS

Q1.4. How does the concept of 'Maxwell's Daemon' relate to the Second law of thermodynamics? How is this relevant to the ways in which some biological molecules appear to use thermal noise as part of their physiological function? Does this mean that these molecules can break the Second law?

Q1.5. Classical thermodynamic potentials are derived from ensemble assumptions, but the ergodic hypothesis maintains that there is an equivalence between ensemble and single-molecule properties. Under what situations might there be differences between derived values of thermodynamic potentials using ensemble and single-molecule experimental methods, and why might this be relevant to real biological systems?

Q1.6. Many molecular processes in biology occur via several discrete steps from a starting molecular conformation to an end molecular conformation, as opposed to being a more simple transition that occurs between start and end states. Are there specific thermodynamic advantages to the former approach?

Q1.7. Explain with reasoning why the Gibbs free energy is in general the most appropriate form of free energy to use in the context of determining whether processes in the cell will occur or not, and under what conditions the Helmholtz free energy might actually be more sensible.

FOR THOSE WHO HAVE NOT MADE UP THEIR MIND

Q1.8. There are many examples of single biological molecules that exist in multiple, heterogeneous states. Is this advantageous or disadvantageous to the cell?

Q1.9. What sets the limits on length and force scales for single biological molecules?

Q1.10. There are some well-known examples in biophysics research of studies of multi-molecular ensemble systems that have led to a knowledge of single-molecule properties. What are these examples, and what special conditions allow ensemble properties to be used in this way and why?

Q1.11. In what ways could we say that cells act as storage vessels of negative entropy? What implications does this have for the Second law of thermodynamics? Where does all the positive entropy go?

Q1.12. If a single 'fluorescent protein' (see Chapter 3), a natural fluorescence emitting biological molecule of effective diameter ~1 nm, is conjugated to the surface of the Earth which (naively!) is modelled as a smooth sphere of radius ca. 6400 km, what is the greatest area over the Earth's surface that the fluorescent protein could light up in principle, neglecting any effects due to diffraction and refraction? What is the equivalent area on a cell's surface if instead the fluorescent protein is conjugated to the surface of a perfectly smooth, spherical cell of radius ca. 5 μm. (Hint: This is in effect the same problem as working out how far away the horizon is.)

2 The molecules of life – an idiot's guide

MOLECULE, n. The ultimate, indivisible unit of matter. It is distinguished from the corpuscle, also the ultimate, indivisible unit of matter, by a closer resemblance to the atom, also the ultimate, indivisible unit of matter. Three great scientific theories of the structure of the universe are the molecular, the corpuscular and the atomic. A fourth affirms, with Haeckel, the condensation or precipitation of matter from ether – whose existence is proved by the condensation or precipitation. The present trend of scientific thought is toward the theory of ions. The ion differs from the molecule, the corpuscle and the atom in that it is an ion. A fifth theory is held by idiots, but it is doubtful if they know any more about the matter than the others.

(THE COLLECTED WORKS OF AMBROSE BIERCE (1911), VOL. 7, *THE DEVIL'S DICTIONARY*, PP. 220–221)

GENERAL IDEA

Here we outline some of the key concepts and terminology in cell and molecular biology to orientate readers from a more physical science background.

2.1 Introduction

The classical biological view is that living organims are typically structured in a hierarchical manner in terms of *physiological function* relating to *length scale*. For example, a complex multi-cellular *organism* is composed of smaller units called *organs* which appear to be dedicated primarily to a subset of biological processes, and these may be further deconstructed into different *tissues*, and these tissues may be further sub-divided into smaller structural features consisting ultimately of individual *cells*, or some structural matrix secreted by cells. Single cells, whether part of a multi-cellular organism as in the human body or simply the organism itself as for unicellular life forms such as bacteria, can in turn be broken up conceptually into smaller subunits. In essence these are structural sub-cellular features which appear to work together to perform a narrow subset of biological functions, for example cell *organelles* such as the nucleus in certain cell types. Ultimately, smaller sub-cellular feautures can be perceived as collections of *single biological molecules*.

This old-school hierarchical, ordered image to life has its place in being simple to understand and intellectually appealing in its reductionist order, however it runs the risk of being misleading at times and potentially artificial. Often in the living cell there are mulitple layers of *interaction* and *feedback* between ostensibly different processes, occurring over a range of *different length and time scales*, and involving multiple regions of the cell. In other words, it is not quite as easy as dividing a cell up into structures which perform unique functions, rather sub-regions of the cell can be involved in many different processes at the same time, some of which may consequently utilize shared elements. It is hard to render such explicit interconnectedness as words and diagrams – perhaps the best

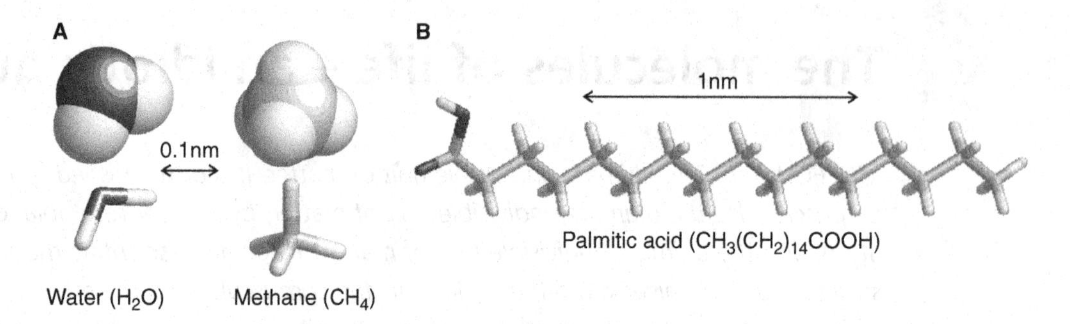

FIGURE 2.1 Carbon chemistry. **(A)** Space-filling (upper panels) models which give a good representation of the volume occupied by constituent atoms, and stick models (lower panels) which give a clearer visual indication of the positions of atoms in relatively complex structures, here showing water (left) and methane (right). **(B)** A stick model of the fatty acid molecule palmitic acid, used as a component in many fats (lipids), all prepared using RasMol (see references).

way is to treat the cell as a grand *network* of connected features then where there is a *clustering of nodes* we can start to ask about what range of biological functions are performed in that particular *hub*. Ultimately, what is more clear is that this interconnectedness involves collections of single biological molecules.

In this chapter you will find a concise outline of the principal molecules of interest in biology, as well as some of the core biological concepts that are relevant to single-molecule biological techniques. There are several books the reader can refer to for more detailed information (Alberts *et al.*, 2008 is arguably the most comprehensive, but for a robust yet compact and clear overview of the essentials the reader is strongly referred to Rose, 1999).

2.2 The atomic components of single biological molecules

If we consider the proportion by number of atoms, then just four atomic elements constitute greater than 99% of the total number of atoms in a living cell: hydrogen (~63%), oxygen (~25%), carbon (~9%) and nitrogen (~1%). Roughly 70% of a cell's mass is water; however, in terms of molecules this equates to almost 99% of all molecules in the cell. It is clear that the most important molecule is water (the chemical structure is shown in Figure 2.1A, left panel). Water is used as a reactant in many biochemical processes, is generated as a product in many others, acts to stabilize many large-scale molecular structures through hydrogen bonding or hydrophobic force effects and is, essentially, the *solvent of all life* as we know it.

However, for the remaining non-water molecules, it is primarily the chemistry of *carbon* atoms that has a central importance to living cells. A whole branch of 'organic' chemistry is dedicated solely to understanding such carbon-containing molecules. There are two prime reasons why carbon is such a good atom for biological purposes, which relate to its *stability* and its ability to form extended chains, or to *catenate*.

KEY POINT

Carbon has a central role in biological molecules because (1) carbon–carbon bonds are relatively **stable**, and (2) carbon atoms can **catenate** to form extended, uniquely defined three-dimensional structures.

The normal valence of a carbon atom is four. This can be satisfied with multiple carbon-bond combinations which result in either linear or planar structural features (a triple plus a single covalent bond, two double covalent bonds, or a double and two single covalent bonds), or it can be satisfied with with four single covalent bonds which results in a three-dimensional *tetrahedral* arrangement (as shown for example in the organic gas molecule methane generated by certain bacteria as a metabolic by-product from the breakdown of the sugar cellulose found in plant cell walls, Figure 2.1A, right panel). The comparative stability of carbon–carbon bonds compared to like–like bonds of other elements in the periodic table, combined with the potential for further carbon bonding to either carbon atom in such a pair, allows *catenation* to occur, that is, chain formation involving multiple carbon atoms, an example of which is the fatty acid palmitic acid (Figure 2.1B) which we will encounter later in this chapter.

This means that carbon-containing molecules can form relatively stable extended three-dimensional structures. As we will see later throughout this book, the shape of single molecules is intrinsically related to function, and thus carbon chemistry allows multiple forms for different biological functionality, an essential feature for cellular processes. However, biological molecules are not so much optimized for greatest stability, but rather occupy a compromise zone of being comparatively stable and yet not *too* stable. For example, one physical form of pure carbon is diamond which is a highly stable structure, but this is not encountered in living cells. It is just as important for the cell to be able to break up biological molecules controllably as it is to be able to make them. Thus, some level of chemical *instability* could also be said to be a feature of biological molecules.

Other important but less abundant atomic elements include nitrogen (a vital component of all proteins among several other types of molecule), phosphorus (for example, found in all nucleic acids such as DNA) and sulphur (present in some proteins, especially those with more *structural* functions), as well as sodium, potassium and chlorine (present in ionic form in all cells), calcium (very important in muscle cells, but also secreted into extra-cellular components such as bone by some cells), magnesium (plays a vital role in manipulating abundant phosphate-containing molecules in the cell), and more trace but still essential elements such as certain transition metals (for example iron and zinc to name only two out of about 30). Beyond this are more elements still found in less abundance in the cell, which are deemed *non-essential* per se but are still found in many cell types.

2.3 Cell structure and sub-cellular architecture

Single biological molecules perform their physiological functions in the context of a living organism: either inside living cells, attached to their outer surface, or outside cells but in their vicinity in the so-called *extra-cellular matrix* which some cells secrete. Some organisms appear to consist just of a single cell, bacteria and yeast for example, whereas other cells are part of a more complex multi-cellular organism, for example the human body which contains over 200 different types of cell as defined by tissue of origin, with $\sim 10^{14}$ human cells in total in each of us. The conventional definitions of uni- and multi-cellular organisms and of cell type in general have recently become a little blurred. Bacteria, for example, generally exist in colonies of many cells which in effect communicate with their neighbours with many features similar to a more complex multi-cellular organism, whereas individual cells from a complex multi-cellular organism can be extracted and cultured and are as 'alive' as they were in their native context of the whole organism. In fact so-called stem cells are in a pre-specialized state and can *differentiate* under the right conditions into any different cell type from a particular organism.

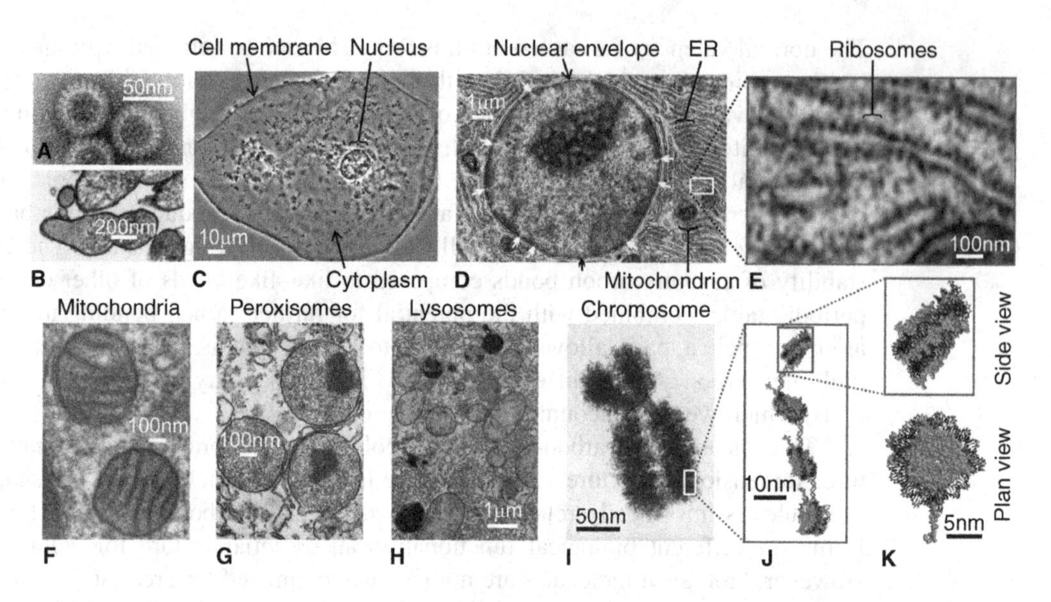

FIGURE 2.2 The inner architecture of cells. **(A)** Negative-stained TEM of human rotaviruses (original data adapted with permission from Cornelia Büchen-Osmond), and **(B)** *Mycoplasma pneumoniae* cells. **(C)** Phase contrast image of human cheek cell (adapted, CIL:12594). **(D)** TEM of nucleus: endoplasmic reticulum (ER), and pores (arrows), **(E)** expanded to show grain-like ribosomes (adapted, Fig. 145, Fawcett, 1981). TEMs of organelles: **(F)** mitochondria (adapted, Howard, 2000), **(G)** peroxisomes (adapted, Fig. 282, Fawcett, 1981), and **(H)** lysosomes (adapted, Fig. 265, Fawcett, 1981). **(I)** Chromosome (adapted from Fig. 129, Fawcett, 1981), with **(J)** the DNA wound tightly around protein bead-like structures of nucleosomes, an expanded section **(K)** of a single nucleosome indicated (PDB ID:1AOI).

Despite this diversity of form and function and the confusion over what an 'organism' really might be, all cells share certain common features. They are essentially the basic structural and functional unit of life. What does this really mean though? Without entering the arena of philosophy, arguably the least bad definition is given as the key point below.

KEY POINT

A cell is a **functionally autonomous**, physically enclosed and structurally independent **unit of life**, from either a single-celled or a multi-cellular organism, which contains biological molecules and which has the capacity to **self-replicate independently**.

The range of length scale for most cells spans a little over two orders of magnitude at roughly 1–200 μm (Agutter and Wheatley, 2000 contains interesting arguments for why this might be so), though there are some exceptions at either end of the range which mean that the actual full scale spans more like eight orders of magnitude. At the small end, a minority of biologists consider *viruses* (Figure 2.2A) to be the minimally sized unit of life, with the smallest known viruses having an effective diameter of ~20 nm. However, although viruses are self-contained structures physically enclosing biological molecules, in that they consist of a protein *capsid* coat which encloses a viral genetic code of a nucleic acid, of either *DNA* or *RNA* depending on the virus type, viruses can only replicate by utilizing some of the extra genetic machinery of a *host cell*, so in other words they do not fulfil the criterion of *independent* self-replication and cannot thus be considered a cell, if we want to pick nits.

The smallest definitive cell belongs to bacteria of the *Mycoplasma* genus (Figure 2.2B); these are membrane bound 200–300 nm diameter enclosed units containing only around 500 genes, which are believed to be the minimal set needed to perform the minimum catalogue of functions which are characteristic of life. This compares with an estimated 23,000 different genes for a human cell. Recently, *Mycoplamsa* cell *ghost* membranes (i.e. with their native genetic material removed) were used as the host *capsule* to store a mix of *synthetic genes* to produce a self-replicating life form in what arguably is the first example of a *man-made cell* (Gibson *et al.*, 2010).

Even the simplest cell is now known to be much more than just a bag of chemicals, but rather contains complex *local architecture* and sub-cellular features. A typical example of this is shown in Figure 2.2C, taken from the inside of a human cheek. Every cell is bounded by some type of cell envelope, which involves at the very least a *cell membrane* composed primarily of *lipid* (i.e. fat) molecules, but may also include in some but not all cells an additional membrane and/or a tougher *cell wall* (in bacterial, fungal and plant cells) made of large, stiff *protein-sugar* (called *glycoprotein*) molecules.

The cell envelope serves essentially as a controlled barrier between the inside cell world and the world outside. The cell membrane itself, as we shall see later (see Chapter 8), is far from a simple structure but rather has several architectural complexities which are likely to be linked to biological functions. Inside the cell is the so-called *cytoplasm* which consists primarily of water and ionic salts, but also includes a multitude of other biological molecules, either singularly or in complexes, as well as local larger length scale ordered structures. For example, the genetic code, which, as we shall see later, is written into the nucleic acid DNA inside cells, is stored nominally in a condensed, tightly packed format. In relatively simple cells called *prokaryotes* (a broad category of unicellular organisms, including bacteria), this tight-packed region of DNA is called the *nucleoid*, and its state of condensation is regulated by several proteins which bind to the DNA. More complex cells called *eukaryotes* (again, a broad category, including all animal and plant cells) encapsulate the genetic material into a separate smaller membrane-bound structure called the *nucleus* (Figure 2.2C).

The nucleus is an example of a cell organelle, a substructure within the cell which performs primarily a set of highly specialized functions (in this case the replication of the genetic code and the initial stage in the generation of proteins by first *transcribing* sections of the genetic code into another nucleic acid, RNA).

Other organelles (three such are indicated in Figure 2.2F–H) include the *centrosome* (an organizing centre for coordinating the construction of stiff structural proteins used in the so-called *cytoskeleton* to maintain cell shape), *chloroplasts* (inside plant cells which can self-replicate by virtue of containing their own DNA separate from the cell nucleus, and which convert light energy from the Sun into chemical energy stored in sugar molecules via *photosynthesis*, see Chapter 8), the *endoplasmic reticulum* or ER (a transport/targeting network for molecules), the *Golgi apparatus* (a tubular system involved in molecular processing and packaging), *lysosomes* (which contain multiple digestive enzymes used to break up old organelles as well as foreign bodies which the cell has engulfed), *mitochondria* (which like chloroplasts possess their own DNA and can self-replicate, and which generate the *universal* biological fuel of *adenosine triphosphate*, or ATP), *peroxisomes* (which are involved in breaking down components of fats as well as eradicating toxic *free-radical* chemicals), *proteasomes* (involved in degrading superfluous or damaged proteins), *ribosomes* (large molecular machines involved in manufacturing protein from transcribed RNA), and *vacuoles* (used in food, waste and water storage).

The alphabetical list of organelles above is a little artificial; not all cells contain all such organelles. Also, if we wished we could move down to smaller length scales and find smaller sub-cellular structures which again appear to be specialized in terms of biological function but are not conventionally deemed to be organelles. For example,

DNA in eukaryotes is packaged into *chromosomes* (Figure 2.2I) which are nominally localized in the cell nucleus organelle. They possess a specialized set of functions, but at a smaller length scale chromosomes themselves consist of bead-like structures called *nucleosomes*, which in turn are composed of a few protein molecules called *histones*, around which the DNA is wrapped (Figure 2.2J,K) when in its quiescent state, i.e. when not being read off by proteins during replication or protein production.

This blurring of the definition of what is an organelle, or a sub-organelle, or a large molecular complex, or a collection of a few molecules bound in a complex, is fortunately largely a semantic issue as opposed to a scientific one. What is clearly the case is that cells possess a great deal of locally defined sub-cellular features over a range of length scales spanning a few nanometres to several hundred nanometres, which perform a subset of specialized biological processes, which ultimately involve the interaction of single biological molecules. To get an idea of the real complexity of cellular processes it is revealing to have a look at so-called *metabolic maps*, which are visual depictions of the complex linked biochemistry that occurs in living cells (see Nicholson, 2003).

2.4 Amino acids, peptides and proteins

Proteins, which are so small that 1 billion would fit on the full-stop at the end of this sentence, carry out most of the vital activities in living cells, sometimes requiring the assembly of multiple proteins into remarkable biological machines, constituting something like half the mass of all organic molecules in the cell. There are of the order of 10,000 different proteins in a typical, small eukaryotic cell which, assuming a diameter of $\sim 10\,\mu m$ might contain around 10^5 molecules of a given typical protein on average; this number seems rather large considering our interest here in 'single' molecules.

If we look at smaller prokaryotic cells of more like $1\,\mu m$ characteristic length scale, these contain perhaps 2000–3000 different types of protein but have the same typical concentrations (i.e. the same average number of protein molecules of that type in a given volume), and this then equates to a smaller value of more like 100 molecules of a typical protein per cell. But this number still appears large compared to the *single* molecule world. However, at this typical protein concentration the mean separation of individual protein molecules is actually around 100 nm. This is 10–100 times larger than the length scale of a single protein molecule or complex of molecules, so although there can appear to be hundreds upon thousands of protein molecules per cell, each given molecule is separated from its nearest neighbour by a distance which, on average, is one to two orders of magnitude larger than its own size – a bit like playing football on a full-sized pitch but with only one person on each team. When viewed in this way it is clear that single-molecule effects are likely to be highly relevant for proteins in the living cell.

A protein is essentially a polymer composed of multiple subunits of molecules called amino acids. An amino acid (Figure 2.3A) is so-called because it possesses both an acid carboxyl (–COOH) group and an amino ($-NH_2$) group at either end of the molecule, though in a water solvent most really exist as a so-called *zwitterion* in which the amino group has a positive charge ($-NH_3^+$) associated with the nitrogen atom, and the carboxyl group has a negative charge ($-COO^-$) largely localized to the oxygen atom associated with a hydroxyl (–OH) group which loses a hydrogen atom. In nature, there are 20 different amino acids which can be manufactured by living cells with an average molecular weight of $\sim 136\,g$ (though note, in human cells eight of these are not

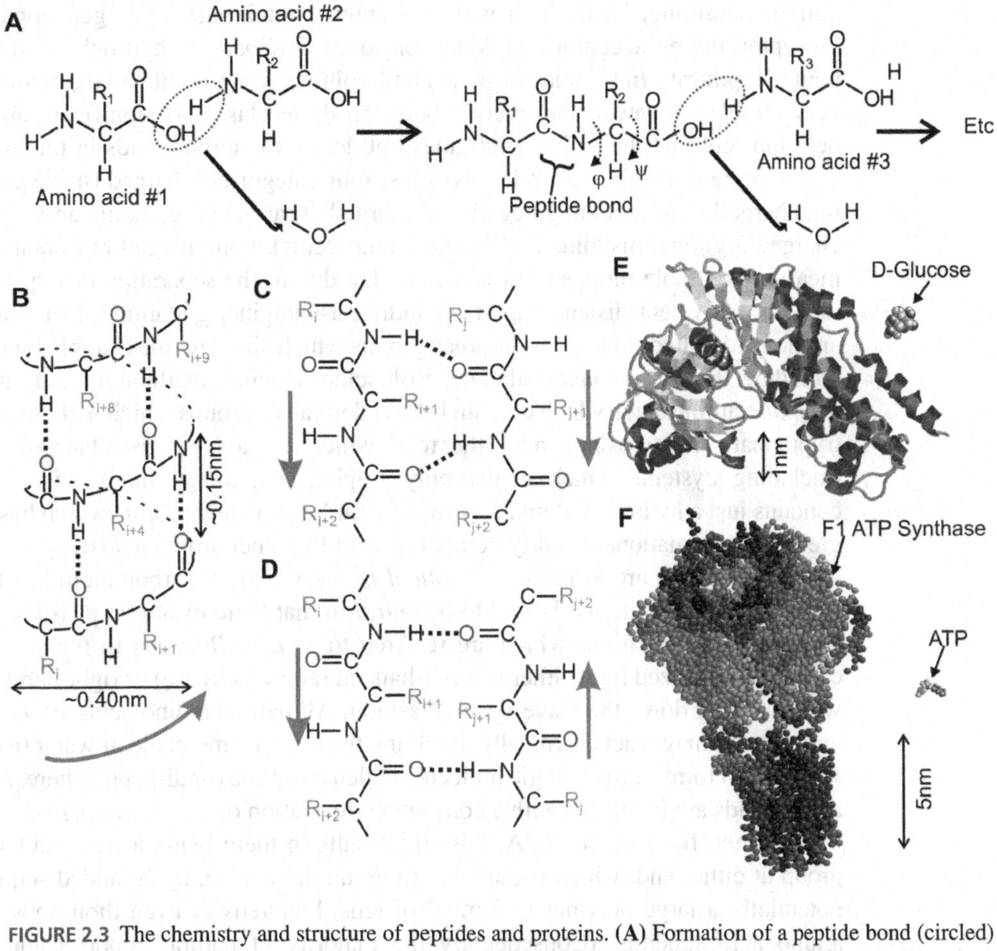

FIGURE 2.3 The chemistry and structure of peptides and proteins. **(A)** Formation of a peptide bond (circled) via loss of a molecule of water, substituent groups labelled 'R'. Dihedral angles φ and ψ are indicated (dotted arrows) relative to the axis of the peptide bond. **(B)** α helix secondary protein structure is stabilized by hydrogen bonds (dotted lines) between electron-dense oxygen O and electron-sparse hydrogen H bound to a nitrogen atom N, substituent groups labelled R. Each helix turn is associated with 3.6 amino acids on average; in principle helices can be left or right handed (as with screw threads), but in nature less than 1% of helix-containing proteins contain left-handed helices because of the comparative energetic instability when formed from L-amino acids. β sheet secondary structures are also stabilized by hydrogen bonding but result in an in-plane 'sheet' orientation, either with **(C)** parallel or **(D)** anti-parallel bonding depending on the relative direction of the protein chain (arrow). **(E)** Example of a real tertiary protein structure of the enzyme hexokinase (PDB ID:1BDG), with α helices, β sheet and random-coil shown in ribbon format, with its substrate glucose illustrated alongside in space-filling format with individual atoms. **(F)** Example of a complex quaternary structure of the enzyme F1 ATP sythase (PDB ID:1QO1, an example of a natural rotary molecular machine) with different protein sub-groups shaded differently, shown alongside substrate ATP.

manufactured at the level of the genetic code and so are described as *essential*), differing only by the substituent group on the central 'α' carbon atom.

The ways in which these amino acids can be categorized into different apparent classes illustrates both the best and worst of the language of biology: up to *eleven different classes* have been formally recognized! At one level this seems sensible, in that these classes include measurable physicochemical parameters: charge (positive, negative or uncharged), chemistry (whether the substituent groups are acidic, aliphatic, aromatic, amide-containing, basic, hydroxyl or sulphur-containing), hydrogen bonding potential (acceptor, donor, acceptor and donor, or none), *hydropathy* (hydrophilic, hydrophobic or neutral), polarity (non-polar or polar) and volume (very small, small, medium, large and very large!). However, the overlap between these classes is significant and one cannot help but feel that the cell as such does not define the amino acids in this way.

A more tractable system involves just four categories: charged (those groups containing charged ions, including negatively charged aspartic and glutamic acids, and positively charged arginine, histidine and lysine amino acids), non-charged but polar (possessing a measurable dipole moment but less than that due to the separation of oppositely charged ions an equivalent distance apart, including asparagine, glutamine, serine and threonine amino acids), hydrophobic (non-polar groups which show no measurable binding to water, including the *non-aromatic* alanine, isoleucine, leucine, methionine and valine, and the *aromatic* amino acids which contain benzyl derivative groups which include phenylalanine, tryptophan and tyrosine), and 'the rest' which are a few miscellaneous special cases (including cysteine which is the only sulphur-containing amino acid, glycine which contains just a hydrogen atom as its substituent 'group', and proline which has significantly greater conformational rigidity compared with the other amino acids).

Amino acids are examples of *optical isomers* – the α carbon atom has four different groups attached to it, and is said to be *chiral* in that there exist two possible mirror-image structural conformations which are referred to as L or R owing to their ability to rotate circularly polarized light either in a left-handed (anti-clockwise) or right-handed (clockwise when looking down the wave-vector) fashion. All natural amino acids are L-isomers. Two amino acids may react chemically involving the loss of a molecule of water (a *condensation* reaction) to form a larger single molecule called a peptide (or di-peptide here, since only two amino acids are involved) with a consequent formation of a covalent *peptide bond*.

As seen from Figure 2.3A, this still results in there being a free –COOH and $-NH_2$ group at either end, which means that more amino acids may be added sequentially until potentially a large polymer is formed of tens, hundreds or even thousands of individual amino acid subunits. (Consequently the majority of natural peptides and proteins are 'linear' in sequence at least in having both an –N and a –C terminus at opposite ends, but note that there do exist rare cases of 'circular' proteins in which the –N and –C termini themselves react with each other via a condensation reaction. These include the well-known 'cyclic peptide' of cyclosporin initially isolated from the fungus *Tolypocladium inflatum* and now used biomedically for its immunosuppressant action in human organ and tissue transplants, as well as some larger cyclic proteins which may have evolved to take advantage of a reduced susceptibility towards enzymatic degradation in the cell by virtue of being circular, see Traibi and Craik, 2002.) Such a large amino acid *polypeptide* polymer which contains greater than at least a few tens of amino acid subunits is called a protein. The conventional definition of when a polypeptide is better termed a 'protein' is a little blurred, but many scientists treat 50 subunits as a sensible cut-off, so for example the first protein to have its chemical amino acid sequence determined was insulin with 51 amino acid subunits (Sanger and Tuppy, 1951a, 1951b; Sanger and Thompson, 1953a, 1953b), and the largest known proteins are expressed in muscle tissue and are different forms of a molecule called *titin* (the largest of which contains ~35,000 amino acid subunits).

The structure of a protein may be characterized to some extent by the relative spatial orientation of each peptide bond in the protein backbone, represented by three so-called *dihedral* or *torsion* angles. These are commonly referred to as phi, psi and omega (φ, ψ and ω). The ω angle is at the peptide bond itself and is generally taken as being 180° because the peptide bond has a *partial* double bond character which results in planarity.

The other two angles φ and ψ correspond to the relative angles of the next two bonds in the backbone of the protein following the peptide bond itself, in the direction of the –COOH group (see Figure 2.3A). This has proved of significant use if the φ and ψ relation is plotted for all peptide bonds in a given polypeptide/protein, resulting in a *Ramachandran diagram* which will generally show several identifiable clusters which relate to specific secondary structures, indicating typically ~3 stable angles for each of φ and ψ (see later Q5.4).

The length scale for individual amino acid subunits is characterized by the typical separation of neighbouring α-carbon atoms at roughly 0.36 nm. However, proteins in general are not linear structures, rather the amino acid chain folds in on itself to form intra-molecular bonds which result in three-dimensional structures. Biologists refer to four levels of *structural hierarchy* for proteins.

(i) *Primary* (the underlying chemical sequence of the individual amino acid subunits).
(ii) *Secondary* (local repeating structural motifs stabilized by hydrogen bonding between different amino acids in the same chain typically involving tens of individual amino acid *residues* or subunits).
(iii) *Tertiary* (the overall three-dimensional shape of the single protein formed by folding of the secondary structure which can be stabilized by a combination of hydrogen bonding, *salt bridge* links made between different parts of the protein via ions, disulphide covalent bonds made via the reduction of two –SH groups from two cysteine amino acids to form a strong S–S link, as well as hydrophobic forces in the core of the folded protein).
(iv) *Quaternary* (the final structure adopted by multiple individual proteins, either of the same or different type, in forming a functional multi-protein complex).

A good example of these four structural hierarchies is seen in an important class of molecule called an *antibody*. An antibody, also known as an *immunoglobulin*, is a large protein complex which is produced naturally in animal cells as part of the *immune response* and contains binding regions of high specificity to so-called *antigen* ligands. Its biological function is primarily to bind to foreign micro-organisms during an infection of the body, with the associated antigens often being located on the outer membrane surface of these invading cells. The antibody–antigen specificity is normally characterized as the equivalent equilibrium dissociation constant K_d of the antigen–antibody complex which, depending upon the antibody, varies over the range 10^{-7}–10^{-11} M (note, here M is the SI concentration unit of moles per litre, and the smaller value of K_d the higher the specificity).

Each antibody has a primary structure embodied in its peptide sequence, which is typically arranged in several *barrel-like* modular structural motifs organized into either *light* or *heavy chains* that are stabilized mainly by β sheet secondary structure interactions. Each antibody has a variable sequence region called *Fab* which acts as the binding site to the antigen. The strength of the binding to a given antigen depends critically on the atomic-level topographical shape of the Fab region, which embodies the tertiary level structure. Finally, multiple antibodies may often aggregate together (for example, *IgM* type antibodies), thus embodying a quaternary level structure (Figure 2.4).

Note, antibodies also have a very useful role in *bio-conjugation* chemistry for single-molecule biophysics research, for example in generating highly specific bonds for a fluorescent tag to a biological molecule of interest (see Chapter 3). Typically, a *primary antibody* binds to the molecule of interest via an Fab binding site, whereas a *secondary antibody* binds to a conserved *Fc* region at the far end of the primary antibody. The most common type of immunoglobulin is IgG which has a characteristic Y shape with two Fab sites at one end of the molecule joined via a hinge region to the Fc stem.

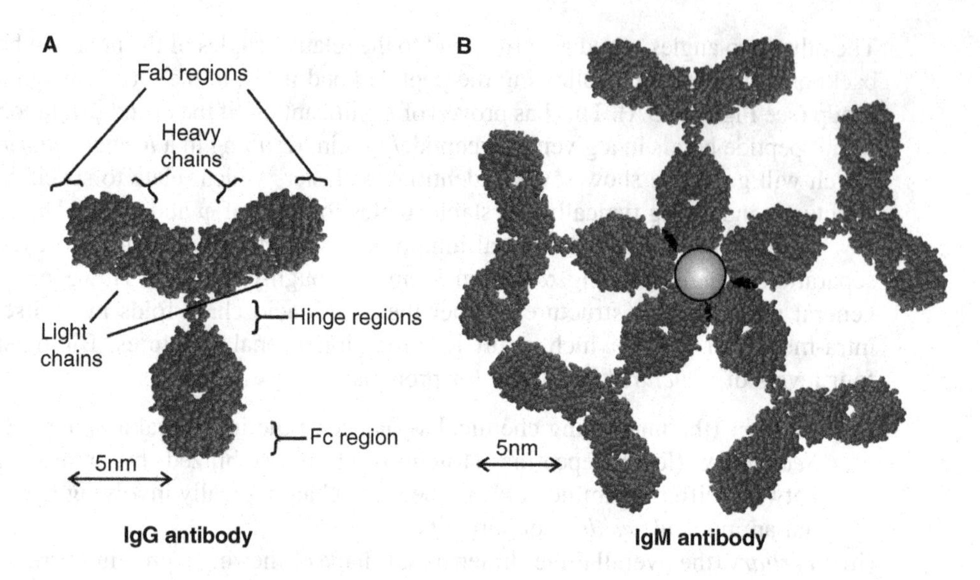

FIGURE 2.4 Structure of antibodies based on PDB ID:2IG2 for (**A**) the IgG antibody, and (**B**) IgM (which contains 5 IgG subunits bound together via the Fc region, thus possessing in effect 10 Fab antigen binding regions).

Proteins are arguably the most important class of biological molecule, the prime reason being that they include a large sub-class of molecules called *enzymes*. The majority of chemical reactions in a cell either would not occur spontaneously over a realistic cell lifetime, or would occur very slowly without additional assistance. This assistance is provided by a variety of enzymes which are catered to catalyse many different reactions, and thus are absolutely fundamental to the viability of any living cell. A significant subset of enzymes is the family of *motor proteins* – protein molecules that in essence undergo energized and directed motion along a variety of different molecular tracks, and are vital to many fundamental biological processes ranging from muscular contraction, cell motility and division, molecular cargo transport inside cells and even *DNA replication* and *gene expression* (see sections 2.6 and 2.9).

2.5 Sugars

Carbohydrates, the technical name for sugars, make up around 15% of the mass of organic molecules in the cell. The name derives from their basic chemical formula of $C_i(H_2O)_j$ where i and j are integers – in other words, these molecules are *hydrates* of carbon. Many biochemists still refer to carbohydrates as *saccharides*, and as is the feature of the language of biology these are often then subdivided into different categories, the most useful being based on the number of equivalent saccharide subunits.

Monosaccharides can exist as an equilibrium mixture of both a straight-chain and a ring-like conformation (illustrated in Figure 2.5A, B for glucose), but in water solution, as inside the cell, at least 99% exists as the ring form. This category can be further divided into classes relating to the number of carbon atoms present (for example those containing the lowest number three of carbon atoms are called *trioses*, four carbon atom monosaccharides are *tetroses*, five carbon atom monosaccharides such as *ribose* and *fructose* are *pentoses*, those containing six carbon atoms such as *glucose* are *hexoses* etc.). Biochemists prefer also to classify according to whether the presence of the residing *carbonyl* C=O group in the straight-chain form of the molecule is an *aldehyde* (meaning the

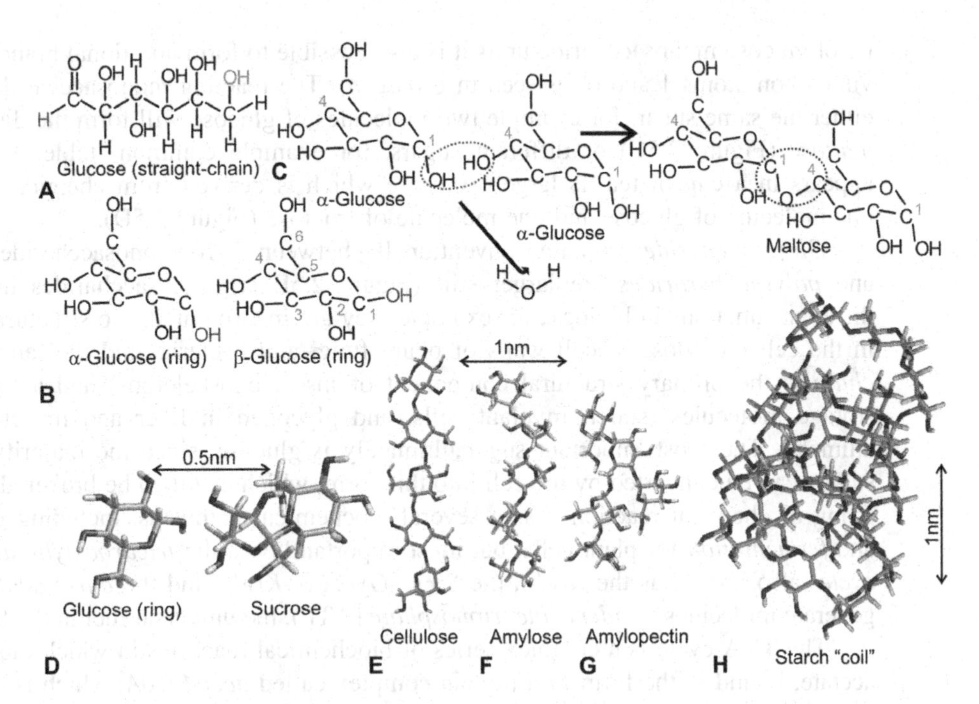

FIGURE 2.5 Common sugars in biology. **(A)** The straight-chain chemical structure of glucose; the hydrogen atom of the carbonyl group, chiral carbon atom and attached hydroxyl group are indicated. **(B)** Ring forms of glucose (carbon-bound hydrogen atoms not shown for clarity), illustrating the α and β conformations dependent on the position of the hydroxyl on the carbon 1 atom, carbon atoms numbered conventionally from the carbonyl atom of the straight chain form. **(C)** Two monosaccharides can react at the hydroxyl groups shown to form a disaccharide involving the formation of a glycosidic bond. **(D)** Stick depiction of D-glucose illustrating its physical length scale, and the disaccharide sucrose composed of monosaccharide subunits of D-glucose and the pentose D-fructose. Common polysaccharides of glucose are shown in **(D–G)** illustrating **(D)** cellulose, a common structural component of plant and many bacterial cell walls, with **(E–G)** starch, an energy storage molecule in plant cells which is composed of a mix of **(E)** amylose and **(F)** amylopectin (structurally similar to glycogen found in animal muscle cells). Amylose often forms **(H)** extended coiled starch structures.

attached substituent group to the carbonyl is hydrogen, see Figure 2.5A), in which case the resulting sugar is an *aldose*, or a *ketone* (the attached substituent group is not hydrogen), in which case the resulting sugar is a *ketose*. There is yet a further classification (historically due to the ability of sugars to rotate the plane of polarized light), which is due to the fact that in the straight-chain form of the molecule there exists a chiral carbon atom furthest away from the carbonyl group (see Figure 2.5A). If the molecule is oriented as shown, then the hydroxyl group attached to this chiral carbon atom above the carbon chain indicates a D form, whereas a hydroxyl group attached below it indicates an L form.

Disaccharides are chemically derived from the bonding of two monosaccharide molecules via two available hydroxyl –OH groups; these groups can be in a downwards-pointing conformation as shown in Figure 2.5B (the 'α' form) or may be upwards-pointing (the 'β' form). Loss of water during the chemical condensation reaction results in the formation of a *glycosidic bond*, with the disaccharide product retaining two free hydroxyl groups at either end allowing further subunits to be added to build larger sugars (Figure 2.5B). As can be seen from the example given in Figure 2.5C, the glycosidic link between glucose monosaccharide units may form between the labelled carbon atoms 1 and 4 (as seen in *cellulose* and *starches*), but in other larger sugars built up of glucose monosaccharide units it is also possible to form additional branching links via carbon atoms 1 and 6 (as seen in *glycogen*). The reacting monosaccharides may be either the same sugar, for example two molecules of glucose will form the disaccharide *maltose* (Figure 2.5C) or different sugars, for example common 'table sugar', as it appears in a cup of tea, is largely *sucrose* which is derived from chemically binding one molecule of glucose and one molecule of fructose (Figure 2.5D).

Oligosaccharides contain conventionally between 2–10 monosaccharide subunits, and *polysacchararides* are larger still (Figure 2.5E–H). Polysaccharides have many different functions in biology, for example they are incorporated into structural features in the cell (cellulose in cell walls of plant, fungi and bacteria, and the 'amino-sugar' *chitin* is the primary structural component of insect exoskeletons) and act as energy storage molecules (starch in plant cells, and glycogen in liver and muscle cells of animals). The most important sugar ultimately is glucose, since the majority of other sugars can be converted by the cell into this form, which can then be broken down via a string of chemical reactions using several biochemical pathways, including *glycolysis*, and *fermentation* for plant cells, but most importantly via the *tricarboxlylic acid (TCA) cycle* (also known as the *Krebs*, the *Szent–Györgyi–Krebs* and the *citric acid cycle*) to generate molecules of *adenosine triphosphate* (ATP), the universal fuel in the living cell.

The TCA cycle is a complex series of biochemical reactions in which the chemical acetate, bound in the form of a protein complex called *acetyl-CoA*, which is one of the intermediate breakdown products of sugar, protein and fats, is converted in nine sequential steps into different organic acids that are all characterized as having three –COOH groups (Figure 2.6). Some of these reactions are coupled to a chemical reduction process for the compound NAD^+ (a *nucleoside* derivative called *nicotinamide adenine dinucleotide*, see section 2.6). The equivalent reduction reaction is simply:

$$NAD^+ + 2[H] \rightarrow NADH + H^+ \quad (2.1)$$

Here, the atomic hydrogen [H] is generated from the oxidation of one of the associated acids in the TCA cycle. This *NADH* is then subsequently used in *oxidative phosphorylation* (OXPHOS) to generate molecules of ATP (this is discussed later in greater detail in section 8.4).

2.6 Nucleic acids

Around 15% by mass of all cellular organic molecules are nucleic acids, however as a proportion of numbers of molecules this equates to less than 0.01%. This illustrates that nucleic acid molecules are generally very massive. The most relevant to the living cell are *ribonucleic acid* (RNA) and *deoxyribonucleic acid* (DNA). A nucleic acid is a biological polymer composed of subunits called *nucleotide bases* (hence the term *polynucleotide* can be used synonymously), but the range in number of subunits is very large across different molecules; at the small end are short fragments of RNA called *small interfering RNA* (siRNA) which have only 21 subunits, whereas at the high end are truly massive molecules of DNA associated with entire chromosomes, for example the largest chromosome in the human cell is chromosome 1 which contains almost 250 million subunits (in the case of DNA in particular these subunits are paired, and so the term *basepair* is generally used instead of subunit).

A nucleotide consists of three chemical components: a nitrogenous *nucleobase*, a phosphate group and a pentose sugar. The pentose sugar can be either ribose, in the case

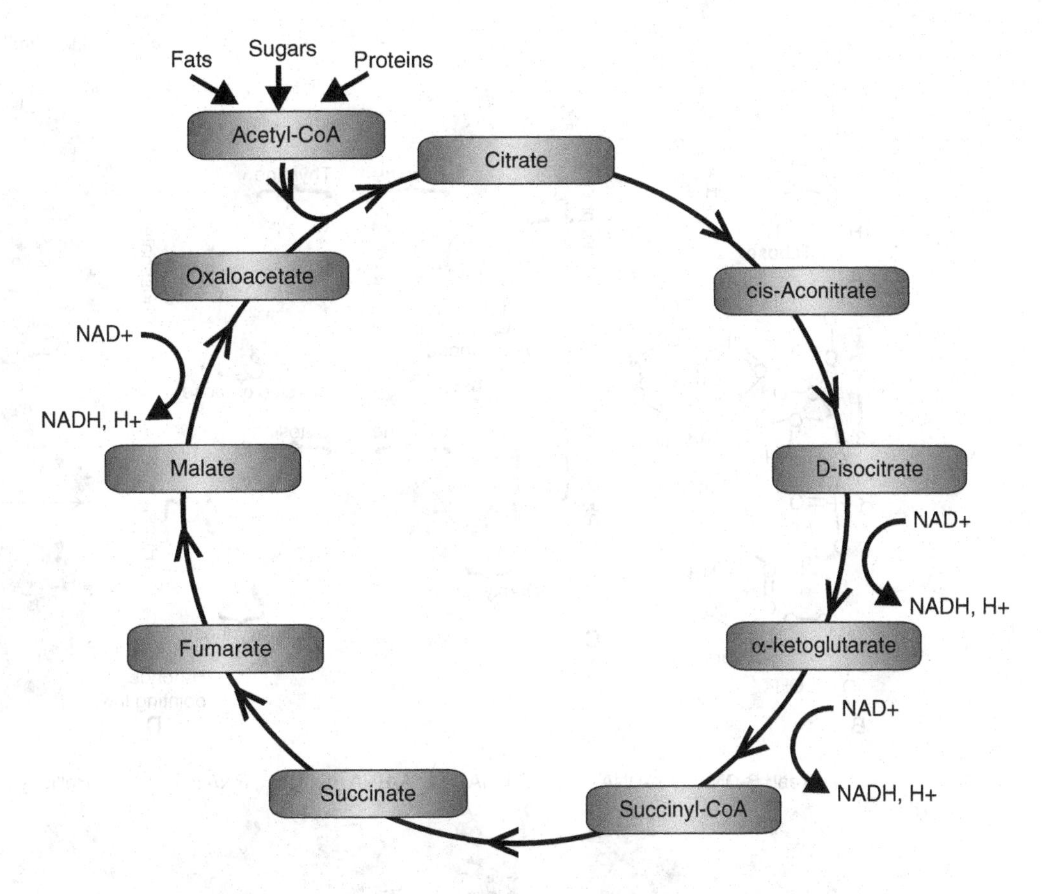

FIGURE 2.6 Simplified version of the TCA cycle. The diagram indicates the entry point of acetyl-CoA as a breakdown product of the degradation of fats, sugars and proteins, as well as the main reactions which are coupled to generating NADH which is used in OXPHOS to generate ATP.

of RNA, or *deoxyribose* (identical to ribose apart from the absence of an oxygen atom on the carbon 2 atom, see Figure 2.7A) for DNA. Nucleotides are classified as either *pyrimidines* (a six-member ring of carbon and nitrogen) including *cytosine* (C), *thymine* (T) and *uracil* (U), or *purines* (a pyrimdine fused to a five-member carbon–nitrogen ring group called an *imidazole*) including *adenine* (A) and *guanine* (G). A, C and G are common to both RNA and DNA, but RNA contains U as its fourth and final base whereas DNA contains T, though note that some *non-standard* bases exist which are most common in *transfer RNA* (tRNA), discussed below.

The combination of pentose sugar and nucleobase is called a *nucleoside*, and a nucleic acid primary sequence is made up of a string of nucleosides which are linked via negatively charged phosphate groups between the 3 and 5 carbon atom positions, such that sugar and phosphate groups alternate to form the nucleic acid *backbone* (Figure 2.7B).

In most natural cases, DNA consists of two backbone strands whereas RNA consists of just one. The double-stranded nature of DNA is due to hydrogen bonding between certain pairs of bases (A with T forming two hydrogen bonds, and C with G forming three hydrogen bonds, Figure 2.7C). This so-called *Watson–Crick base-pairing* results in a regular, uniform structure which is energetically most stable as a double-helix conformation (see Watson and Crick, 1953 for the original description), such that the hydrogen bonds point in towards the axis of the helix and the phosphate groups point outwards (Figure 2.7D), with some

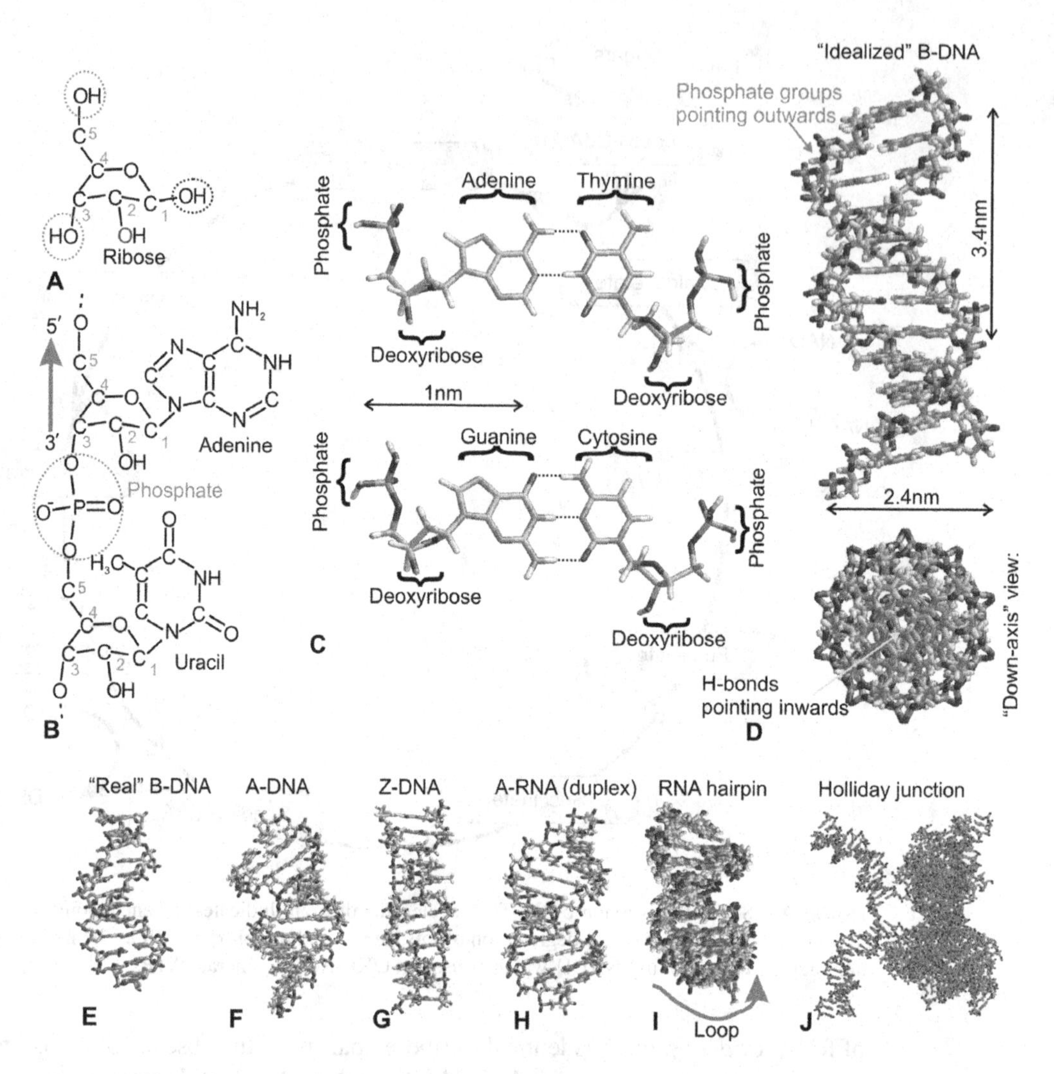

FIGURE 2.7 The molecules of the genetic code. **(A)** Chemical structure of ribose (deoxyribose has an O atom missing), with phosphate and nucleobase binding points indicated. **(B)** A short section of RNA consisting of just 2 bases (here with nucleobases uracil and adenine) which are joined by a negatively charged phosphate backbone; the '3′–5′ directionality' relative to the 3 and 5 carbon atoms of the sugar is shown (arrow). **(C)** Stick models of DNA nucleotides, showing the points of H-bonding between base-pairs (dotted lines). **(D)** 'Idealized' B-DNA, side-on and down-axis. **(E)** 'Real' DNA, and **(F–H)** other less common helical structures found for DNA and RNA (**(D–F)** prepared using RasMol based on coordinates kindly provided by Jason Kahn, University of Maryland). **(I)** The single-stranded RNA hairpin, loop region indicated (arrow) (PDB ID:1ATV). **(J)** Four-way Holliday junction (PDB ID:3CRX).

further *stacking interaction* between bases above and below. Averaged over long sections of natural DNA, the numbers of each nucleotide base are roughly the same.

The most common double-helix DNA structure in nature is *B-DNA*, which in a uniform 'idealized' form (for example, consisting of repeats of just the same base in one given strand, with the Watson–Crick base-pair on the other strand) is a *right-handed* helix (i.e. if you put the thumb of your right open hand parallel to the helix axis pointing in the 3′–5′ direction of the backbone then the bases will coil in the same direction as the motion of your fingers closing your hand). It has a diameter of ~2.4 nm and pitch of ~3.4 nm, depending upon solvent conditions, with one complete turn of the helix using

10 consecutive bases. In practice, most DNA is non-ideal, in that there are not long sequences of the same bases on each strand, which results in deviations from the ideal B-DNA structure (Figure 2.7E), manifest as localized heterogeneity along the helix such as localized variations in diameter and pitch.

Also, *mismatch* base-pairs are possible (i.e. incorrect bases appear opposite each other in the double-helix, stabilized only weakly by one hydrogen bond instead of two or three) though unlikely (measured errors are typically 1 in 10^5 base-pairs from test tube experiments, though in the cell there are additional *error-checking* enzymes which reduce this much further to more like 1 in 10^9 base-pairs in human cells for example).

In addition to the two helices generated by the sequence of base-pairs there are also two helices in effect generated by the *absence* of base-pairs (i.e. the *grooves* between the strands). In B-DNA there is a *major groove* of width ~2.2 nm, and a *minor groove* of width ~1.2 nm. These grooves, especially the more accessible major grooves, are important features in providing accessible sites for proteins which bind to DNA and regulate its function.

Other less common double-helical structures of DNA exist (Figure 2.7F, G) including *A-DNA* (a wider, shallower right-handed double-helix, only thus far observed under relatively non-physiological dehydrated conditions) and *Z-DNA* (a tighter, steeper left-handed double-helix which may be advantageous for providing torsional strain relief to the DNA compared to the B-DNA form during periods when the DNA code is converted into RNA during the manufacture of proteins, discussed below). In addition, there is a double helical form of RNA which exists, A-RNA, also known as the *RNA duplex* (Figure 2.7G), but this is only stable in short sections. A more common RNA secondary structure is due to looping of the same single-stand of RNA back upon itself to form a *hairpin* type structure (Figure 2.7I). RNA also forms other non-helical specialized structures concerned primarily with the manufacture of proteins, which is discussed later in this chapter.

Although a DNA molecule in the cell will typically assume the B-form structure when relaxed, it can undergo phase transitions to other more exotic structures when external forces, for example provided by protein-based molecular motors bound to the DNA, cause it to stretch and twist. A common stable intermediate structure of this type is a *Holliday junction* (Figure 2.7J). This meeting of four individual strands associated with two double helices is actually mobile and can migrate relative to the relaxed DNA structure axis and is biologically important in the transfer of genetic material between two different DNA double helices, for example during so-called crossing-over as occurs in meiosis during which germ cells are made, or in cases of repair of breaks in DNA.

BIO-EXTRA

Eukaryotic cells divide either via **mitosis**, which preserves the number of chromosomes, or via **meiosis**, which halves the number of chromosomes to produce **germinal cells**, such as spermatozoa or ova. During meiosis, paired chromosomes in the cell derived ultimately from the father and the mother line up in the mid-plane of the cell and it is at this stage that exchange of genetic material from either chromosome source can occur, via a process called **crossing-over**, with the resulting **heteroduplex** DNA structure called a Holliday junction. This general DNA exchange process is called **homologous recombination**, and is also utilized in simpler prokaryotic cells.

As previously mentioned, the DNA in the cell is generally tightly packed for most of its existence into condensed regions, either in the nucleoid region of prokaryotic cells or in the nucleus of eukaryotes. In simple prokaryotes such as *Escherichia coli* (a type of bacterium found in the bowels of animals) this DNA is often in the form of a single complete, closed circle which coils in on itself during compaction, but may also contain several smaller circles of DNA called *plasmids* (which also feature in chloroplasts and mitochondria separate from the cell). In more complex prokaryotes there may be more than one of the large circles of DNA (with some complex prokaryotes possessing linear as opposed to circular sections of DNA), and some biologists refer to these as prokaryotic chromosomes. Such predominantly circular chromosomes can have a range in the number of base-pairs present from 160,000 up to over 12 million.

In the more advanced eukaryotic cells, DNA is packaged around bead-like protein nanostructures called nucleosomes, which are then packed into fibres, and coiled again with the assistance of a variety of DNA condensing proteins (biologists refer to this complex of DNA and protein as *chromatin* fibres which, prior to further coiling of the fibre itself, are ~30 nm wide) into linear chromosome structures. For example, a normal human cell generally contains 23 pairs of chromosomes, with each member of the pair derived genetically from the father and mother. The minimal number of chromosomes per cell in any species is two (found in *nematode* worms), whereas the highest is over 1000 (found in a species of *fern*). Eukaryotic cell chromosomes are generally localized in the cell nucleus (unless the cell is in a state of division in which case the nuclear envelope is no longer present). They contain larger numbers of base-pairs than circular prokaryotic chromosomes, for example in human cells the range is from 57 to 247 million.

Nucleic acids have two principal functions in the cell. Firstly, the genetic code is written into the DNA of every cell, but also in the separate DNA of certain cell organelles such as chloroplasts (in plant cells) and mitochondria. This DNA can undergo replication during cell/organelle division by a process which involves unzipping the Watson–Crick base-pairs to form two single-strand templates, and then reforming Watson–Crick base-pairs on each separate strand to generate two almost identical new double helices, which are then each incorporated into the two daughter cells or organelles. This process of replication is somewhat more involved than this short description suggests, but has been explored using single-molecule methods, discussed later in greater depth (see Chapter 9).

The second primary function of nucleic acids is their involvement in *transcription* and *translation*, which are the two processes by which the DNA code is converted into actual proteins. This is discussed in more detail below in section 2.9 outlining the *central dogma of molecular biology*.

2.7 Lipids

Around 90% of all organic molecules in the cell are lipids, though by mass this is more like 10% indicating their relatively low molecular weight compared to the macromolecules such as proteins, nucleic acids and polysaccharides. In its most basic form, that of the so-called *glycerolipid*, a lipid is composed of two components: *fatty acids* (between one and three constituent molecules per lipid molecule), and a *glycerol* molecule (Figure 2.8A), the resulting mono-, di- or triglyceride product resulting from a condensation reaction between the fatty acids and the glycerol. Fatty acids are long chain hydrocarbons containing 4–24 carbon atoms, terminating in an acid carboxyl group; they can be classified as *saturated* (containing only single C–C bonds) or *unsaturated* (containing double or triple bonds), and may also possess additional functional groups containing different atoms such as sulphur and nitrogen.

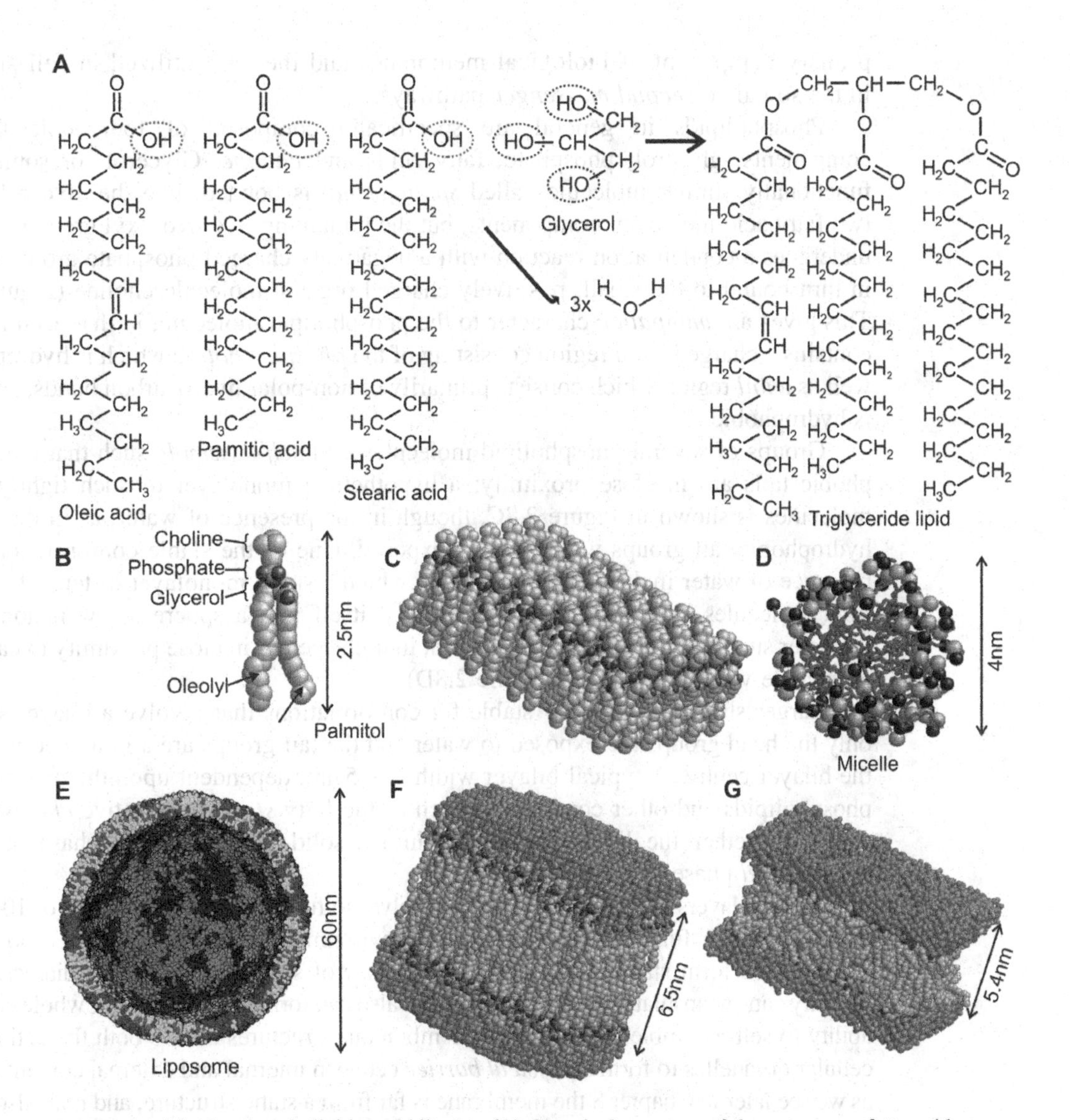

FIGURE 2.8 Lipids, phospholipids and bilayers. (A) Chemical structure of three common fatty acids, reacting with glycerol, reacting hydroxyl groups shown. (B) Space-filling model of the common phospholipid phosphotidyl choline. (C) Two-dimensional tight-packed monolayer array of several phosphotidyl choline molecules. (D) A micelle composed of 65 molecules of phosphotidyl choline, for clarity only carbon, nitrogen and phosphorus are shown (coordinates from Tieleman *et al.*, 2000). (E) Liposome comprising di-C16-PC, di-C18:2-PC and cholesterol, based on coarse-grained MD simulation by Daniel Patterson, Oxford University, rendered using VMD indicating microdomains of saturated lipid, unsaturated lipid and cholesterol. (F) Two-dimensional tight-packed bilayer array of several phosphotidyl choline molecules in a static 'crystalline' phase with surrounding water molecules, bilayer width indicated, and (G) similar arrangement but in a dynamic 'gel' phase (coordinates from Heller *et al.*, 1993).

A multiple carbon–carbon bond in an unsaturated fatty acid is physically more constrained than the single carbon–carbon bonds and is thus manifest as a kink in the chain of carbon atoms, shown in Figure 2.8B. This is an example of a so-called *phospholipid*, which resembles a glycerolipid but with some additional features. While glycerolipids have biological functions limited to fat energy storage and thermal insulation in animal tissues, phospholipids have more extensive applications as the primary component of biological membranes and they are utilized in cell signalling in the so-called *second messenger* pathways.

Phospholipids in general are specifically composed of four molecular subcomponents: glycerol, phosphate, fatty acids and *choline*. Glycerol (or sometimes a functionally similar molecule called *sphingosine*) is bonded, like the glycerolipids, to two fatty acid molecular components, but the remaining free hydroxyl group of glycerol undergoes a condensation reaction with a negatively charged phosphate group, which is in turn bound to the small, positively charged organic molecule choline (Figure 2.8B). This gives an *ampipathic* character to the phospholipid molecule, in that each molecule contains a charged *head* region consisting of the *phosphocholine* which is hydrophilic, as well as a *tail* region which consists primarily of non-polar hydrocarbon bonds, and which is hydrophobic.

Groups of several phospholipid molecules will *self-assemble* such that their hydrophobic tails are in close proximity; a hypothetical monolayer of such tightly packed molecules is shown in Figure 2.8C, though in the presence of water, as in the cell, the hydrophobic tail groups would not be exposed. One of the stable conformations in the presence of water includes the *micelles* in which a small monolayer of tens of phospholipid molecules is in effect wrapped around itself into a sphere a few nanometres in diameter such that the tail groups all point in to the centre in close proximity to each other to exclude water completely (Figure 2.8D).

Larger structures are only stable for conformations that involve a bilayer such that only the head groups are exposed to water and the tail groups are all directed inwards to the bilayer centre. A typical bilayer width is ~5 nm, dependent upon the precise mix of phospholipids and other constituents such as the fatty *steroid* derivative *cholesterol* but also on whether the lipids are packed into a solid-like *crystalline* phase or a more dynamic *gel* phase (Figure 2.8E–G).

These bilayers can also fold in on themselves to form spherical *liposomes* of 10–100 nm diameter, which fully enclose an aqueous environment. More importantly, phospholipid bilayers also form stable larger scale *membranes* of several micrometres diameter which not only can encapsulate whole organelles but also can form the boundary of whole cells. This ability to self-assemble into enclosed membranous structures allows both the cell and subcellular organelles to form a *physical barrier* between internal and external contents, though as we see later in Chapter 8 the membrane is far from a static structure, and may also contain significant heterogeneity in terms of *microdomain* substructure.

Other less common lipids involve sugars reacting with one or two of the available hydroxyl groups of glycerol, to form *glycolipids* which are generally expressed on the outer surface of cell membranes and are utilized in cell-type recognition (for example, by cells involved in the immune response in higher organisms).

Lipid bilayers make highly effective barriers to charged molecules, especially to small ions. The free energy ΔG required to take such an ion of charge e and effective radius a from water-based solution, either from the inside of the cell or from outside the cell, into the centre of a lipid bilayer, can be calculated using:

$$\Delta G = \frac{e^2}{8\pi\varepsilon_{\text{free-space}}a}\left(\frac{1}{\varepsilon_{\text{lipid}}} - \frac{1}{\varepsilon_{\text{water}}}\right) \tag{2.2}$$

Here, the ε values are electrical *permittivity* constants which measure the ability of the surrounding medium to store charge. Lipid has a typical permittivity ~40 times smaller than that of water which results in a ΔG value of ~$65k_BT$ (see Chapter 1 for an explanation of energy units). Simple thermodynamics predicts that the probability for making such a transition is the associated *Boltzmann factor* $\exp(-\Delta G/k_B T)$, which here equates to $\sim 10^{-28}$. Thus, the chance of spontaneous partition of an ion into a cell membrane is exceptionally small.

PHYSICS-EXTRA

There is in addition a **screening effect** of the charge of an ion in water by orientation of the polar water molecules, which are electric dipoles. A more accurate calculation for the ΔG above would account for the microscopic structure of the water molecules close to the ion; most of the energy associated with screening can be attributed to this first **hydration shell.**

Lipid bilayer membranes form an essential component of the bio-energetics system in cells because in effect they allow energy derived chemically from a chemical breakdown of food (for example in glycoloysis, section 2.5), or from light (as in photosynthesis in plants) to be stored as electrical potential energy in much the same way as a charged capacitor does in an electrical circuit. Cells can convert energy from light and nutrients into immediately usable energy stored in the form of ATP using a mechanism localized to lipid bilayer membranes. This energy transduction is called *chemiosmotic coupling* as it links chemical ATP synthesis to the transport of protons across a lipid bilayer. The energy is then stored as *electrical capacitance charge*.

For prokaryotic cells such as bacteria this *chemiosmosis* takes place at the cell membrane itself, but for more complex eukaryotic cells ATP generation occurs at the lipid bilayer membranes encapsulating either mitochondria or chloroplast organelles (see section 2.3). In mitochondria, free energy is transferred in a series of coupled chemical reactions largely via a series of *electron transport* processes, resulting in a chain of electron-transporting protein complexes being sequentially reduced then oxidized as electrons are transferred to and then lost from each complex in the chain. In chloroplasts the free energy input is from sunlight absorbed by molecules of chlorophyll, and the electrons are taken *from* water to generate oxygen and are used to synthesize carbohydrate from carbon dioxide. This all occurs in association with a system of lipid bilayers called *thylakoid* membranes inside chloroplasts and is very similar to the invaginated *cristae* membranes found inside mitochondria. The free energy derived from the sunlight is ultimately converted into high energy chemical bonds of sugar molecules in the process of photosynthesis, but the complex electron transport chemistry involved at the thylakoid membranes is very similar to the electron transport chemistry associated with OXPHOS in mitochondria (see Chapter 8).

Each electron reduction–oxidation, or *redox*, transfer reaction can be modelled as the sum of two separate *half-reactions* involving the individual reduction and oxidation processes, each of which has an associated *reduction potential*. The reduction potential is a measure of the equivalent electrode voltage potential which would be measured if the chemical half-reaction were electrically coupled to a *standard hydrogen electrode*. In biochemistry textbooks, this reduction potential is often described as being in the *normal standard state*, i.e. the *standard* reduction potential V_0. There is a lot of confusing jargon here, but essentially the term *standard* implies that all components are at concentrations of 1 M. Just to complicate things further, this should not be confused with the *biochemical* standard state electrode potential which is the same as the normal standard state electrode potential apart from the pH being 7. Since the pH is defined as $-\log_{10}[H^+$ concentration] this indicates a concentration of H^+ of only 10^{-7} M for the biochemical standard state.

For example, the reduction half-reaction for the electron-acceptor NAD^+ is:

$$NAD^+ + H^+ + 2e^- \rightarrow NADH \qquad V_0 = -0.315\ V \tag{2.3}$$

An oxidation half-reaction involving the electron-donor *malate*$^-$ is the reverse of the following reduction half-reaction of the electron-acceptor *oxaloacetate*$^-$:

$$\text{oxaloacetate}^- + 2H^+ + 2e^- \rightarrow \text{malate}^- \qquad V_0 = -0.166\ V \tag{2.4}$$

Therefore, we can combine these two half-reactions (in effect Equation 2.3 minus Equation 2.4) to give:

$$NAD^+ + \text{malate}^- \rightarrow NADH + H^+ + \text{oxaloacetate}^- \qquad V_0 = -0.149\ V \tag{2.5}$$

The free energy change per mole associated with a general redox process is given by $-nFV_0$ where F is *Faraday's constant*, with n electrons being transferred in the process. Note, this also allows the molar equilibrium constant K to be calculated as $K = \exp(nFV_0/RT)$where R is the molar gas constant, at absolute temperature T. The negative reduction potential in Equation 2.5 indicates that the free energy change for this reaction is positive, and so it will not occur spontaneously under standard conditions. However, in the cell, concentrations are not 'standard' (i.e. they are not all at a concentration of 1 M). The presence of the enzyme acetyl-CoA used in the TCA cycle depletes the oxaloacetate component and builds up the citrate component, and the resulting concentration difference (which equate to *chemical potentials*) drives each reaction in the cycle in the direction suggested by Equation 2.5 (see Q2.6).

The free energy released as the electrons move to states with ever lower potential energy is ultimately used to *pump* protons across the membrane and, at the end of the chain of electron transfer reactions, electrons are transferred to molecular oxygen which reacts with protons to produce water. This *proton gradient* established across the lipid bilayer membrane, created by the proton pumps, is used to drive the *rotary molecular machine* called *ATP synthase* (Figure 2.3F) which couples the energy released in *discharging* the lipid bilayer capacitor to synthesize molecules of ATP. This rotary molecular machine has itself been the target of several single-molecule investigations, with a seminal study using a stiff *actin* protein filament as a marker for the molecular rotation (Noji *et al.*, 1997) which was tagged to allow it to be visualized using fluorescence microscopy (see Chapter 3).

Newly synthesized ATP is then free to diffuse to different regions of the cell and will couple in highly coordinated ways to a multitude of different biological processes. The trapped chemical energy in each molecule of ATP may be released by an *exothermic* hydrolysis reaction to yield a molecule of the diphosphate ADP and free inorganic phosphate, along with the equivalent of 20–25 k_BT of free energy per ATP molecule, or 80–100 pN nm, which can used to energize a huge variety of different subsequent biochemical reactions which may be parts of completely unrelated cellular subsystems.

But why does ATP contain so much locked chemical energy? Well, firstly there is a *resonance* effect involving partially *delocalized electrons* which are then free to move between the oxygen atoms bridging the three phosphate groups. This results in a *shared* oxygen atom in the terminal phosphate bond which therefore reduces the number of arrangements of these electrons compared to the hydrolysis products

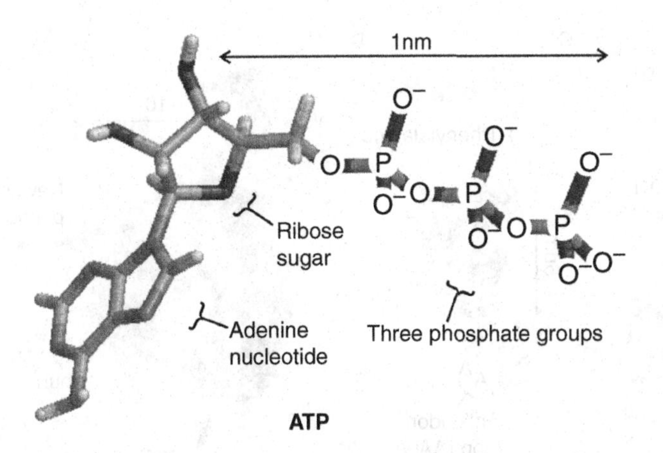

FIGURE 2.9 Stick depiction of the chemical structure of adenosine triphosphate (ATP), positions of charged phosphate groups indicated.

so there is a favourable entropy gain resulting from hydrolysis (Figure 2.9). Secondly, there is *charge repulsion* between adjacent phosphate groups, which is reduced upon hydrolysis. The most important factor, however, is due to the *hydration free energy* – this is more negative for the products (i.e. more hydration energy is *released*), partly due to entropy and partly due to better hydrogen bonding and better *electrostatic charge screening*, again resulting in a significant free energy release upon hydrolysis.

There is actually only ~100 g of ATP in a typical human body, and yet this in principle is the molecular energy currency of all cells. This is possible because ATP undergoes very rapid *turnover* in cells, and is replaced very quickly; typically the lifetime of a single molecule of ATP in a single cell is ~1 min, and in a whole human body ~40 kg of ATP will be used and subsequently replaced in a given 24 hour period.

2.8 Miscellaneous 'small' molecules

The class of 'other' biological molecules makes up around 10% in the cell and involves the organic molecules which do not assemble into large-scale and/or macromolecular structures. These include several *hormones* (a signalling chemical released in small quantities by a cell or group of similar cells in one region of the body in a higher organism which can be detected by cells in another region of the body to initiate some form of cellular response), and many different cell *metabolites* (any chemical intermediate involved in the biochemistry of a functioning cell, many of which will be derived from the other classes of molecules described earlier in this chapter).

Also, there are various *neurotransmitter* molecules. These are chemicals which can transmit a nerve signal across a *synapse* gap between neighbouring nerve cells in animals, the commonest being *acetylcholine*.

Furthermore, other types of molecule include the steroids and the associated alcoholic-steroids called *sterols* such as cholesterol which is used as a stabilizing element in most cell membranes, and a variety of cholesterol-derived hormones such as *oestrogen*.

And finally there are the *vitamins*. These are a varied group of small chemicals which are required for several different cellular processes but cannot be synthesized in significant quantities by the host organism and so need to be ingested.

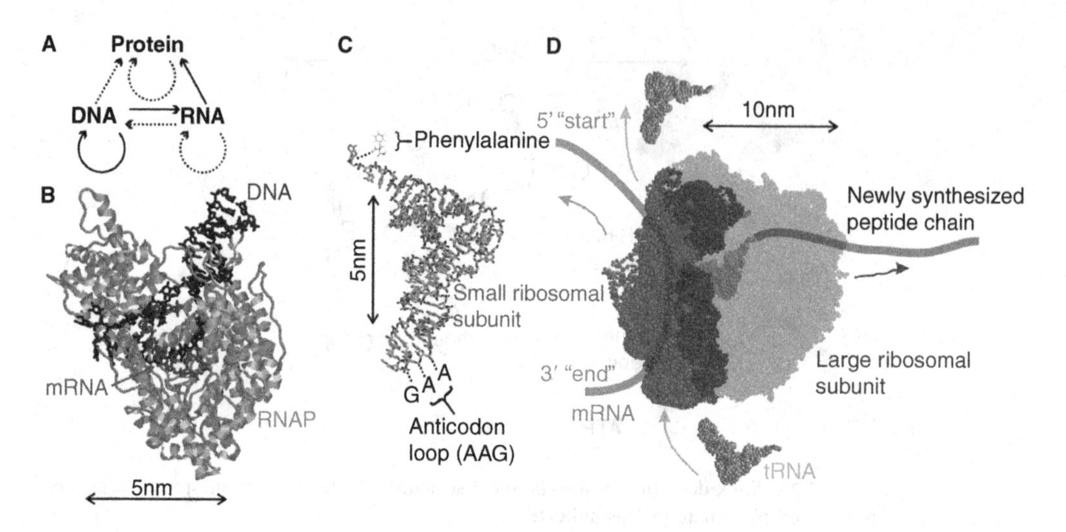

FIGURE 2.10 DNA → RNA → Protein. **(A)** Flow-diagram for 'genetic information'. The 'general' flow routes are indicated in solid lines, with dotted lines 'special' cases (many biologists prefer not to include the protein–protein route, but this is known to exist with prions). **(B)** Molecular structure of the RNA polymerase captured 'mid-translation', protein, mRNA and DNA (based on PDB ID:1MSW). **(C)** Structure of tRNA with amino acid phenylalanine attached (based on PDB ID:4TNA). **(D)** Schematic of mRNA translation at a ribosome (structures based on PDB IDs:1FKA (small subunit) and 1FFK (large subunit)) and tRNA (PDB ID:4TNA), note the rotation of the tRNA at the active translation site.

2.9 The 'central dogma' of molecular biology

The phrase 'central dogma of molecular biology' was coined first in a symposium in 1958 by Francis Crick, later more formally described by him in 1970 with some of the experimental gaps filled in. Crick acknowledged that the use of the term 'dogma' was imprudent since many took this to imply that it was in some way an immutable belief, whereas in fact it is just like any other hypothesis in biological science, but now the experimental evidence in support of it is significantly more substantial than when it was first proposed. Essentially it is a description of the *flow of information* in the cell between the nucleic acids DNA and RNA, and proteins. In its modern form it is primarily used to describe how information written into the DNA of the genetic code is first transformed into information written into RNA (in the form of *messenger RNA* or mRNA) by a process called transcription, and then this mRNA is transformed into polypeptides and protein by a process called translation (Figure 2.10A).

As the figure suggests, there is also a flow of information between DNA to itself (DNA replication) and special cases of direct *RNA replication* which occurs in many viruses. In addition, there is a special case of genetic information flowing back from RNA to DNA; this is a process called *reverse transcription* and occurs in so-called *retroviruses* (these are RNA-containing viruses which convert their RNA to DNA, which is then incorporated into the DNA of the host cell, an infamous example being *HIV*). Direct conversion of DNA into protein again is classed as a special case, since it can be made to occur in the test tube under special conditions using cell-free bacterial extracts which contain so-called *ribosomes*. There is also the rarer possibility of direct transfer of genetic information from protein to protein. An example of this is involved in so-called *prions*, which are proteins that can replicate themselves directly without a nucleic acid intermediary. This occurs in fungi from one generation to the next, but was also recently discovered in humans in the disease *CJD*, the human equivalent of *mad cow's disease*.

So, although many biologists tend to deny its relevance, possibly for historical reasons, it is indeed another special case of the central dogma. The primary flow of genetic information however is *DNA→RNA→protein*.

The central dogma does not make any direct comment about control mechanisms, but simply about the *directionality* of genetic information transfer. That being said, DNA replication, translation and transcription are all clearly very fundamental biological processes and have been investigated using a variety of single-molecule techniques to elucidate the underlying molecular mechanisms, which are described later (see Chapter 9). To summarize here, we now know that the double-helix of DNA is first *unzipped* and converted into single-stranded mRNA by an enzyme belonging to a family called *RNA polymerase*, or RNAP (Figure 2.10B).

Subtle alteration (so-called *post-transcriptional modification*) of the mRNA molecule may then occur, usually in eukaryotic cells as opposed to prokaryotes. This modification takes the form of *splicing* as follows.

(i) Sections of the mRNA are cut out enzymatically.
(ii) The cut ends then re-anneal. This stage is usually performed by a small RNA–protein complex called the *spliceosome* in the nucleus, and may occur with different permutations of the cut sections joining back together to result in several *alternative splicing* forms of the original mRNA molecule.
(iii) *Capping* is then performed which involves the addition of a modified G base to the leading 5′ end of the mRNA.
(iv) *Polyadenylation* then occurs which involves the addition of around 250 A bases to the trailing 3′ end of the mRNA called a *poly(A) tail* which is believed to protect the mRNA from degradation in the cell by enzymes called *exonucleases*.
(v) Sometimes further *editing* is then performed, in which relative minor alterations are made.

This mRNA molecule, modified or otherwise, then localizes to a ribosome. In eukaryotic cells these are located outside the nuclear envelope and so mRNA molecules are first trafficked through nuclear pores to the cytoplasm (see Chapter 9). In prokaryotes there is no nuclear envelope and so it is possible for transcription and translation to be physically linked. Once at the ribosome, the information stored in the mRNA molecule is then read off in consecutive units of three base-pairs called a *triplet codon* by the transient binding of transfer RNA (tRNA). tRNA is often described as having a cloverleaf type three-dimensional structure (Figure 2.10C), with the attachment of a specific amino acid at one of the arms (there are 20 different tRNA molecules for each of the 20 different amino acids, each of which is pre-matched with its correct amino acid using a collection of enzymes called *aminoacyl-tRNA synthetases*), and an *anti-codon* site on the opposite arm of the molecule which binds to the mRNA codon by complementary base-pairing.

The ribosome is a large 20–30 nm diameter complex composed of both proteins and nucleic acid in the form of *ribosomal RNA* (rRNA) and, although differences exist between prokaryotes and eukaryotes, the general structure consists of a large and a small subunit which assemble onto the start region of the mRNA molecule, sandwiched together (Figure 2.10D), with the three-dimensional structure of each subunit stabilized by base-pairing within and between rRNA molecules. The sandwich interface acts as the site of mRNA translation, with molecules of tRNA binding transiently to the active codon site of translation and fusing their attached amino acid residues with the nearest *upstream* amino acid coded by the previous three base-pairs of the mRNA molecule, ultimately permitting a long polypeptide chain to be created. In other words, the DNA has regions of consecutive three base-pairs each of which can code for a different amino acid in the final protein produced from translation of the transcribed mRNA.

Other ribosomes may bind to the same mRNA once the start region has been processed, generating a *polysome* or *polyribosome* cluster which in effect manufactures copies of the same protein molecule from just one molecule of DNA with each individual ribosome outputting protein molecules in parallel with the other ribosomes in the polysome cluster.

In principle since there are four possible bases in mRNA, with three possible positions for each base in each triplet codon, this would imply 4^3 or 64 possible combinations. However, since there are only 20 naturally expressed amino acids, it is unsurprising that most of the triplet sequences are *degenerate* (i.e. there is more than one three base-pair combination, typically between 2–6, that will code for the same amino acid, primarily by variation in the third base-pair), and in fact only two amino acids (methionine, or formylmethione in bacteria, and tryptophan) are uniquely specified by just a single codon. The mRNA sequence for methionine (AUG) is also a special case since it acts as the *start codon* in most organisms, including humans.

This start codon, in combination with several protein *initiation factors*, directs the start of translation for a sequence of mRNA, and so signifies the start of the polypeptide chain produced. Some organisms use a different codon sequence to act as the start codon (either GUG or UUG). Similarly, three base-pair triplets exist as *stop* codons (UAA, UAG and UGA). Stop codons do not code for any amino acids but act to terminate mRNA translation (and are thus also called *nonsense* or *termination* codons). The region between the start and the nearest upstream stop codon is generally referred to as the *open reading frame* or ORF.

An ORF which actually codes for a protein (not all do) is then designated as a *gene*. In other words, sections of a cell's DNA are divided up into coding regions which can ultimately code for a certain protein. The number of genes in a cell depends upon species. As we saw earlier, the smallest set of genes (called the *genome*) is ~500 from the *Mycoplasma* genus, whereas in comparison humans have ~23,000 genes, composed of $\sim 3 \times 10^9$ base-pairs. Also, the variability in the human genome, i.e. the measure of how different a complete set of genes is from one cell in one human to those of another cell in another human, is only roughly 1%. However, the number of different human protein *sequences* which actually exist is estimated to be 1–2 million. The discrepancy is most likely due to post-transcriptional modification of the original mRNA – the corresponding DNA regions which are ultimately cut out of the mRNA are 'non-coding' in that they do not result in polypeptide manufacture, and are called *introns*, whereas the corresponding DNA regions which are re-annealed back into a modified mRNA molecule are 'coding' and called *exons*. On average therefore the mRNA from each human gene can be spliced/edited in ~50 different ways to generate the final protein.

2.10 Molecular simulations

A significant point to note concerning the structure of 'macromolecular' structures in biology, including not just proteins (section 2.4) but also polysaccharides (section 2.5) and nucleic acids (section 2.6), is that the actual, real molecule in the living cell can often in general be very dynamic. Standard ensemble-average structural determination techniques of x-ray crystallography, NMR and EM (see Chapter 1) all in effect produce experimental outputs which are biased towards the *least dynamic* structures. Experiments are often performed using anomalously high local concentrations of the molecule in question, divorced from the native physiological range, which may force a tightly packed state which does not naturally exist in the live cell, and the largest average measurement signal generated will be related to the most static structural state which may not necessarily be the most probabilistic state in the functioning cell. In addition, thermal fluctuations in the real living cell can result in considerable variability about any mean-average apparent structure.

Furthermore, the presence of different ions and of water molecules in the solvating medium can result in important differences in structural states which are not recorded in the standard structural models for a given biological model. This is not to say that these determined structures are no good, far from it, but we have to be suitably cynical as to whether or not a structure displayed truly represents what happens inside a living cell.

To model the effects of thermal fluctuations and electrostatic forces due to ions and water molecules, many researchers perform *molecular dynamics simulations* on biological structures. These are characterized by in effect digitizing the structure in a sensible way, placing each digitized element in a potential energy field at a given point in time and space, often with randomized elements of water molecules and ions (the term *Monte Carlo simulation* is often used synonymously to indicate this randomized feature), and then solving physical equations to predict the direction and magnitude of the local forces acting on each given element (the force equates to the negative gradient of the local potential energy function), indicating where the particle will move to after a small incremental time. The process is then repeated on all digitized elements in the system to result in a new predicted 'structure' at the next time point, very marginally different from the previous one. This is then iterated for subsequent time points over all digitized elements in the structure.

The largest problem with this process is computational, since small length scales require small incremental times to follow robustly the fate of digitized elements with sufficient precision, and with multiple elements in the system the permutation problem of computation of forces can become immense, and so in modern research requires multi-processor supercomputer clusters working in parallel continuously over several months. Even so, the maximum time scale that can be sensibly simulated can be very limited, with typical current 'good' simulation rates running at the equivalent of $\sim$1 ns of molecular simulation for every day of real computation time.

To give a better idea, at the smallest length scale are quantum mechanical simulations, which can be used to predict the effect of mechanical force on chemical bond instability, and so must use raw molecular orbital wave function predictions. For a typical single protein structure a total maximum quantum mechanical simulation time of $\sim 10^{-10}$ is achievable. However, this is an order of magnitude smaller than the rotational diffusion time scale of a typical protein molecule, so in other words there is a time scale gap between molecular motion and predicted bond behaviour in simulations.

To overcome some of the time scale issues it is therefore common to apply *coarse-graining* both to the length scale and to the potential energy functions, essentially increasing the length scale of the digitized elements by pooling smaller elements together as one and modelling the potential energy field with a simpler shape. So, atomic scale simulations do not in general consider individual molecular orbital wave functions but rather assume primarily *Van der Waals* potential energy functions (the *Lennard–Jones potential*, also known as the *6-12* or *12-6 potential*, discussed in the section below, between pairs of atoms separated by distance r made up of an *attractive* element varying as $\sim 1/r^6$ and a *repulsive* element varying as $\sim 1/r^{12}$) with additional refinements due to electrostatic interactions modelled by the so-called *Boltzmann–Poisson* equation.

This allows time scales up to $\sim 10^{-5}$ s to be realistically simulated. However, again there is a time scale gap since large molecular conformational changes can be much slower than this, perhaps lasting hundreds to thousands of microseconds. Further coarse-graining can allow access into these longer time scales, for example by pooling together atoms into functional structural motifs, and modelling the connection between the motifs with, in effect, simple springs, resulting in a simple *harmonic* potential energy function.

These *mesoscopic* scale simulations can model the behaviour of macromolecular systems potentially over a time scale of seconds, but clearly what they lack is the fine detail information as to what happens at the level of specific single atoms or molecular orbitals. A recent novel approach has involved attempts to combine *hybrid* elements of mesoscale simulation with smaller length scale simulation on smaller specific sub-regions of the mesoscale structure, and these increasingly utilize information from single biomolecule experiments to feed in local parameters for mechanical properties of the mesoscale model.

Powerful as they are, of course, computer simulations are only as good as the fundamental data, and the models, that goes into them, and it often pays to take a step back from simulation results to see whether simulation predictions really make intuitive sense or not. As Eugene Wigner, one of the founders of modern quantum mechanics, is attributed to have said, 'It is nice to know that the computer understands the problem. But I would like to understand it too.'

2.11 Importance of non-covalent forces

In practice, the tertiary and quaternary structures of proteins, and the equivalent higher order structures of other biological macromolecules, are affected by a variety of *non-covalent* interactions. One such is the presence of electrostatic layers of polar water solvent molecules and solvated ions, which holds true for the real live-cell structures of other classes of macromolecules possessing surface charges. Most commonly there is an additional *electrical double layer* (EDL), also known as the *Gouy–Chapman layer*; the first layer is composed of ions directly adsorbed onto the molecular surface due to a variety of chemical interactions, while the second layer is normally relatively diffuse and due to weak electrostatic binding of counter charges to this first layer.

Other relevant non-covalent forces include the following.

(i) *Van der Waals* or *dispersion-steric repulsion forces* – these are characterized by a short range steric repulsive potential energy having a $\sim 1/r^{12}$ distance dependence and a longer range 'dispersive' attractive component to the potential energy due to the interaction between non-bonding electrons in different molecular and atomic orbitals which results in *induced electrical dipoles*, and has a $\sim 1/r^6$ dependence (this is the combination referred to as the 'Lennard–Jones potential'). The short range repulsion is due to volume-exclusion or steric force; essentially molecules occupy a given volume of space and there is a significant free energy barrier to moving them too close together due to consequent overlapping of electron orbitals which is disallowed by the *Pauli exclusion principle* of quantum mechanics, manifest as a steric force keeping the molecules separate.

(ii) *Hydrophobic forces* – this is due to a tendency for non-polar molecules to pool together to exclude polar water molecules. It is relevant to stabilizing the core of many protein molecules from which water is often excluded, and also to the structure of lipid substructures such as membranes and vesicles (see section 2.7).

TABLE 2.1 **Approximate distance and bonding energy properties of the main forces relevant to biomolecules**

Force		~Distance (r) dependence for the potential energy	~Bonding energy (kJ mol^{-1})	~Bonding energy (k_BT)
Attractive	Covalent bonding	r^2	C-N: −275	−110
			C-C: −370	−150
			H-O:−455	−185
			ADP-P:−50	−20 *
	Electrostatic	r^{-1} monopole	−20 to −50	−8 to −20
		r^{-3} dipole	−4 to −8	− 1 to −3
	H-bonding	r^{-10}	− 5 to −10	−2 to −4
	Hydrophobic	complicated, due to electrostatic and H-bonds	−5 to −50	−2 to −20
Repulsive	Dispersive	r^{-6}	−0.1	−0.04
	Electrostatic	r^{-1} monopole	+20 to +50	+8 to +20
		r^{-3} dipole	+4 to +8	+1 to +3
	Steric repulsion	r^{-12}	>+100	>+40
	Helfrich	r^{-2}	<+0.1	<+0.04
	Langevin	r^0	<+0.1	<+0.04

* The hydrolysis of the terminal phosphate bond in ATP results in a release of ca. −30 kJ mol^{-1} free energy in vitro, however in the cell this is modified by the presence of Mg^{2+} ions and is closer to −50 kJ mol^{-1}.

(iii) *Hydrogen* (or H-) *bonding*, as already referred to for its key role in determining the secondary structure of proteins, is a relatively short range force over a length scale of ~0.2–0.3 nm due to a small polar attraction between slightly electronegative atoms on biological molecules such as nitrogen and oxygen and the orbital electron of a nearby hydrogen atom.

(iv) *Electrostatic forces* – these can be attractive or repulsive depending on the charge pair, but for proteins are not particularly important owing to the double layer effect resulting in significant screening at length scales larger than single molecules.

(v) *Helfrich forces* – these are relevant to molecules in the vicinity of lipid membranes, but in essence thermal fluctuations of the membrane due to random collisions of solvent water molecules generate mechanical undulations in the lipid membrane that produce an apparent short range repulsive force that can counter the attractive component of the dispersion forces.

(vi) *Fluctuation forces* – this is the so-called *Langevin force* (see section 1.4) due to the random thermal fluctuations of the surrounding water solvent molecules.

Each of these forces is the negative gradient of a potential energy function which has a different shape, and thus different associated total integrated bonding energy associated with it. The properties of the principal forces are listed in Table 2.1.

THE GIST

- Aside from water, the most important biological molecules in the cell are characterized by organic molecules of carbon chemistry.
- Cells have complex sub-cellular architecture in which organic molecules often perform highly specialized functions.

- Most organic molecules can be classed as protein, nucleic acid, sugar or lipid.
- Lipids are mainly composed of glycerol and fatty acid components, but in the cell are more common as phospholipids which have a polar head and hydrophobic tail region allowing stable bilayers to self-assemble in the presence of water, constituting a biological membrane.
- Lipid bilayers are essential components in energy transduction in all cells by acting as electrical capacitors in the 'chemiosmosis' process.
- Sugars can be low molecular weight monosaccharides, such as glucose, or may be polymers of monosaccharide subunits forming large polysaccharides such as starch.
- Proteins, nucleic acids and the polysaccharides are macromolecules.
- Proteins are composed of amino acid subunits linked by covalent peptide bonds, with three-dimensional structures stabilized mainly by hydrogen bonding.
- Protein molecules can bind together stabilized by a combination of weak forces to form multi-protein complexes, as often occurs in enzymes.
- Nucleic acids in the cell are either DNA or RNA, which are polymers of nucleotide bases with a backbone of alternating phosphate groups and pentose sugars.
- DNA normally forms a double-helix structure stabilized by hydrogen bonding between complementary base-pairs as well as base-stacking interactions; RNA is normally single-stranded with weakly stabilized hairpin loops as in mRNA, but may adopt complex three-dimensional structures in other forms such as tRNA and rRNA.
- The central dogma of molecular biology describes the flow of genetic information between DNA, RNA and protein, and primarily consists of sections of DNA called genes being transcribed into mRNA, and mRNA being translated into protein.

TAKE-HOME MESSAGE Cells contain DNA which can ultimately make protein, and proteins in the form of enzymes catalyse thousands of different biochemical reactions which permit all of the other molecular components of the cell to be made. DNA also permits the cell to replicate.

References

GENERAL

- Alberts, B. *et al.* (2008). *Molecular Biology of the Cell.* Garland.
- **CIL** (Cell Image Library) (2011). An open-access repository of annotated images, videos, and animations of cells from a variety of organisms: http://cellimagelibrary.org/
- Fawcett, D. W. (1981). *The Cell.* American Society for Cell Biology open-access online atlas of cell EM images: http://bioeducate.ascb.org/FawcettTheCell.html
- Howard, L. (2000). *Dartmouth College EM Facility Image Database.* An open-access EM image online database: http://remf.dartmouth.edu/imagesindex.html
- Nicholson, D. E. (2003). *Metabolic pathways.* Sigma Chemical Co., St. Louis, MO. Downloadable version also available at: www.iubmb-nicholson.org
- **PDB** (RSCB Protein Data Bank). The worldwide repository of biological macromolecular structural data. Originally initiated by the Brookhaven National Laboratory in 1971, the archive contains information about over 70,000 experimentally determined structures of proteins, nucleic acids, and complex assemblies, and provides a variety of tools and resources: www.pdb.org/

- RasMol. An intuitively designed open-access program for molecular graphics visualization originally developed by Roger Sayle, which can be used to display molecular structures whose atomic coordinates are available from the PDB (reference above), software freely downloadable from: www.openrasmol.org/OpenRasMol.html
- Rose, S. (1999). *The Chemistry of Life*. Penguin.

ADVANCED

- Agutter, P. S. and Wheatley, D. N. (2000). Random walks and cell size. *BioEssays* **22**: 1018–1023.
- Crick, F. (1958). On protein synthesis. *Symp. Soc. Exp. Biol.* **XII**: 139–163.
- Crick, F. (1970). Central dogma of molecular biology. *Nature* **227**: 561–563.
- Gibson, D. G. *et al.* (2010). Creation of a bacterial cell controlled by a chemically synthesized genome. *Science* **329**: 52–56.
- Heller, H., Schefer, M. and Schulten, K. J. (1993). Molecular dynamics simulation of a bilayer of 200 lipids in the gel and in the liquid crystal-phases. *Phys. Chem.* **97**: 8342–8360.
- Noji, H., Yasuda, R., Yoshida, M. and Kinosita, K. J. (1997). Direct observation of the rotation of F_1-ATPase. *Nature* **386**: 299–302.
- Sanger, F. and Thompson, E. O. (1953a). The amino-acid sequence in the glycyl chain of insulin. I. The identification of lower peptides from partial hydrolysates. *Biochem. J.* **53**: 353–366.
- Sanger, F. and Thompson, E. O. (1953b). The amino-acid sequence in the glycyl chain of insulin. II. The investigation of peptides from enzymic hydrolysates. *Biochem. J.* **53**: 366–374.
- Sanger, F. and Tuppy, H. (1951a). The amino-acid sequence in the phenylalanyl chain of insulin. I. The identification of lower peptides from partial hydrolysates. *Biochem. J.* **49**: 463–481.
- Sanger, F. and Tuppy, H. (1951b). The amino-acid sequence in the phenylalanyl chain of insulin. 2. The investigation of peptides from enzymic hydrolysates. *Biochem. J.* **49**: 481–490.
- Tieleman, D. P., van der Spoel, D. and Berendsen, H. J. C. (2000). Molecular dynamics simulations of dodecylphosphocholine micelles at three different aggregate sizes: micellar structure and lipid chain relaxation, *J. Phys. Chem. B* **104**: 6380–6388.
- Traibi, M. and Craik, D. J. (2002). Circular proteins – no end in sight. *Trends Biochem. Sci.* **27**: 132–138.
- Watson, J. D. and Crick, F. H. C. (1953). A structure for deoxyribose nucleic acid. *Nature* **171**: 737–738.

Questions

FOR THE LIFE SCIENTISTS

Q2.1. All natural amino acids are L-isomers but all natural sugars are D-isomers. Why is this? (Hint: No one really knows, but it may be instructive to look into theories of how life began and the importance of clay...)

Q2.2. The average and standard deviation of heights measured from a population of adult men of the same age and ethnic background were 152 cm and 8 cm respectively. Comment on how this compares to the expected variation between the sets of genes from the same population.

Q2.3. A roughly spherical bacterium possessing a single chromosome was measured as having a radius of 1 μm, and it was estimated that 1% of its mass was taken up by DNA. (a) What is the mass of the bacterium's chromosome? (b) Assuming that the molecular weights of the raw nucleotides A, C, G and T are 267 g, 243 g, 283 g and 257 g respectively excluding phosphate groups, and that the molecular weight of a phosphate group is 80 g, estimate the number of base-pairs in the bacterium's genome.

Q2.4. How true is it to say that the chemistry driven by light in chloroplasts is the reverse of that in mitochondria?

Q2.5. (a) What is the hydrophobic effect, how does it originate and how do hydrophobic forces compare with other attractive and repulsive non-covalent molecular and inter-atomic forces in relation to magnitude and distance dependence? (b) Why and how is the hydrophobic effect important in the formation of the structure of (c) cytoplasmic proteins, (d) integral membrane proteins and (e) ion channel proteins. (f) How does the effect depend on temperature? At room temperature the standard enthalpic energy change for the association of two protein subunits was estimated from AFM spectroscopy to be $+15$ kJmol^{-1}, with a corresponding change in standard entropy estimated at $+60$ kJmol^{-1}. (g) What is the significance of the signs of the changes in enthalpy and entropy? (h) Is association of the subunits favoured at room temperature? (i) What would you predict to happen if there were (1) a large rise or fall in temperature, or (2) a small rise or fall in temperature?

Q2.6. (a) What is the meaning of standard electrode potential in the normal and the biochemical standard states, and why can they take different values? The (normal) standard electrode potential for the reduction of acetic acid to acetaldehyde $CH_3COOH+2H^++2e^-\rightarrow CH_3CHO+H_2O$ is -0.19 V at room temperature. (b) What is the biochemical standard electrode potential of this reaction? In the biochemical standard state the reduction of NAD^+ to NADH has an electrode potential of -0.32 V at room temperature. (c) Is this electrode potential likely to be more or less sensitive to pH than for the reduction of acetic acid. (d) What is the molar equilibrium constant for the oxidation of acetaldehyde by NAD^+ and the associated free energy change per molecule in units of k_BT? (e) What extra information would you need to determine what the corresponding enthalpic and entropic changes are in this reaction?

FOR THE PHYSICAL SCIENTISTS

Q2.7. (a) Find expressions for the total Gibbs free energy required to bend a symmetric planar phospholipid bilayer of thickness 2w into a spherical liposome of radius r, which is much greater than w, assuming that the energy required to change the packing area by a small amount ΔS of a single phospholipid molecule head group is equal to $\frac{1}{2}a\Delta S^2$, where a is a constant. In an experiment, the total bending free energy of a liposome was measured at ca. 150 k_BT and the polar heads tightly packed into an area equivalent to a circle of radius of 0.2 nm each, with the length from the tip of the head group to the end of the hydrophobic tail measured at 2.5 nm. (b) Estimate the free energy in joules per phospholipid molecule required to double the area occupied by a single head group.

Q2.8. A typical hydrogen-bond requires an energy equivalent to ca. $5k_BT$ to break. (a) Assuming a roughly equal mix of A, C, G and T nucleotide DNA bases, estimate the probability for generating a 'mismatch' base-pair in the DNA double-helix, stating any assumptions you make. (b) Compare this against the observed error of ca. 1 in 10^5 measured in the test tube? In a living cell there is typically one error per genome per generation (i.e. per cell division). (c) How does this compare with the value obtained above?

Q2.9. (a) Assuming there are ca. three thousand million base-pairs in the human genome, how many raw bits of information is this equivalent to? (b) Assuming the base-pairs are primarily stored in the form of B-DNA, how does the storage capacity compare against that of a man-made 1 Tb hard drive of size 11.7 × 8.0 × 1.6 cm? (c) How many copies of a single genome could in principle be fitted into a single cell nucleus of diameter 10 μm?

FOR THOSE WHO HAVE NOT MADE UP THEIR MIND

Q2.10. Consider the definition of a cell as being the minimal structural and functional unit of life which can self-replicate and can exist 'independently'. In reality any given cell is not isolated, for example there are channels and pores in the membrane that can convey molecules into and out of the cell. Does this alter our notion of 'independence', and are there more robust definitions of a cell?

Q2.11. There are ca. 10^{15} cells in the human body, but only ca. 10^{14} of them are human. What are the rest, and should this alter our view of the definition of an 'organism'?

Q2.12 Most cells, barring a few exceptions (for example certain nerve cells, and some unicellular algae), have a characteristic length scale which is of the order of 100 μm or less. (a) It was previously thought that random diffusion considerations of single molecules sets this upper limit; in what ways is this a sensible theory? We now know that cells contain many tightly packed substructures which hinder simple random diffusion, such that any given diffusing single molecule is likely to hit some such feature after diffusing a distance of only ca. 20 nm. (b) How does this alter the simple diffusion theory for maximum cell size, and what other explanations could there be?

Q2.13. (a) Go to www.pdb.org (the Protein Data Bank). Download the PDB coordinates for the following PDB IDs: 1AOI and 1QO1. Install RasMol and open the pdb files. (b) Where appropriate, display separate strands of the DNA double-helix red and blue, with non-polar amino acid residues yellow and polar amino acid residues magenta. (c) What is the maximum separation of any two atoms in either structure?

Q2.14. (a) What are the attractive and repulsive non-covalent forces relevant to single biomolecule interactions, and how (and why) do they differ in terms of the relative distance dependence and magnitude. (b) What forces are most relevant to the folding of a protein, and why?

3 Making the invisible visible: part 1 – methods that use visible light

It is very easy to answer many of these fundamental biological questions; you just look at the thing! ... Unfortunately, the present microscope sees at a scale which is just a bit too crude. (FEYNMAN, 1959)

GENERAL IDEA

In this chapter we discuss the techniques which are available to the experimental scientist who wishes to visualize or detect single biological molecules primarily using visible light, both in the test tube and in the living cell.

3.1 Introduction

How can we 'see' something like a single biological molecule, which is of the order of a thousand million times smaller than a typical object in the macroscopic world that we visualize with our naked eyes? The region of the human eye responsible for detecting light, the *retina*, consists of two types of cells called rods and cones (rods differ by being 100 times more sensitive than cones but they respond more slowly, have less spatial resolution and do not discriminate colour), both of which can convert detected photons of light into electrical signals, conveyed via ion channels and nerve fibres into the brain. The resolving power of the human eye, the *visual acuity*, is a measure of the smallest angular separation that the eye can resolve, which for humans has a theoretical limit equivalent to ~0.01°, about 20 milliradians, determined by the limit of optical diffraction set by the wavelength of the incident light and the diameter of the aperture in front of the 'imaging device' (here set by the pupil of the eye), as shown in Figure 3.1A.

Individual waves of light from objects which are greater than a few wavelengths distance away (for example, greater than approximately 1 μm for visible light) diffract, which results in an apparent blurring of the light source such that light originally emitted from a very small region in space (a 'point', approximated as a single molecule for example) appears to spread out and occupy a larger space, defined by the *point spread function* of the imaging system detecting this light. When two such blurry point spread functions are too close together they cannot be resolved as two separate intensity distributions but rather appear as a single intensity function. The minimal separation of two such point spread functions, when they are just discerned as being two separate intensities, is defined as the optical resolution limit of that imaging system. For an ideal light microscope, the theoretical limit is roughly half a wavelength of light.

The closest an object can be viewed by the naked eye, the *near distance*, is around 20 cm, and at this distance the length subtended by this small angle can be calculated simply from the product of the minimum angular separation in radians with the near distance, which turns out to be around 40 μm. The *angular resolution limit* of the human

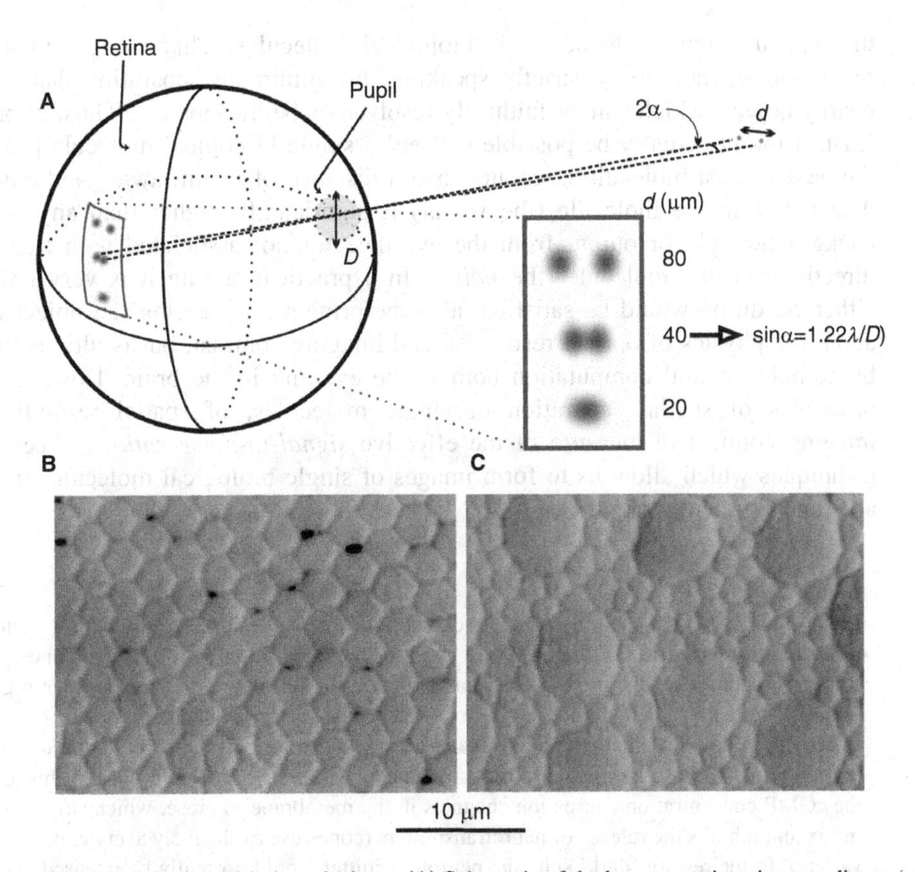

FIGURE 3.1 Seeing with the naked eye. **(A)** Schematic of the human eye imaging two distant 'point source' objects (for example, approximated by single biological molecules) separated by a distance d and imaged through the pupil aperture (diameter D) onto the retina, with the effect of three different values of d illustrated in the expanded region of the retina (rectangle). The cut-off point for being able to resolve the two objects as separate is when they are separated by roughly 40 μm apart, assuming they are at the near distance from the eye, at which point the angle α is the angular resolution and obeys the 'Rayleigh criterion' formula as shown, where λ is the wavelength of light. Histological 'plan-view' sections of the human retina (original data used with permission from Christine A. Curcio, University of Alabama School of Medicine, for full details see Curcio, 1990) visualized by dye staining followed by light microscopy are shown in **(B)** for the central 'fovea' region consisting only of tightly packed cone cells associated with high visual acuity and **(C)** a region just outside the fovea consisting of both cone (larger) and rod (smaller) cells. The rods, unlike the cones, multiplex by combining the photodetection output from roughly 15 neighbouring rod cells.

eye is in practice worse than the theoretical diffraction limit by a factor of ~2 as a result of *spherical aberration*. To produce an electrical impulse the most sensitive retinal cells need a minimum input signal of just one photon of light. However, 50% of photons incident upon the photodetection layer of cells in the retina make it through to a photoreceptor molecule, and only 10% of incident photons at the outer surface of the eye, the *cornea*, are transmitted through the eye as far as the retina surface. Whether or not an object is 'seen' is determined not only by the photons detected, but also by the detection noise of both scattered light inside the eye and *dark noise* due to thermally induced changes in the photoreceptor molecules.

This limit of resolution is the size of certain large cells (for example, amoebae), however, as we saw from the previous chapter, it is roughly 100,000 times larger than the typical length scale of single biological molecules. That being said, the spatial resolution of the eye is, strictly speaking, the minimum separation distance of two nearby objects which can be faithfully resolved as being separate. Thus, in principle at least, it might actually be possible to 'see' a single biological molecule provided that the next nearest molecule is greater than a distance of 40 μm away, and that the light detected from the molecule (the *signal*) is signficantly higher than any surrounding background light or output from the eye detector not associated with light emission directly from the molecule (the *noise*). In a practical setting it is very unlikely that either condition would be satisfied; also the principle of 'seeing' an object is not just about the physics of optical resolution and imaging contrast, but is ultimately affected by complex neural computation both in the eye and in the brain. However, the core principles of spatial separation of single molecules, of spatial resolution and of imaging contrast (a measure of the effective *signal-to-noise ratio*) are central to the techniques which allow us to form images of single biological molecules in a practicable and efficient manner.

BIO-EXTRA

Light in the eye is detected by a process called **phototransduction**, starting with a photo-induced ***cis–trans* isomerization** of the molecule **rhodopsin** expressed in both rods and cones. Rhodopsin comprises the protein opsin integrated in the plasma membrane, and the organic molecule retinal which undergoes a shape change, causing a series of changes in the opsin that lead to activation in the regulatory protein **transducin**, causing downstream activation of the enzyme **cGMP phosphodiesterase** which results in the conversion of cGMP to 5′-GMP. This lowering of the cGMP concentration causes ion channels in the membrane to close, which stops sodium ion influx and inhibits the release of neurotransmitters (cones use exclusively acetylcholine, rods use a variety). In the resting 'dark' state the neurotransmitter would normally be released across a synaptic junction to a neighbouring bipolar cell situated adjacent to a **ganglion cell**. The neurotransmitter causes **hyperpolarization** in the bipolar cell, which prevents it from releasing neurotransmitter into the bipolar-ganglion synapse, thereby stopping conduction of the nervous signal to the brain.

3.2 Magnifying images

The most obvious way to improve our chances of seeing a single molecule is to magnify its image before it reaches our eye. This can be done most efficiently using a combination of lenses that constitute a high-power optical microscope. In its most simple form such a microscope consists of a high numerical aperture (NA) objective lens placed very close to the sample, with a downstream imaging lens focusing the sample image onto a highly sensitive light detector (such as a high-efficiency CCD camera, or sometimes a photomultiplier tube in the case of a scanning system like confocal microscopy), as depicted in Figure 3.2A.

The NA of an objective lens is defined as $n\sin\theta$, where n is the refractive index of the imaging medium (in air this is 1 so to increase the NA, high-power objective lenses are generally *oil immersion* lenses which means they have a blob of special imaging oil placed in optical contact between the glass microscope slide or coverslip and the objective lens which has the same high value of refractive index as the glass (~1.52). The angle θ is the maximum half-angle subtended from the sample of a ray of light which can be captured by the objective lens, so in other words higher NA lenses can capture more light from the sample.

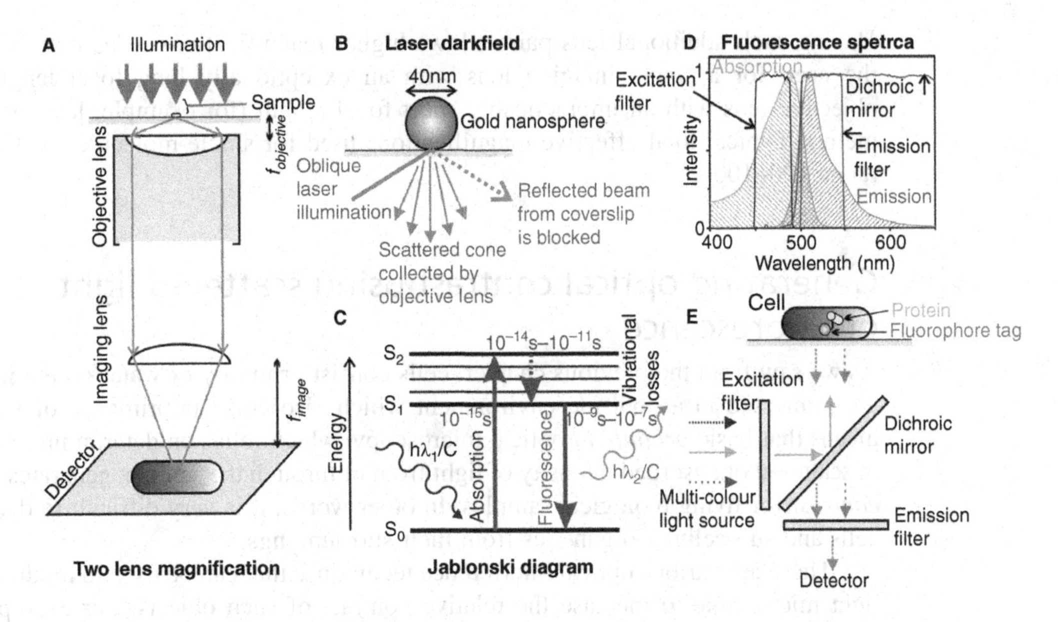

FIGURE 3.2 Magnifying images and enhancing contrast. **(A)** Schematic of a very simple optical microscope in which illumination results in scattered light from the sample which is captured by a high numerical aperture objective lens with the collimated light then focused by a higher focal length imaging lens onto some form of either widefield detector (such as a CCD camera) or point detector such as a PMT (with the sample then being scanned by relative movement of the stage and the illumination as occurs in confocal microscopy), resulting in a magnification of the final image. **(B)** In laser darkfield imaging, an oblique incident laser beam illuminates the sample; the strong 'forward' scatter beams from a nano-gold particle are captured by the objective lens but the back scatter reflection from the glass coverslip is at a sufficiently oblique angle to be physically blocked from reaching the detector. **(C)** A simplified 'Jablonksi diagram' which illustrates the energy level transitions and corresponding time scales involved during fluorescence absorption and emission. **(D)** Absorption and emission spectra for the commonly used fluorescent protein eGFP, often just referred to as 'GFP', illustrating the approximate mirror-image appearance due to the Stokes shift. Typical transmission spectra for real excitation and emission are shown, as is that of a dichroic mirror. **(E)** Schematic of a typical filter set used in a fluorescence microscope, indicating how a potentially multi-coloured light source (e.g. from a mercury-arc or tungsten-halogen light if a single-wavelength laser is not used) can be first filtered then reflected onto the fluorescently tagged biological sample, indicated with a schematic cell which has a protein tagged with a fluorescent dye. The fluorescent dye generates emissions but components in the cell may also generate natural 'autofluorescence', both of which might be transmitted by the dichroic mirror but ideally only the fluorescence signal is transmitted by the emission filter.

In its simplest form as reduced to Figure 3.2A, the magnification M is given by the ratio of the focal length of the imaging lens to that of the objective lens (the latter typically being a few millimetres):

$$M = f_{\text{image}}/f_{\text{objective}} \tag{3.1}$$

In practice, there is likely to be a series of further lens pairs placed between the imaging lens and the detector, arranged in effect as telescopes, for example with focal lengths $f_1, f_2, f_3, f_4, \ldots, f_n$ corresponding to lenses placed somewhere between the objective lens and the camera detector. The total magnification of such an arrangement would then be equal to:

$$M = (f_2/f_1)(f_4/f_3)\ldots(f_n/f_{n-1})(f_{\text{image}}/f_{\text{objective}}) \tag{3.2}$$

Having such additional lens pairs allows higher magnifications to be obtained without the need for a single imaging lens with an exceptionally long focal length and an objective lens with an impracticably short focal length (for example, less than a millimetre). Typical total effective magnifications used for single-molecule work are in the range 100–1000.

3.3 Generating optical contrast using scattered light or fluorescence

As we saw from the previous chapter, cells consist primarily of water on the inside, and are immersed in an outside environment which also consists primarily of water. This means that basic *brightfield* optical microscopy, which relies on determining differences in scattered or absorbed intensity of light from or through the sample, generates very poor *contrast* for living biological samples. In other words, it is very difficult to discriminate cells and sub-cellular organelles from their surroundings.

There are various optical interference techniques that can be used to modify a visible light microscope to increase the relative contrast of such objects. For example, *phase contrast microscopy* allows subtle differences in refractive index to be accentuated. The inside of a cell has a refractive index which can vary not just from cell to cell, but also in different regions of the same cell, over a range of 1.35–1.45 compared to the water environment outside the cell of more like 1.33. This method utilizes the fact that the phase of scattered/diffracted light from a sample is shifted by roughly *a quarter of a wavelength* relative to that of the incident light, and by utilizing matched optical phase plates above and below the sample this can be converted into a *half wavelength phase shift* which is then the condition for *destructive* interference. The scattered light from the sample then appears to have a much lower intensity than the non-scattered light from the surroundings, thereby increasing the imaging contrast. In other words, this process in effect transforms phase information at the sample into amplitude information in the intensity of the final image, which is tremendously powerful yet remarkable in its simplicity.

Also, there is *differential interference contrast microscopy* (DIC). DIC utilizes differential *polarization excitation* to accentuate differences in the *spatial gradient* of the refractive index of a sample. It is therefore an excellent technique for identifying the boundaries of cells, but also of cell organelles.

However, much as these approaches assist in our attempts to visualize cells and tissues, they do not help in actually seeing single molecules. The proportion of incident light which in general is scattered from any given single biological molecule is extremely small compared to the intensity of the incident light itself, which means that other techniques which enhance the intensity of the signal from a single molecule compared to its local surroundings are required. The most practicable approach is to tag the molecule specifically with some form of reporter probe which is designed to generate a high local signal upon excitation with light. One such technique uses a probe which is coated in a metal such as gold which generates a high scatter signal for certain wavelengths of visible light, generally obtained using laser excitation (Figure 3.2B).

The principal photon-scattering process here is *elastic* and so the wavelength of scattered light is the same as that of the incident light, thus it is vital to block any scattered light not from the tagged molecules themselves (for example, back reflections from the glass microscope slide). As a result, the regions of the sample not containing tagged molecules appear dark on the camera detector, hence the name *laser darkfield*. By careful selection of the size of the probe (typically tens of nanometres) and the wavelength of light, an additional enhancement to the signal can be achieved by the generation of so-called *surface plasmons*. This technique can generate an exceptionally high signal-to-noise ratio and has been used to great success on certain in vitro (i.e. 'test tube') biological samples to measure nanometre length scale movements of single molecules and complexes with extremely high time resolution of the order of a few microseconds.

One issue with laser darkfield is its relatively limited use with living samples. The main reason is a technical one in that the relatively large size of a cell compared to the probe can result in a measurable scatter signal from the cell body itself, even though it is not metallized, which can swamp that of the probe. Also, it is not a trivial task to label molecules of choice specifically in the living cell with a metallized nano-probe without generating some non-specific labelling of other cellular substructures. In addition, the size of the probe is relatively large compared to single molecules themselves, implying likely steric hindrance effects with subsequent impairment of normal biological function. And finally, for certain cell types (for example, prokaryotes), it is difficult to introduce such a relatively large probe into the cell whilst keeping the cell intact. An alternative approach, which circumvents many of these problems, is to utilize *fluorescence microscopy*.

Fluorescence is the process by which energy from one, or (rarely) more than one, photon of light is first absorbed by the lower energy electron shells of an atom causing an electron to jump to a higher energy state. The electron then undergoes some slower vibrational energy losses before jumping back down to its ground state, resulting in the fluorescence emission of a photon of light of slightly lower energy (and hence longer wavelength) than the excitation light (Figure 3.2C).

PHYSICS-EXTRA

The simplified **Jablonksi diagram** of Figure 3.2C omits the first excited **triplet state** energy level, which can be reached from the excited state via **intersystem crossing** involving a classically forbidden transition from a net spin zero to a spin one state and therefore has a much lower rate constant that the fluorescence transition. The transition from the triplet state back to the ground state results in emission of a lower energy **phosphorescence** photon than for the fluorescence process, over a range of characteristic times of microseconds up to several hundreds of seconds, depending upon the dye and the chemical environment. One effect of the fluorescence emission is to cause **photon bunching** over these time scales (instead of a continuous emission), which is often not observed for fluorescence imaging time scales of milliseconds or higher as the effect is averaged out, but may be seen with other fluorescence detection techniques performed over faster time scales.

This difference in wavelength between excitation and emission light, the so-called *Stokes shift* (Figure 3.2D) can be utilized by using a combination of optical filters and specially coated mirrors such that the excitation light incident upon and scattered from the sample can be blocked out with a high attenuation of typically 10^6–10^8 (typical fluorescent samples will have a ratio of emitted fluorescence intensity to back scattered excitation light of 10^{-4}–10^{-6}, and as a rule of thumb a factor of 10–100 smaller than this will allow sufficient contrast for single-molecule applications). This allows essentially only the fluorescence emissions to be detected (Figure 3.2E). Thus, if a fluorescent reporter probe is specifically attached to a single molecule of interest, this can generate sufficiently high local contrast to permit the position of the reporter to be detected above the surroundings, provided the other tagged molecules in the vicinity are at a distance greater than the optical resolution limit, as is discussed in more detail later (see section 3.9).

The main sources of 'noise' in fluorescence imaging are either natively autofluorescent molecules in the biological sample contaminating the fluorescence signal from the specific fluorescently tagged molecules, or scattering of the incident excitation light 'bleeding through' into the fluorescence detector channel. This scattering is primarily of two types, either *Rayleigh* or *Raman scattering*. Rayleigh scattering is an elastic process in which the emergent scattered photon has the same wavelength as the incident photon. Raman scattering is inelastic and results in either the incident photon losing some energy prior to being scattered (*Stokes* scattering) or, more rarely, gaining some energy (*anti-Stokes* scattering), due to vibrational and/or rotational energy changes in the scattering molecule. The anti-Stokes peak generally appears at a constant energy difference relative to the energy of the incident excitation photon. In biological samples the largest Raman contribution is due to the anti-Stokes scattering of water molecules.

A Raman peak position is normally described in terms of wavenumbers ($2\pi/\lambda$ with typical units of cm^{-1}) and in water this is generally $\sim$3400 cm^{-1} lower or higher than the equivalent excitation photon wavenumber, depending on whether the peak is Stokes or anti-Stokes (typically higher or lower wavelengths by $\sim$20 nm for visible light excitation), which for some dim fluorophores can have a comparable amplitude to the fluorescence emission peak itself (i.e. this Raman peak is then the limiting noise factor). This Raman effect can also be put to good use in *Raman spectroscopy* (see Chapter 4).

KEY POINT

Fluorescence microscopy in general is an **invaluable biophysical tool** for probing biological processes in living cells. It results in exceptionally **high signal-to-noise ratios** for determining the localization of molecules tagged with a fluorescent dye but does so in a way that is relatively **non-invasive** compared to other single-molecule methods. This **minimal perturbation** to the native physiology makes it a **probe of choice** in many single-molecule studies in the live cell.

3.4 Organic dyes, FlAsH/ReAsH, fluorescent amino acids and quantum dots

Although there are many naturally fluorescent compounds or *fluorophores*, much progress has been made in designing bright, photophysically optimized fluorescent organic dyes. There is a large range of different organic dyes which are typically manufactured by chemically modifying just a few different originator dyes such as *cyanines* and *xanthenes*. They are generally characterized by having delocalized electron systems which in effect allows the whole molecule to operate as an electric dipole. They are designed to be bright and relatively photostable and optimally excited either at common peak wavelengths from a *broad-spectrum* white-light source such as a *mercury-arc* or *tungsten-halogen* lamp, or at common *laser* emission wavelengths.

Common organic dyes include variants such as *fluorescein* (green) and *rhodamine* (orange) derivates (Figure 3.3A). Historically, such dyes were first used to label single biological molecules specifically using a series of two antibody *adapters* (Figure 3.3B), a *primary antibody* that would bind to the biological molecule of interest, and a *secondary antibody* chemically labelled with typically up to $\sim$10 dye molecules that would bind to the primary antibody.

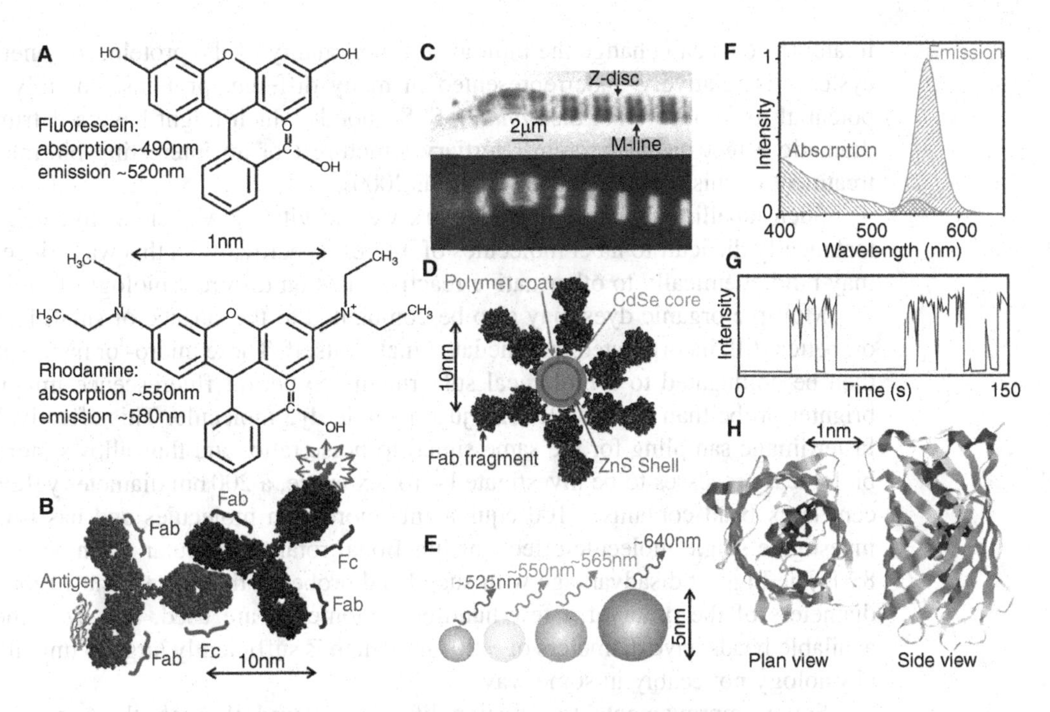

FIGURE 3.3 Fluorescent tags. **(A)** Chemical structures of two common organic dyes, fluorescein (upper panel) and tetramethyl rhodamine (lower panel). **(B)** Primary IgG binding via one of its Fab groups (PDB ID: 2IG2) to an example antigen molecule of interest (PDB ID:1BPV), with a secondary IgG with fluorescein dye molecule attached binding the Fc region (PDB ID:1FC2) of the primary IgG. **(C)** Example of immunofluorescence from a single myofibril showing brightfield (upper panel) and fluorescence (lower panel) probed using a specific antibody that binds to the N2-A region of the muscle protein titin (original data, author's own work, for full details see Leake, 2001). **(D)** Schematic of a functionalized QD with Fab fragment bioconjugating layer. **(E)** Schematic illustrating relative core QD sizes with peak emission wavelength. **(F)** Absorption and emission spectra for a common commercial QD 'QD565' (data from PubSpectra). **(G)** Schematic depiction of intensity versus time traces for a surface-immobilized QD, illustrating 'blinking'. **(H)** Structure of GFP (PDB ID:1GFL), β strands, α helices and chromophore of three amino acids are indicated.

The specificity of such *immunofluorescence* labelling can be very high, but the main difficulty lies in introducing the antibody labels into living cells. Some exceptional cases such as muscle cells called *myofibrils* can be permeabilized relatively easily without significantly impairing sub-cellular features, allowing the relatively large antibodies to bind to specific filamentous features in the cell (Figure 3.3C). Each antibody has a typical effective viscous-drag radius of ~10 nm, so a primary–secondary complex is relatively large compared to the molecule which is being tagged, and so functional impairment is highly likely.

An alternative method for introducing an organic dye tag onto a biological molecule of interest is to attach a dye directly using covalent chemical conjugation. This is sometimes done directly to existing reactive groups in the molecule, but is more often done by some prior genetic manipulation of the molecule of interest to engineer in one or more chemically reactive groups. For example, in some proteins one of the amino acids in the structure can be replaced with a cysteine amino acid residue which contains an –SH, or *sulphydryl*, group that can be used to conjugate to a dye molecule or, more typically, *biotin* tags. Note, however, if the protein already contains cysteine residues this can lead to problems due to multiple labelling, and there is always a potential issue that a change in amino acids can change the biological functionality of the protein. In general though, cysteine is relatively underrepresented in many different proteins, possibly due to its potential to form internal *disulphide* –S–S– bonds which might have a detrimental and dramatic effect on the protein's tertiary structure (for an interesting bio-mathematical treatment of this see Miseta and Csutora, 2000).

Such labelling techniques often work well in vitro, however for live-cell work it is technically difficult to label molecules of interest specifically in this way, since the labels may bind chemically to other native reactive sites on different biological molecules.

Multiple organic dyes may also be conjugated to the surface of small latex spheres or, better still, incorporated into the latex matrix itself. These micro- or nano-spheres may then be conjugated to a biological substructure to permit fluorescence imaging with a brighter probe than is possible with just a single dye molecule. This effectively permits faster image sampling for the same signal-to-noise ratio, and thus allows more dynamic biological processes to be investigated – for example, a 200 nm diameter yellow fluorescent latex bead contains ~100 equivalent fluorescein molecules and has been used to investigate single-molecule effects of functional rotary molecular machines (see section 8.3). The biggest disadvantage with such bead probes is their size – most typically, bead diameters of the order of a few hundred nanometres are used, and even the smallest available beads have diameters of ~20 nm, which is sufficiently large to impair the native physiology noticeably in some way.

Some improvements to functionality are offered through the use of the green *fluorescein arsenical helix binder* (FlAsH) and the pink equivalent using the organic dye *resorufin* (ReAsH), both of which come under the general category of *tetracysteine-tags*. This technology utilizes a genetically encoded *tetramic* arrangement of four cysteine amino acid residues in a specific protein under study in a cell. External incubation of the cell with the non-fluorescent membrane-permeable FlAsH/ReAsH reagent then results in eventual binding to the four cysteines and conversion to a fluorescent form. The prime advantage of this method is that the FlAsH/ReAsH reagents are relatively small (1–2 nm) and the binding site in a protein can be engineered from as few as six amino acids in total (four for the cysteines, and two more to generate the necessary tetramic-shaped binding pocket), thus offering potentially minimal steric hindrance to functionality. However, it can be technically difficult to introduce the reagents into certain cell types, and *non-specificity* of binding combined with *cellular toxicity* has limited its current use.

Fluorescence in situ hybridization (FISH) is a common labelling technique for use in probing specific regions of nucleic acids. Originally developed using radioactive probes (see for example Gall and Pardue, 1969), now fluorescence probes are the reporter molecules of choice. Here, a probe consists of a short ~10 nucleotide base sequence of either singled-stranded DNA or RNA, which will bind to a specific region of nucleic acid sequence from a cell extract or a thin, *fixed* (i.e. dead) tissue sample via complementary base-pairing (see Chapter 2), following suitable incubation protocols normally of at least 10 hours but sometimes involving a few days. Either a fluorophore is chemically attached directly to the FISH probe via one or more bases or, more commonly, a secondary labelling method is used employing either a fluorescent secondary antibody or a biotin tag that binds to the probe.

FISH has been used to isolate the position of individual genes on chromosomes, and is a commonly used clinical assay in probing for a range of disorders in a developing foetus in the womb. Although the long incubation time and in vitro nature of the sample make this technique disadvantageous, it can be applied to single-molecule precision.

BIO-EXTRA

Biotin is a small, natural molecule of the B-group of vitamins, which binds with very high affinity to two structurally similar protein molecules called **avidin** (found in egg white of animals) and **streptavidin** (found in bacteria of the genus *Streptomyces*, a notorious family of microbe which produces at least 100,000 different types of natural antibiotics, several of which are utilized in common clinical practice), typically with a K_d of 10^{-14}–10^{-15} M. As a result, biotin–avidin or biotin–streptavidin systems are commonly used by biochemists in conjugation chemistry. In proteins, the biotin component is most commonly introduced by chemically labelling one of the 'primary' (free) amine groups found in the amine acid lysine with a coupling molecule called ***N*-hydroxysuccinimide** (NHS) to biotin, or using derivates of the chemical biotin bound to the conjugating chemical **maleimide** which can react with a free sulphydryl group in the protein. Separately, either avidin or streptavidin can be chemically labelled with a variety of fluorescent tags and used to probe for the 'biotinylated' sites on the protein following incubation with the sample. The most common primary labelling molecule for immunuohybridization chemistry in general involves the class of small steroid-derived molecules called **haptans** which are known to be highly immunogenic (i.e. animal cells develop highly specific antibodies in response). Biotin is a haptan, as is **digoxigenin** (DIG), which is commonly used in FISH assays. DIG is normally covalently bound to a specific nucleotide triphosphate probe, and the fluorescently labelled IgG secondary antibody anti-DIG is subsequently used to probe for its location on the chromosome.

Some recent successes have been made using genetically encoded *synthetic fluorescent amino acids* as reporter molecules. These are artificial amino acids which have a covalently linked fluorescent tag engineered into the substituent group. The means of tagging is not via chemical conjugation but rather the amino acid is genetically coded directly into the DNA which codes for a given protein of interest to be labelled, essentially by modifying one of the *nonsense* codons that normally do not code for an amino acid (see Chapter 2). At present, the brightness and efficiency of such fluorescent amino acids is still relatively poor for general application in single-molecule studies, but there may be significant scope for development in the future.

Organic dyes, FlAsH/ReAsH tags and fluorescent amino acids all undergo irreversible *photobleaching*. A single fluorophore molecule will emit an average characteristic number of fluorescence photons before bleaching to a permanent 'dark' state, most likely as a result of non-specific chemical damage from the local generation of highly reactive free-radical groups from the surrounding water solvent. This ultimately limits their application in that a given reporter tag can only be observed for a certain time before it vanishes.

Quantum dots (QDs) have a photostability which is at least 100 times greater than that of the most stable organic dyes, and so for most experimental applications they can be considered not to undergo significant irreversible photobleaching, and thus have many advantages for monitoring single biological molecules. They are manufactured crystalline alloy spheres typically of two components (*cadmium selenide*, CdSe, being the most common) or sometimes three components containing of the order of 100 atoms, and which are typically 3–5 nm in core diameter (Figure 3.3D, E). They have semiconductor properties and can undergo fluorescence due to an *exciton* resonance effect within the whole crystal, with the energy of fluorescence relating to their precise physical size. An exciton is a correlated particle pairing composed of an electron and a *positive-hole* (the name physicists use for the net positive space left in a material after an occupying electron has moved elsewhere). It is analogous to the excited state of traditional fluorophores but has a significantly longer lifetime of $\sim 10^{-6}$ s.

They are characterized by a very *broad* absorption spectrum and very *narrow* emission spectrum – this means that they can be excited using a range of different lasers whose wavelength of emission does not necessarily correspond to an absorption peak in the fluorophore as is the case for organic dyes; the tightness of the spectral emission means that emissions can be relatively easily filtered without incurring significant loss of signal (Figure 3.3F).

In addition, this means that several different coloured QDs can be discriminated in the same sample on the basis of their spectral emission, which is useful if each differently coloured QD is tagging a different biological molecule of interest. QDs are brighter than corresponding organic dyes at similar peak emission wavelength, however their relative brightness is often overstated (for example, a single QD emitting in the orange-red region of the visible light spectrum at corresponding excitation wavelengths and powers is typically just 6–7 times brighter than a single molecule of rhodamine).

A recent application of QDs has been in their development as *fluorescent nanothermometers* (Maestro *et al.*, 2010). The spectral emission peak of QDs is temperature sensitive since the population of high-energy excitons is ultimately governed by a temperature-sensitive *Boltzmann distribution* (see later in section 5.5 on *detailed balance*), but this sensitivity is significantly more for two-photon laser excitation compared to the standard one-photon excitation process (see section 3.6). This is manifest as a drop in QD brightness when measuring over a wavelength window close to the peak of a factor of ~2 when changing the local temperature from 30 °C to 50 °C, and thus potentially is a good probe for investigating temperature changes relevant to biological samples. It has been tested as a proof-of-principle to measure the local temperatures inside human cancer cells.

There are two principal weaknesses of QDs. One is that they undergo a photophysical phenomenon of *blinking* (Figure 3.3G). Several single-molecule fluorophores also undergo blinking of some sort, which is essentially a reversible transition between photoactive (*light*) and inactive (*dark*) states, so in other words the tag appears to be bright then momentarily dark, but in a *stochastic* manner (Nirmal *et al.*, 1996). In general, for organic fluorophores these blinks are more prevalent at excitation intensities higher than those which would normally be used for fluorescence imaging, with typical dark state dwell times of ~tens of milliseconds or less, which is often sufficiently fast to be in effect averaged out during the typical imaging capture times used for fluorescence microscopy of ~100 ms or more. QDs, however, blink more appreciably at lower excitation intensities with longer dwell times of dark states more comparable to the time scale of typical fluorescence imaging; this can make it difficult to assess whether what you see from image to consecutive image in a continuous acquisition is the same tagged biological molecule or a different molecule that has diffused into the field of view from elsewhere.

The second disadvantage is that the real diameter of QDs in practice can be more like 15–20 nm if, as is generally the case, the core is further coated with a solvent-protective shell (typically of *zinc sulphide*) and a polymer-matrix *functionalization layer* (to permit chemical conjugation to a given biological molecule). The size can increase further still to ~30 nm if, for example, an antibody label is also attached to the QD to allow more specific binding to a given biological molecule. This is then an order of magnitude larger than a typical biological molecule in a cell. This results in the same type of steric impairment of function as found in laser darkfield, and suffers similar difficulties of getting the QD into the cell in the first place, especially with prokaryotes.

3.5 Fluorescent proteins, SNAP/CLIP-Tags and HaloTags

The most useful fluorescent tag for current single-molecule research is the so-called *fluorescent protein* (FP). In general, these are photophysically inferior to the other tags discussed in the previous section (for example, they are less bright, absorb excitation light less efficiently, and are less photostable and so will photobleach after emitting fewer photons than typical equivalent organic dye molecules). Despite this, significant insight has been gained into the behaviour of protein molecules inside living cells since the early 1990s using FPs (for a good collection of reviews on structure, photochemistry and application of FPs see the issue of *Chemical Society Reviews*, 2009 listed in the references).

The formal discovery of FPs was made in the 1960s when it was found that a species of jellyfish called *Aequorea victoria* produced a naturally fluorescent molecule called green fluorescent protein (GFP) (Shimomura *et al.*, 1962), which is now known to be utilized by many different marine organisms. The first written account of such *bioluminescence* was actually ca. 2000 years ago in 77 AD. The Roman historian Pliny the Elder was describing various medicinal remedies derived from aquatic animals, writing that the 'pulmo marinus' creature (this translates from Latin as 'aquatic lung', a small jellyfish which is now known to be the 'mauve stinger' we call *Pelagia noctiluca*, an apt Greek/Latin phrase meaning 'night light of the sea') appeared to produce a glowing substance:

> If wood is rubbed with the pulmo marinus, it will have all the appearance of being on fire; so much so, indeed, that a walking-stick, thus treated, will light the way like a torch (Bostock and Riley, 1855).

A crucial breakthrough in the use of GFP came when the gene was sequenced in the early 1990s, and researchers were able to use genetics techniques to introduce its DNA code into several other organisms from different species, including bacteria and nematode worms, which demonstrated that no proteins specific to the jellyfish were actually required to generate this bioluminescence. Using further techniques of DNA manipulation, the gene encoding GFP has been modified to make it brighter and to glow with different colours (Tsien, 1998). It can then be fused directly to the DNA of a gene encoding a completely different protein from a completely different organism and when the genetic code is read off during transcription (see Chapter 2) the protein encoded by this gene will be fused to a GFP molecule, meaning that each of these protein-fusion molecules will in principle possess a single-molecule GFP fluorescent tag.

Several variants of GFP have now been discovered in many other classes of organism, including corals and crustaceans, and in the past ten years or so the use of the GFP family as *molecular reporters* for the location of tagged proteins in living organisms has increased enormously (see Yuste, 2005 for a brief review). They are widely used as non-invasive probes to study different biological systems, from the level of whole organism tissue-patterning down to single individual cells, including monitoring of protein–protein interactions and measurement of a cell's internal environment including the concentration of protons (i.e. pH) as well as ion-sensing.

The basic structure of a fluorescent protein is a barrel shape of 2–4 nm characteristic length scale, with the actual *electric dipole* which is responsible for the fluorescence formed from the reaction of three neighbouring amino acids to generate a *cyclic chromophore* which is enclosed by 11 β strands (Figure 3.3H). Genetic modification of the chromophore groups and the surrounding charged residues inside the core of the protein has resulted in a huge range of synthesized variants having different absorption and emission peaks, with the excitation wavelength spanning not only the visible light spectrum but now extending into the infrared. In addition, mutation of some of the outer residues in the barrel itself has resulted in mutants that mature into a fully functional folded conformation faster in the living cell and also have less risk of aggregating together via hydrophobic forces, and so are definitively *monomeric* tags.

An FP is larger than a single organic molecule; usually the DNA code for a short linker region of a few amino acid residues is inserted between the FP gene and that of the protein of interest to allow some degree of rotational flexibility to the protein–FP fusion construct, thus reducing some of the functionality issues relating to steric hindrance effects. In many biological systems the protein–FP construct can be inserted at the same location as the original protein gene, deleting the native gene itself (for example, this can be done relatively easily in prokaryotic cells), and thus the tagged protein is manufactured at roughly physiological concentration levels in the cell, typically the same as that of the native protein in equivalent untagged cell strains. Since the FP is fused at the level of the original DNA code, the tagging is in effect 100% efficient, which is a significant advantage when compared to the other fluorescent tagging technologies discussed previously.

However, there are disadvantages to the uses of FPs. The principal ones are related to the relatively *poor photophysics* compared to typical organic dyes of similar emission wavelengths at corresponding excitation intensities – they are relatively dim and bleach relatively quickly. For example, the most common variant of GFP used today, so-called enhanced GFP, or *eGFP*, is 2–3 times less bright and will photobleach after emitting typically $\sim 10^6$ photons, ~ 10 times fewer than for comparable organic dyes.

Also, although improvements have been made recently in developing faster maturing FPs, this *maturation* is still an issue for experiments performed in the living cell. Essentially, when the FP–protein fusion construct is transcribed from mRNA translated from the DNA code, it still needs to fold into its final three-dimensional structure and undergo several chemical modifications until its final photoactive conformation is reached (the slowest, and thus *rate-limiting* step, is usually an *oxidation* stage in most FPs). The amount of time this maturation process takes depends on both the FP and the cell type, but typically for a population of freshly made enhanced GFP molecules it will take ~ 40 min for 50% of them to reach photoactive maturity. This means that there is always likely to be a small proportion of immature GFP present in a cell during a given fluorescence imaging experiment, which is thus 'dark'.

In addition, the FP itself results in some potential steric hindrance of function of the native protein which is tagged. For example, GFP contains 238 amino acids with a physical size which is comparable to that of the protein under study. Although much research time is invested in experimenting with different types and lengths of linkers between the protein and the FP to minimize such impairment, it is not unusual for functional activity to be reduced by $\sim 50\%$ (as assessed by the speed of movement of an FP-tagged 'motor' protein compared, if feasible to do so, against the untagged version, as discussed later in Chapters 8 and 9). In addition, in all but a few exceptional cases, the FP tag can only bind to the –N or –C terminus of a protein (see Chapter 2) as opposed to being expressed somewhere in the middle of a protein. This can limit its application in terms of understanding conformational changes within a molecule as part of its biological function.

And finally, many research groups using FP–protein technology perform their experiments using artificial *over-expression* conditions (for example, in bacteria this would involve using FP–protein encoded not into the native chromosome but onto a small engineered circular plasmid of DNA, separate from the cell's chromosomal DNA, which has been delivered into the cell from outside typically by creating holes in the cell membrane using a process called *electroporation* and allowing the plasmid to diffuse through from the external culture medium). This is primarily because it is often difficult to obtain a functional cell without having unnaturally high levels of the FP–protein construct (possibly due to steric hindrance effects reducing the level of integration into large molecular complexes, and so having a locally higher concentration of the construct makes successful integration more likely). But such over-expression can result in non-physiological behaviour elsewhere in the cell, including the generation of artifactual *aggregates* of the FP–protein construct.

In relatively simple prokaryotic cells and some single-celled eukaryotes (for example yeast) it is technically possible to delete a native gene coding for a given protein entirely and replace it with a suitable FP–protein fusion with reasonably high efficiency, thereby producing close to native levels of expression of the FP–protein (and in general relying on modifications in linker properties to increase functionality). However, in more complex eukaryotic cells (such as those in mammals) this is technically much harder to achieve since the efficiency of generating a full deletion mutant is low, so in practice the numbers of proteins which have been investigated using this *genomic encoding* of FPs is much larger for prokaryotes and simple eukaryotes. Therefore, it arguably has more limited biomedical relevance.

Some of the disadvantages of FPs have been overcome by combining the same type of fluorescent protein fusion technology to generate a highly specific primary probe fused covalently to the DNA of a protein of interest, which is not fluorescent itself, as used in the so-called *CLIP-Tag* or the closely related *SNAP-Tag* technology. Here, the primary tag is expressed with the protein itself inside a living cell and in most applications consists of a modified DNA repair protein called *O6-alkylguanine-DNA alkyltransferase* (AGT). Then, the cell can be incubated with a secondary probe that will bind with very high specificity to the primary probe.

HaloTags employ a similar strategy to SNAP/CLIP-Tags (Los *et al.*, 2008). Here, a genomically encoded protein tag (the HaloTag) is fused to the DNA of a protein of interest. The HaloTag is a modified haloalkane dehalogenase enzyme of molecular weight ~33 kDa, which is designed to bind covalently to synthetic ligands which comprise a chloroalkane linker attached to a variety of useful molecules, including bright organic fluorescent dyes, which can be incubated with the cells to be labelled fluorescently. The covalent bond formation between the HaloTag and the chloroalkane linker is highly specific, occurring rapidly under physiological conditions, and is essentially irreversible.

The advantage of SNAP/CLIP-Tags and HaloTags compared to other secondary labelling approaches is that the protein itself is not directly labelled – which might otherwise significantly affect biological functionality. There are still disadvantages in that secondary probes may themselves be large. Also, although the AGT has a 20% smaller molecular weight than a typical FP, the effective radius is only smaller by ~5% and so it actually suffers similar steric hindrance issues, and in fact the HaloTag is marginally larger than a typical FP. The most significant advantage is that they permit much brighter, more photostable organic dye fluorophores to tag the protein of interest via a secondary probe, which in some cases allows certain biological questions in living cells to be addressed which would be difficult using the dimmer FP equivalents (most notably when probing for protein–protein interactions, as discussed in the section 3.9 on 'FRET' below).

However, despite the flaws in fluorescent protein technology, their application to single-molecule imaging has dramatically increased our understanding of a large number of fundamental biological processes in living cells (which is reflected in their appearance in several chapters of this book). In addition, highly pH-sensitive variants of fluorescent proteins have now been developed, for example a derivative of GFP called *pHlourin* (see Chapter 7 for its application in studying the cellular process of *exocytosis*). This molecule has increased brightness sensitivity to pH at long excitation wavelengths but is insensitive to pH change if excited at shorter wavelengths, and so is used as a powerful *ratiometric live-cell pH indicator* at a highly localized level (the fluorescence intensity at shorter wavelengths can be used to normalize the measured signal at the longer excitation wavelength for the total amount of dye present).

A particularly useful tool for researchers utilizing fluorescent proteins in live cells is the *ASKA library* (Kitagawa *et al.*, 2005). This stands for '*A* complete *S*et of E. coli *K*-12 ORF *A*rchive' and is a collection, or library, in which each open reading frame (ORF), i.e. the region of DNA between adjacent start and stop codons which contain genes as well as certain non-coding sequences (see Chapter 2), in the well-characterized bacterium *Escherichia coli* has been fused with the DNA sequence for the yellow variant of GFP, YFP. The library is stored in the form of plasmids of DNA which can be inserted into *E. coli* bacteria and which will express in the presence of an inducer chemical called IPTG (see Chapter 9).

Thus, in principle each protein product from most bacterial genes is available to study using single-molecule fluorescence microscopy, and many of them have equivalent very similar proteins in more complex eukaryotic cells. The principal weakness with the AKSA library, however, is that the resultant protein fusions are all expressed at cellular levels which are far more concentrated than those found for the native non-fusion protein. This is a result of the IPTC expression system employed, which may produce non-physiological behaviour as discussed previously and, as a consequence, many research groups are now attempting to develop non-plasmid genomic fusions in which the expression levels are close to the *wild-type* native values.

3.6 Illuminating and detecting fluorescent tags

There are several methods available in practice which allow fluorescently tagged single biological molecules to be excited and detected. These include camera-imaging methods using *widefield* illumination modes, such as *epifluorescence* and *oblique epifluorescence*, as well as narrower illumination modes such as *slimfield*, used normally in combination with a high quantum-efficiency camera detector. They also include spectroscopic approaches such as *fluorescence correlation spectroscopy* and *scanning confocal microscopy*.

3.6.1 Widefield modes of epifluorescence and oblique epifluorescence

Widefield is so called simply because it excites a wide field of view of the sample. Epifluorescence is the most standard form of fluorescence illumination and essentially involves focusing a light beam (either from an incoherent light source such as a mercury-arc lamp or from a coherent laser source) onto the *back focal plane* of an objective lens, centred on its *optical axis*, resulting in an excitation field which is roughly constant with height into the sample from the microscope slide or coverslip surface, but which has, typically, a radially symmetrical two-dimensional Gaussian

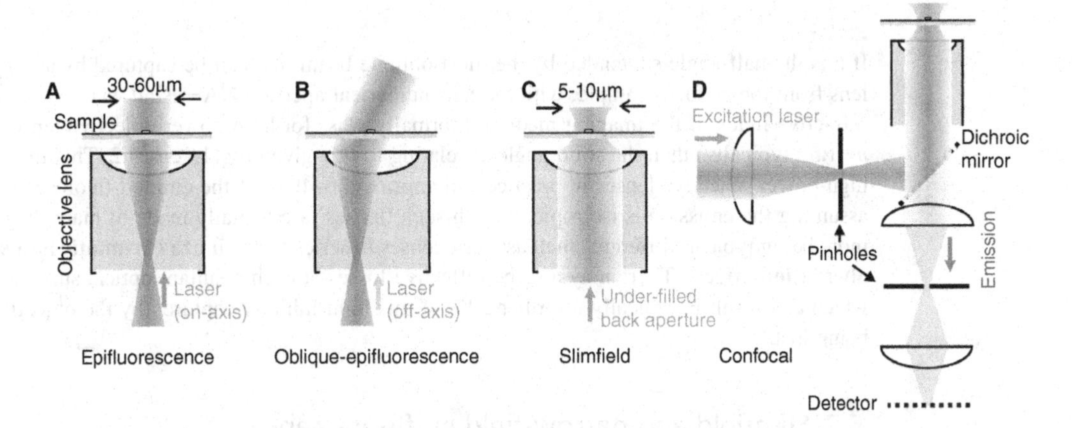

FIGURE 3.4 Fluorescence illumination. Schematics illustrating common modes of laser excitation for fluorescence: **(A)** epifluorescence, **(B)** oblique epifluorescence, also referred to as 'variable-angle epifluorescence' and 'pseudo/quasi TIRF' illumination, **(C)** slimfield and **(D)** confocal illumination.

profile laterally in the sample plane, with a typical full width at half maximum in the range 30–60 μm (Figure 3.4A).

'Epi' illumination refers to the excitation and emission light being channelled via the same objective lens in effectively opposite directions. In principle it is possible to have 'trans' fluorescence illumination, in which the excitation and emission beams travel in the same direction, by illuminating the sample from the opposite side relative to where the objective lens is. This method is not used significantly in practice since the proportion of unblocked excitation light entering the objective lens is much higher, resulting in a higher effective background noise signal.

Since a laser beam can be tightly focused into the objective lens *back aperture* there is room to move the focus laterally away from the optic axis, with the effect that the angle of the emergent beam becomes more inclined, generating oblique epifluorescence – this is essentially synonymous with other approaches described as *variable-angle epi-fluorescence*, *oblique epi illumination*, *pseudo-TIRF*, *quasi-TIRF* and *highly inclined and laminated optical sheet illumination* known as HILO (which selectively combines in-focus high spatial frequency information with low frequency components to generate thin optical sections from deep within a relatively thick sample) when applied to generate fluorescence – an arrangement which has similarities to laser darkfield illumination (Figure 3.4B).

This approach still generates a uniform excitation intensity parallel to the direction of the emergent beam (so typically will uniformly excite a whole cell), but arguably has marginally lower back scatter reaching the camera detector from back reflections of non-biological structures such as the coverslip or microscope slide itself (though note, modern laser *notch rejection* filters are exceptionally good at blocking these back reflections anyway so it is not clear how advantageous this approach is in this case). If the focus in the back focal plane is moved laterally beyond a critical limit then instead of the emergent beam being transmitted through the sample it is *totally internally reflected* – this is utilized in so-called *TIRF* which is discussed with the super-resolution techniques in section 3.9. Fluorescence emissions are then captured by the objective lens (the larger the NA the more fluorescence emissions are captured, typical NA values for high-power objective lenses being 1.2–1.6).

PHYSICS-EXTRA

If θ is the half-angle subtended by the most oblique beam that can be captured by an objective lens from the in-focus sample, with the lens numerical aperture NA=$n\sin\theta$ where n is the refractive index of the imaging medium (normally glass for high-power objective lenses, and so n=1.52 typically) then the solid angle Ω relating to θ is given by $(1-\cos\theta)/2$. This means that high-power objective lenses in practice can **capture 20–40%** of the emitted fluorescence, assuming the emission is isotropic, though such 'lenses' are actually made of many (e.g. five or more being typical) different small aspheric lenses in series to minimize **chromatic and spherical aberration** effects. There may also be reflective losses at each resultant optical surface interface which can result in typically a further 20% of this light initially captured by the objective lens being lost.

3.6.2 Slimfield and narrow-field epifluorescence

By reducing the incident laser beam diameter prior to focusing onto the objective lens back aperture, the emergent epifluorescence excitation field can be shrunk down to a width of roughly 5–15 μm diameter, to yield *narrow-field* epifluorescence, which can be used both to improve the signal-to-noise ratio in large cells, for example by delimiting the excitation field laterally so only sub-cellular features of interest are illuminated (such an approach was used to monitor the organic dye-tagged nuclear pore complexes in the cell nucleus, see Yang and Musser, 2006) and to increase the local excitation intensity to permit faster sampling, as has been applied to monitor the diffusion of single lipid molecules in artificial lipid bilayer membranes with a time resolution of 5 ms (see Schmidt *et al.*, 1996).

A similar technique called slimfield epifluorescence shrinks down a collimated laser beam diameter so as to *under-fill* the back aperture of the objective lens, which results in an *expanded* laser focus at the sample (Figure 3.4C). This can be used to generate very small areas of high local excitation intensity of just a few micrometres diameter and has been applied in monitoring single fluorescent proteins at the millisecond level (see Plank *et al.*, 2009), and for investigating DNA replication in living cells at the single-molecule level (see Chapter 9).

3.6.3 Confocal microscopy

Confocal fluorescence microscopy can dramatically reduce the contribution from fluorescence intensity outside the sample focal plane by using a combination of two *pinholes* in *conjugate* image planes (i.e. they both appear to be in focus at the same time), in effect one in front of the laser light source and the other in front of the detector (Figure 3.4D). For fluorescence imaging of relatively large cells, this can result in a significant increase in the signal-to-noise ratio for the detected fluorescence.

In general, the laser excitation light is focused onto the sample, and either this focused beam is *raster-scanned* across the sample using upstream optics, or the sample stage is raster-scanned relative to a fixed laser focus, with the typical size of the confocal excitation volume being $\sim 10^{-18}$ m^3 or 1 femtolitre (fl), roughly the same size as a small bacterial cell. Fluorescence emissions are ultimately focused onto a highly sensitive single channel detector, typically a photomultiplier tube, and the one-dimensional time-series signals are reconstituted to form the two-dimensional image.

The primary disadvantage of this technique is the lack of speed, in that in scanning even just a typical single cell the speeds are limited to ~100 frames per second. This can be improved by using *high-speed spinning disc* (also known as *Nipkow* disc) confocal microscopy in which two mechanically coupled spinning discs are used to scan ~1000 laser foci across a sample in parallel. In principle this can permit imaging rates of up to ~1000 frames per second. The primary disadvantage here is that the incident excitation light is being split ~1000 ways so even though the laser is focused onto the sample there may potentially be too little intensity for use at the single-molecule level to overcome background noise from the detector readout and autofluorescence in a live-cell assay. However, for performing dynamic experiments on relatively bright, multiply labelled molecular complexes across a large sample area, this is often a technique of choice.

For single-molecule studies, confocal excitation is more often employed in a *non-imaging* capacity to excite fluorescently tagged molecules in *solution-phase*, such that they diffuse transiently into the confocal volume, and are excited into fluorescence resulting in a measurable photon burst on the detector, before the dye-tagged biomolecule diffuses out of the confocal volumes typically a few tens of milliseconds later. This technique may also be combined with FRET approaches (see section 3.9.8).

Recent developments have also been made in the use of *Bessel beams* as a scanning source. These have a potential advantage in obtaining greater information from above and below the focal plane, since a Bessel beam is not limited spatially in its axial dimension in the same way as a confocal volume is (see section 5.2).

3.6.4 Multi-photon excitation

A confocal excitation volume (i.e. a focused laser beam) is applied for multi-photon fluorescence excitation. Here, the initial elevation of a ground-state electron happens following the absorption of more than one photon of light. For visible light fluorophores this implies that the frequency of the excitation light will be reduced by a factor of two (for two-photon excitation) or three (for, more rarely, three-photon excitation), and so typical wavelengths used extend into the *near* infrared of ~1 μm.

The likelihood for two or more photons being in near enough proximity to allow this transition is much lower than for normal single-photon excitation because it is the product not only of two single-photon absorption cross-section areas, but also of the probability that two such photons will be coincident over a very narrow range of time. This means that the incident excitation intensity needs to be significantly higher, hence the use of a focused laser beam. This results in significantly less fluorescence excitation away from the centre of the laser which can improve image contrast in a large sample (for example, in large cells or even deep into tissues penetrating to depths of up to a few millimetres). In addition, infrared radiation is scattered less than shorter wavelength visible light, which again reduces the effective noise, and also the photodamage at these higher wavelengths is correspondingly less.

However, because the effective absorption cross-sectional area for this multi-photon process is significantly less than that for the single-photon excitation process, the input laser intensity may need to be much higher at its peak to produce a measurable fluorescence from the fluorophore used, which in itself can counteract the beneficial longer wavelength effect of less photodamage. The technique typically requires reasonably expensive pulsed laser sources, which has limited its more general application, but it has been used for imaging single biological molecules (Chirico *et al.*, 2001).

3.6.5 Optical lock-in detection

Optical lock-in detection (OLID) is an emerging contrast-enhancement technique which is approaching single-molecule precision and is likely to be highly relevant to single-molecule living cell studies in the near future (Marriott *et al.*, 2008). Fluorescent proteins in live cells are often expressed in small numbers but must be detected over a comparatively large background owing to a combination of camera noise and native autofluorescence from the cell. OLID involves lock-in detection of modulated fluorescence excitation on a class of *optically switchable* dyes through controlling the fluorescent and non-fluorescent states by suitable laser activation and de-activation which is wavelength dependent. Neither the native autofluorescence nor the camera noise responds to this optical switching activation and so only the specific tagging dye molecules will register a modulation signal of the same characteristic driving frequency as the excitation light.

By locking-in on these detected modulating signals in the fluorescence image, a *cross-correlation analysis* can be performed on a pixel-by-pixel basis to estimate a correlation coefficient for a fit to the original laser excitation *reference waveform*. Thus, a pixel-by-pixel two-dimensional reconstruction of this correlation coefficient provides a very high contrast image of the localization of the switchable OLID dye tags, with minimal contamination from background noise. The original OLID tags were organic dyes (for example *nitrospirobenzopyran*-based probes) and so suffered problems with delivery into living cells and non-specificity of binding. However, recent live-cell studies have used the genetically encoded *Dronpa* fluorescent protein, and have been able to generate exceptionally high contrast images of specific sub-cellular structures in cultured nerve cells, as well as in live mammalian and fish embryos.

3.6.6 Light sheet microscopy – SPIM

Light sheet microscopy, also known as *selective plane illumination microscopy* (SPIM), involves orthogonal-plane fluorescence optical sectioning, essentially lighting up a sample from the side. It was first formally put into practice to obtain three-dimensional images of a fixed *cochlear* sample from the ear of a hamster (Voie *et al.*, 1993). Here, the sample is optically sectioned using a very thin sheet of light generated from a *cylindrical* lens. The light is projected perpendicularly to the optic detection axis onto a single plane of a transparent specimen which has been labelled fluorescently.

Because only one plane of the specimen is illuminated, no light is emitted from regions outside the focal plane and, therefore, there are minimal out-of-focus components in the image. Recent improvement to SPIM include the ability to image live and relatively opaque samples, exemplified by live embryos of the fruit fly *Drosophila* (Huisken *et al.*, 2004), and the subsequent ability to track sub-cellular organelles such as nuclei in live zebrafish embryo samples over a duration of several tens of hours (Keller *et al.*, 2008). The most exciting recent development has been the ability to track particles at the single-molecule level; this has been used to image single mRNA molecules labelled with the red dye ATTO647 in the nuclei of living cells in salivary glands, at a depth of some ~200 μm from the surface (Ritter *et al.*, 2010).

3.6.7 Adaptive optics

One issue with attempting to image deeper into tissue samples (as in the SPIM method discussed above) is that the image quality degrades as a result mainly of *heterogeneity* in the refractive index of the multiple cellular layers. To a certain extent this image degradation can be corrected using adaptive optics. Adaptive optics is a technology designed to reduce light wavefront distortion, originally developed for use in astronomical imaging through the distorting effects of the Earth's atmosphere. In essence the techniques use some form of either *deformable mirror* or *micro-lens array* to correct an image by comparing it against some gold-standard image reference. These approaches are starting to be applied to real biological samples (see Ji *et al.*, 2010), with the techniques likely to be highly relevant to future single-molecule research, especially so for deep-tissue biomedical imaging applications.

3.7 Fluorescence correlation spectrosopy (FCS)

Here, fluorescently tagged molecules are detected as they diffuse through a confocal laser volume, generating a short pulse of fluorescence intensity at a given time t, $I(t)$, before diffusing away out of the excitation zone. The distribution of dwell times τ_D between separate pulses is a measure both of concentration of the tagged molecules and of their effective diffusion coefficient (Magde *et al.*, 1972), and the analysis performed looks at the time correlation in the fluctuations of the fluorescence signal due to random molecular diffusion through the confocal volume. In principle this is in one sense an ensemble average technique since the analysis requires a distribution of dwell times to be measured from the diffusion of many molecules through the confocal volume; however, each individual pulse of intensity is in general due to a single molecule, so FCS is rightly described as a single-molecule technique.

The optical setup is similar to that for confocal microscopy, utilizing a high NA objective lens both to generate the confocal laser excitation volume and to capture fluorescence emissions. However, here there is a fast real time acquisition card attached to the fluorescence detector output which can sample the intensity fluctuation data at typically tens of megahertz to calculate an *autocorrelation* function, $G(\tau)$, which is a measure of the correlation of the time-series of intensity pulses with itself when shifted by some amount of time τ such that:

$$G(\tau) = \frac{\left(I(t) - \langle I(t)\rangle\right)\left(I(t+\tau) - \langle I(t+\tau)\rangle\right)}{\langle I(t)\rangle^2} \tag{3.3}$$

Here, $I(t)$ is the intensity at time t and angular brackets '$\langle\rangle$' imply the 'expected' (or mean average over a long period of time) value of the expression inside the brackets. For most applications, the confocal volume can be approximated as a three-dimensional Gaussian shape roughly like a rugby ball, with the vertical axis parallel to the optic axis of the objective lens being longer than each lateral width w in the same focal plane as the sample by a factor a of typically ~2.5, whose approximately femtolitre volume is given by $a\pi^{3/2}w^3$. For diffusion in N spatial dimensions with effective diffusion coefficient D, the general equation relating the average square of the distance travelled, the so-called mean square displacement $\langle R^2\rangle$, for a diffusing particle after a time t is:

$$\langle R^2\rangle = 2NDt^{\alpha} \tag{3.4}$$

Here, the parameter α is the anomalous diffusion coefficient and varies in the range 0–1, such that a value of 1 represents *free* (or *normal*) Brownian diffusion. The micro-environment in many regions of the cell is often tightly packed with several different types of molecules (for example, certain parts of the cell membrane have a protein crowding density of 30–40%), which results in hindered mobility termed *anomalous diffusion* or *sub-diffusion*, with a typical α value of 0.7–0.8. The full form of the autocorrelation function for one given type of molecule diffusing in three dimensions through a roughly Gaussian confocal volume can be modelled as:

$$G(\tau) = G(\infty) + \frac{G(0)}{\left(1 + (\tau/\tau_D)^\alpha\right)\sqrt{\left(1 + (\tau/\tau_D)/a^2\right)}} \tag{3.5}$$

Usually, the experimental autocorrelation function is calculated from Equation 3.3 and fitted by Equation 3.5 to determine the parameters $G(0)$, $G(\infty)$, τ_D and α· $G(0)$ here is simply the mean number of diffusing molecules inside the confocal volume. For fluorescence emissions to be detected, any diffusing molecules need to be reasonably close (within a few hundred nanometres) of the focal plane, which means that the mean dwell time for a detected pulse can be approximated as the time taken to diffuse in the two-dimensional focal plane a mean square distance which is equivalent roughly to the lateral width w of the confocal volume (typically 200–300 nm). Using Equation 3.4, this implies:

$$\tau_D = \left(\frac{w^2}{4D}\right)^{1/\alpha} \tag{3.6}$$

Thus, by using the optimized value of τ_D from the fit of Equation 3.5 to the experimental autocorrelation data, and by rearranging Equation 3.6, the diffusion coefficient can be calculated. In addition, if there is more than one type of diffusing molecule in the system, so-called *polydisperse diffusion*, then Equation 3.5 generalizes such that the total observed autocorrelation function is simply the sum of the individual autocorrelation functions for the separate diffusing molecule types.

FCS can also be used to measure molecular interactions if the interacting molecules are tagged with different coloured fluorophores (most usually dual-colour labelling) whose spectral emissions can be separated optically and subsequently detected in separate channels, known as *fluorescence cross-correlation spectroscopy* (FCCS). A modification of this technique uses dual-colour labelling but just one detector channel which captures intensity only when the two separately labelled molecules are close enough to be interacting, known as FRET-FCS (see section 3.9).

By scanning the sample through the confocal volume, FCS can also generate a two-dimensional map of mobility parameters across a sample, which has been utilized to measure the variation in diffusion coefficient across different regions of large living cells. As with scanning confocal microscopy, speed is a limiting factor, but these constraints can be largely overcome by using a spinning-disc system. The main weakness of FCS in general is its relative insensitivity to changes in molecular weight, M_w. Different types of biological molecules can differ relatively marginally in terms of M_w, however the outputted parameter of dwell time τ_D varies typically as $\sim M_w^{1/3}$, and so FCS is relatively poor at discriminating different types of molecules unless the difference in M_w is at least a factor of ~4.

FCS measurements can also be combined with simultaneous AFM imaging (see Chapter 4). For example, researchers have been able to monitor the formation and dynamics of so-called *lipid rafts* (see Chapter 8) in artificial lipid bilayers using such approaches (Chiantia *et al.*, 2007).

3.8 Fluorescence lifetime imaging (FLIM)

The actual lifetime of the excited fluorescence state (Figure 3.2C) of 10^{-9}–10^{-7} s can be measured by synchronizing fluorescence detection with the excitation using a fast pulsed laser source that can generate pulses of at least tens of megahertz frequency. This is normally done using a confocal-type imaging arrangement, with the sample being scanned laterally through the confocal volume resulting in two-dimensional fluorescence lifetime images. This is useful since the fluorescence lifetime of many fluorophores is dependent upon the local chemical environment, for example the pH and the presence of various ionic species, and therefore FLIM can potentially map out these changes in microenvironment across a sample.

This method has been applied at the single-molecule level in the test tube (for example, see Tinnefeld *et al.*, 2000) but has limited current application in living cells. Molecular interactions may also be monitored using a variant of the technique called FLIM-FRET which utilizes the fact that the lifetime of a fluorophore may change if energy is transferred to or from it from another fluorophore in very close proximity (see section 3.9 for a more detailed discussion of FRET).

3.9 'Super-resolution' techniques

As alluded to at the start of this chapter, when objects are viewed using light at a distance which is greater than several wavelengths away (the *far-field* regime), either scattered or emitted from the object, then optical diffraction of the light becomes significant. As a result, the light intensity from 'point source' emitters (as approximated for example by very small fluorescent tags such as single molecules or even QDs and small fluorescent nanospheres) appears to spread out due to diffraction by convolution with a point spread function, which is determined by the imaging system. The effect is that in a typical high magnification optical microscope such a point source of light will be seen as a much larger blurry spot of light with effective radius w typically 200–300 nm. This is usually limited by the NA of the objective lens and the wavelength λ of the light, such that $w = 0.61\lambda/\text{NA}$, with visible light having a wavelength range of roughly 400–700 nm and typical NA values being 1.2–1.6 (Figure 3.5A).

This formula is well known to physicists as the lateral diffraction optical resolution limit, or the *Abbe limit*, and is typically two orders of magnitude larger than any single biological molecule we actually want to detect, which may appear to limit the application of light to probing molecular processes in the cell. The axial resolution is worse still, with the point spread function stretched further than the lateral extent by a factor of about 2.5 for typical high-power objective lenses as we saw for FCS, and is more typically 500–700 nm. However, this is not as bad as it seems. There are in fact many so-called 'super-resolution' techniques which in effect allow us to break this resolution limit (for a review see Chiu and Leake, 2011).

PHYSICS-EXTRA

The first formal diffraction limit was described by **Abbe** in 1873, using an angular resolution of $\sin\theta = \lambda/2nD$ where θ is the first-order diffraction angle from a standard rectangular aperture diffraction-grating of aperture width D in a medium of refractive index n for light of wavelength λ. If the aperture is a circle (for example, a lens) the diffraction pattern is an **Airy disc**, a circular intensity function having a central peak containing ~84% of the intensity, with multiple outer rings containing the rest of the diffracted light interspaced by zero-intensity minima. The formula

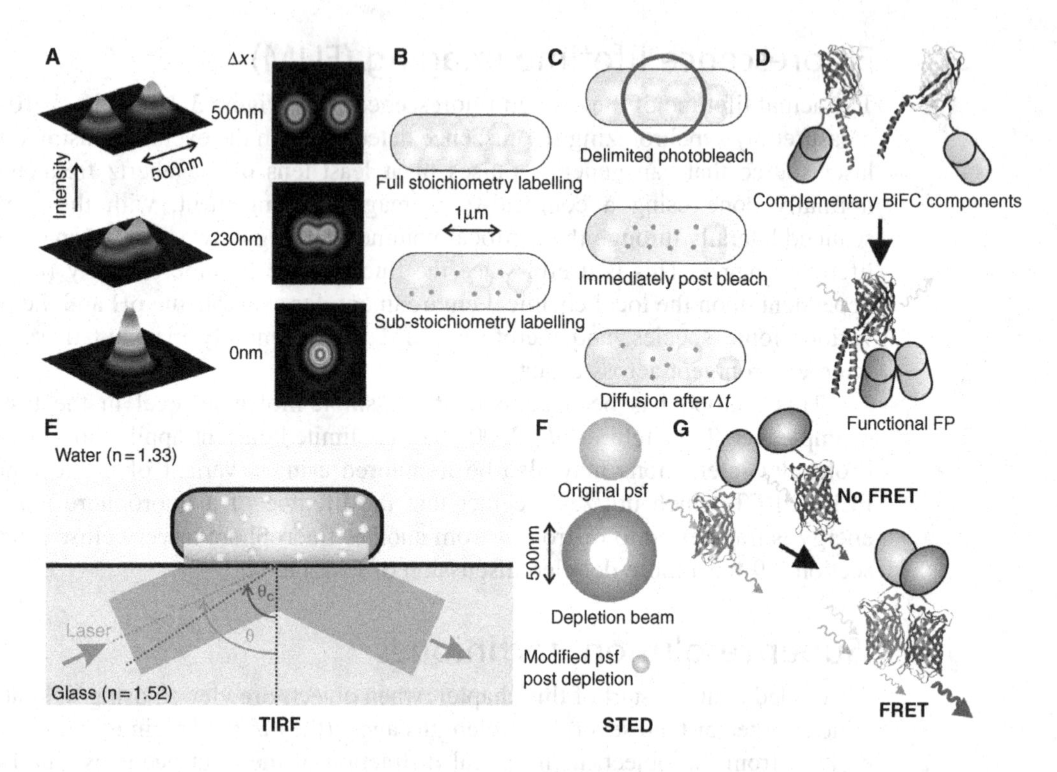

FIGURE 3.5 Super-resolution. **(A)** Contour (left panel) and plan-view (right panel) simulations of two juxtaposed point spread function (psf) images, with separation of intensity centroids Δx shown (here the resolution limit is given by 230 nm). **(B)** The effects of sub-stoichiometric labelling with the psf of labelled molecules (large, bright circles) and the unlabelled molecules (small, dark circles) shown. **(C)** Pre-bleaching a sub-population of fluorophores, followed by diffusion after a time Δt of unbleached fluorophores into the delimited bleach zone (circle). **(D)** Two complementary BiFC components indicated, with the 'missing' components of each structure shown. Also shown is the leucine α helical zipper, which fuses when the tagged molecules are very close. **(E)** Schematic of a cell in water-based medium on a glass microscope slide, illustrating the principles of laser excitation TIRF (here using the objective lens method), with fluorophores inside the depth of penetration of the evanescent field being excited. **(F)** Schematic illustrating the relative sizes of the STED beam and pre- and post-depleted psf images. **(G)** Cartoon illustrating two proteins (ellipsoids) tagged with donor and acceptor FPs (for example, GFP and mCherry respectively) for use in FRET, excitation and emission photons indicated by arrows of varying size.

is modified to give $\sin\theta \approx 1.22\lambda/nD$ where θ is now the angle of the diffracted beam to the first-order zero-intensity minimum, D is the aperture diameter. The astronomer **Rayleigh** subsequently suggested a semi-arbitrary criterion for optical resolution that two optically magnified point objects (stars in his case) could be resolved using circular optics if the central maximum of the Airy disc of one object was in the same position as the first-order minimum of the other object. For optical microscopy, a sample at a focal distance f from the objective lens will be resolved from a similar nearby object a small distance ΔL away according to the **Rayleigh criterion** if $\Delta L \geq f \sin\theta/2$, θ relatively small. The NA of the objective lens is given by f/nD, implying the **Abbe limit** for the minimum ΔL of $\sim 0.61\lambda/\text{NA}$. Another resolution criterion calls for there being no local dip in intensity between two neighbouring Airy discs. This leads to the **Sparrow limit** of $\sim 0.47\lambda/\text{NA}$ which some argue is a more sensible condition (the experimental limit was first measured by the astronomer **Dawes** who concluded that two stars of similar brightness could be just resolved if the dip in intensity between their images was at least 5%, which is closer to the Sparrow limit than to the value predicted from the Rayleigh criterion by ~20%).

3.9.1 Iterative fitting (FIONA-type) approaches

The simplest approach involves fitting the exact theoretical point source function, or a good approximation to it, to the experimentally measured blurry spot of light. In doing so the centre of the light intensity (the *intensity centroid*, which is generally where the single molecule emitting the light actually is) can be estimated to a very high spatial precision, Δx, which is superior to the optical resolution limit (this is analogous to knowing with reasonable confidence where the peak of a mountain is even though the mountain itself is very large):

$$\Delta x = \sqrt{\frac{s^2 + a^2/12}{N} + \frac{4\sqrt{\pi}s^2b^2}{aN^2}} \tag{3.7}$$

This equation (see Thompson *et al.*, 2002, though the first experimental application of Gaussian fitting of diffraction-limited spots of light can be found in Gelles *et al.*, 1988, where it was used to determine the centre position of 190 nm diameter kinesin-coated microspheres conjugated to microtubules from non-fluorescence DIC images to within 1–2 nm precision) models how the spatial precision for pinpointing the centre of the spot of light in an image varies with the effective width s of the spot when approximated by a Gaussian function (roughly the same as the width w above), pixilation and background noise parameters a and b largely due to the camera detector, and number of sampled photons of light N captured from the point-source object. What this equation shows is that, if N can be made very large, Δx varies roughly as $s/\surd N$ – in other words, not only is Δx less than s (i.e. 'super' resolution), but if N is greater than ~250^2 or ~60,000 photons then Δx can in principle be less than 1 nm. This is the basis for so-called *fluorescence imaging with one nanometre accuracy* (FIONA) (see Park *et al.*, 2007).

The actual magnification of the point spread function image on the camera detector must be set such that each equivalent pixel edge length is smaller than the standard optical resolution limit. *Nyquist–Shannon information theory* indicates that the pixel size should be set to at least the same length as the standard deviation width of a typical point spread function resolution – anything greater than this and the point spread function will be under-sampled by the array of pixels on the camera. In practice, Equation 3.7 can be used with realistic estimates on the parameters s, N and b to determine an optimal value of a which minimizes Δx and thus generates the best spatial precision. Typical equivalent values of pixel edge sizes when divided by the total imaging magnification (i.e. the effective size of the pixel if it were imaged onto the original sample) are around 50 nm.

Although now routinely employed by several research labs, this technique is still technically challenging – the *overall* efficiency of most high-magnification microscopes is typically around 10% or less, in other words nine photons out of every ten emitted are not detected by the camera. This means that to achieve sub-nanometre precision a fluorescence emitter needs to emit ~600,000 photons, which in practice can be achieved for a typical bright single organic dye molecule imaged under normal conditions for a total duration of ~1 s. To image faster than this requires dividing up the intensity signal into smaller and smaller time windows, hence sampling fewer photons in each unless the emission flux from the dye is increased in proportion.

For example, to maintain the same sub-nanometre value of Δx in each image but to sample at a faster 'video-rate' of ~25 frames per second would require ~600,000 photons to be emitted every 40 milliseconds instead of every second. For many organic dyes and QDs this is feasible, but for less stable fluorophores such as FPs there is an issue in that irreversible photobleaching limits the total number emitted to typically ~10^6

photons for GFP (for yellow variants such as yellow fluorescent protein, YFP, this is more like $\sim 10^7$ photons, though for some FPs which emit in the blue such as cyan fluorescent protein, CFP, this can be more like $\sim 10^5$ photons). Thus, the total number of frames that can be acquired before a typical FP molecule is photobleached is $\sim 10^6$/600,000 or ~15–20 frames. If these frames are acquired consecutively this means that less than 1 s of information is available, which potentially limits the investigation to reasonably fast processes. If a faster sampling time is required, such as ~5 ms per frame, then as few as 2–3 frames can be acquired before photobleaching.

The total time scale for an experiment can be expanded by in effect *strobing* the illumination so that the available frames are spread out over several seconds if required (see Chapter 9), but it still places a fundamental limit on the amount of information that can be gathered from FPs at the nanometre-precise level relevant to the length scale of single biological molecules. Many real experimental studies involving FIONA-type iterative fitting for single particle tracking sacrifice some of the precision in favour of faster sampling times, so for example there are several video-rate membrane protein diffusion studies which have a spatial localization precision of closer to ~10 nm.

FIONA-type single-molecule imaging approaches have been applied to multi-colour fluorescence imaging, for example where one protein of interest in a cell is labelled with one colour fluorophore while a different protein in the same cell is labelled with a different colour fluorophore, and the two emission signals are split optically on the basis of wavelength to be detected in two separate channels. This has led to the generation of far too many acronyms to be sensible, including *single-molecule high-resolution co-localization* (SHREC) which can measure relative separations of the different coloured fluorophores larger than ~10 nm (see Warshaw *et al.*, 2005 for the invention of the technique, and Churchman *et al.*, 2005 for the invention of the name). There are also related techniques called *single-molecule high-resolution imaging with photobleaching* (SHRImP) (see Gordon *et al.*, 2004) and *nanometer-localized multiple single-molecule fluorescence microscopy* (NALMS) (see Qu *et al.*, 2004), which were invented independently but are essentially the same technique. They use photobleaching to localize two closely placed fluorophores to nanometre accuracy. Another iterative Gaussian-fitting technique has emerged recently called TALM (single particle tracking localization microscopy) which combines iterative fitting of the intensity centroid with single particle tracking of mobile tagged proteins (Appelhans *et al.*, 2012).

These various 'localization of point spread function' approaches work provided the local concentration of fluorescently tagged molecule or complex of interest is sufficiently low so that their average separation is greater than the optical resolution limit – in other words, these approaches allow you to pinpoint the position of a single biological molecule very precisely provided there are not very many of them about. The value of the limiting concentration when averaged over a whole cell depends upon certain geometrical constraints (for example, whether particle movement is in three spatial dimensions, or is confined to two as in the cell membrane) and the real spatial distribution (for example, the existence of localized *hotspots* of activity in the cell). As a rough guide it is equivalent to $\sim 10^{17}$ particles per litre for the case of tagged particles inside a cell, which corresponds to a molar concentration of ~100 nM (for the interested reader, an elegant mathematical treatment for modelling the nearest-neighbour distance can be found by one of the great mathematicians of modern times in Chandrasekhar, 1943).

As seen previously, a typical bacterium a few micrometres in length has a volume of ~1 fl, and at this limiting concentration each cell contains of the order of a hundred such molecules in total. There are many biological systems of interest that have total numbers of molecules in the cell of a given protein within that system in excess of this, in addition to those cases of above normal expression levels of FP-fusion labelling. This potentially presents a problem, since if the concentration of fluorescently tagged molecules is too high then all that can be seen from a fluorescence image is effectively diffuse fluorescence across the cell which cannot be analysed using a FIONA-type approach.

In practice, what often happens is that single fluorescently tagged molecules, most commonly fluorescent proteins, integrate into multi-molecular complexes in living cells (see Chapter 8) – these complexes therefore appear brighter than a single fluorophore tag and can be distinguished as distinct spots of light at higher concentration levels of the local surrounding single fluorophore tagged molecules than would be expected from the simple analysis above. Even so, there are still upper threshold limits to concentration that are ultimately encountered. Also, in some test tube experiments the tagged molecule of interest is relatively tightly bound to some structure which is firmly immobilized to the microscope slide, either fully immobilized itself or constrained to move just along that structure whilst still being bound to it (for example, certain motor proteins running along a filament track).

This means that any unbound molecule can be washed away by using a simple flow-cell arrangement in the microscope. This works well for certain so-called *processive* enzymes, often motor protein enzymes which stay bound to their substrate (a filamentous track in the case of *linear* motor proteins), for prolonged time periods (for example, see Chapter 9). However in the general case many different types of biological molecule perform some of their functions by binding to and dissociating from sub-cellular structures at a noticeable rate. These are therefore *non-processive* and difficult to investigate using the immobilized-structure test tube assay above. For these systems it may be important from a physiological standpoint to have a high local concentration of molecule in the vicinity of certain structures to ensure significant molecular turnover, which thus presents the nearest-neighbour problem when imaging optically using standard far-field methods.

FIONA methods are now routinely applied to two-dimensional tracking. In many cases two spatial dimensions are adequate, either because of the significant prevalence of TIRF illumination and/or the use of high NA objective lenses for imaging which have reasonably narrow *depths of field* (a measure of the region above and below the focal plane in which particles are still reasonably in focus), and/or for studying biological processes which are intrinsically limited to the cell membrane. However, with improvements in cytoplasmic imaging techniques and the desire to investigate biological processes deep inside relatively large cells, there has been a greater need for developing three-dimensional tracking methods. There are three broad classes of techniques for achieving three-dimensional tracking.

(i) *Multi-plane imaging*. Here the light emitted from a tracked particle is split to image it separately on two image planes which each have a slight distance offset, so that the particle will come into sharp focus on each plane at different relative distances above or below the sample focal plane. In practice this is often done using just two different planes and can be achieved by using separate pixel regions of the same camera detector as the respective planes, and hence doing a combination of Gaussian fitting of each to extrapolate the full *xyz* coordinates of the particles. This works well for bright particles, but is limited by having to reduce the equivalent photon flux incident on any individual detection plane as well as suffering from some uncertainty as to whether a tracked particle close to the focal plane is diffusing above or below it.

(ii) *Astigmatism.* Here, a cylindrical lens is used to image tracked particles on the camera, which has a different focal length corresponding to x and y axes. The result is that particles above or below the focal plane have an asymmetric point spread function, and so by measuring the different Gaussian widths in x and y the distance of the particle from the focal plane can be calculated. The prime drawback is that this method is relatively insensitive to changes in z less than ~100 nm.

(iii) *Double-helical point spread function method.* This technique uses phase optics which result in a lobed double-helix shaped point spread function at the camera whose size and shape can be used to estimate z, as well as pinpointing x and y. Although this method has a downside of requiring more expensive phase optics, the sensitivity in z is more than twice as good as the other two competing methods (for a good review of all three approaches see Badieirostami *et al.*, 2010).

3.9.2 Sub-stoichiometric labelling

There are several methods for overcoming the 'nearest-neighbour problem'. The simplest, and arguably least satisfactory, is to label your molecule population of interest *sub-stoichiometrically*. This involves labelling only a sub-population of all of the molecules of a given type, so that the total photoactive population is under the limiting maximum concentration for FIONA-type tracking (Figure 3.5B). This can be achieved relatively easily if the labelling method involves conjugation chemistry, namely by tinkering with concentrations of reactants and incubation time and temperature.

However, as a result there is a dark unlabelled population that may well be behaving in a different manner to the labelled population and affecting the experimental measurements as a result, and it becomes very difficult to demonstrate whether or not this is the case (essentially involving tedious controls at several different sub-stoichiometry labelling levels and monitoring the biological functionality of each). In addition, it is sometimes difficult to achieve consistent, reproducible results owing to experimental variables which are hard to control (for example the incubation temperature and the chemical endpoint of the conjugation reactions).

For FP-fusion over-expression systems it may be possible to control expression levels by changing the concentration of inducer chemicals (common chemicals for plasmid expression systems in bacteria are IPTG and arabinose, see Chapter 9), but the sensitivity of response to inducer concentration in these systems is often quite sharp (the range of inducer concentration over which the expression levels can be controllably varied reliably is relatively small) making this approach unreliable in general. Also, this approach cannot be applied to cases of *genomic* FP-fusion labelling.

A related approach involves labelling all of the molecules of interest, but then irreversibly photobleaching some proportion of the tagged molecules, by illuminating a cell, or part of a cell, with excitation light for a given duration prior to FIONA-type analysis (Figure 3.5C). This method is superior to sub-stoichiometric labelling in that there are not significant numbers of unlabelled molecules of interest in the cell, which would potentially have different physical properties from the tagged molecules such as mobility and rates of insertion into a complex so forth, and also it has the advantage of being applicable to cases of genomic FP-fusion labelling. This method has been used to monitor, for example, the diffusion of labelled proteins in the cell membrane of bacteria by using a high intensity focused laser bleach at one end of the cell to bleach locally a ~1 μm diameter region, and then to observe fluorescence subsequently at lower excitation intensity as labelled molecules diffuse stochastically from the unbleached regions back into the bleached zone, often allowing these individual particles to be tracked using iterative fitting based methods (see section 8.3).

A novel sub-stoichiometric labelling approach, applicable to studying biological processes which involve pairs of interacting proteins, is *bifunctional fluorescence complementation* (BiFC). Here, one of the interacting proteins is labelled with a truncated non-fluorescent part of an FP structure using the same type of DNA fusion technology as for standard FP-fusion labelling, while the other protein in the interacting pair is labelled with the complementary FP structure (essentially the bits which are missing from the label on the first protein). Each truncated FP complement tag also has a short α helical attachment made predominantly of leucine amino acids – two such α helices when near to each other will interact strongly to form a so-called *leucine zipper* (Ghosh *et al.*, 2000) which then facilitates the binding of the two complementary FP truncated peptides to form a fluorescent FP tag (Figure 3.5D).

Thus, even though the nearest-neighbour separation of the two proteins may be less than the optical resolution limit, a functional fluorophore is only formed when a pair of interacting proteins are sufficiently close (i.e. less than the length scale of a few nanometres of the FP itself). The formation of such photoactive BiFC complexes is stochastic and so often the nearest-neighbour distance between them is greater than the optical resolution limit, thereby permitting FIONA-type super-resolution fitting analysis to be performed on their localization.

Sub-stoichiometric labelling may also be used to monitor turnover and movement in protein assemblies, in a method called *fluorescent speckle microscopy* or FSM (Waterman-Storer *et al.*, 1998). Here, a cellular sub-structure is sub-stoichiometrically labelled resulting in a speckled appearance in conventional fluorescence imaging. This is in effect an identifiable pattern, and movement of the assembly as a whole results in a shift of this pattern which can therefore be measured relatively accurately without the need for any iterative Gaussian-type fitting procedure. This method has been applied to the study of microtubules in live cells.

3.9.3 Total internal reflection fluorescence (TIRF)

Another common approach to overcoming the nearest-neighbour problem is to delimit the excitation field in fluorescence microscopy to length scales which are less than the optical resolution limit. The most frequently employed method in bio-optical research labs is to use *total internal reflection fluorescence* (TIRF) microscopy (Figure 3.5E). This can technically also be classed as a widefield method (since laterally, in the plane of the sample, it generates a similar sized field to those of standard epifluorescence) but also utilizes a *near-field* effect axially (i.e. at a length scale range of less than a few wavelengths of light) to generate significant image contrast (for a rigorous discussion of the physics of TIRF see Axelrod *et al.*, 1984).

In its modern form this usually involves a laser beam of wavelength λ directed at an oblique angle of incidence θ to the interface between a glass microscope slide of coverslip and the water-based medium which is surrounding a biological sample. At an angle of incidence above a critical angle θ_c, total internal reflection of the incident beam occurs, sending the beam back from the microscope slide or coverslip away from the sample:

$$\theta_c = \sin^{-1}(n_{water}/n_{glass}) \quad (3.8)$$

Here, n refers to the refractive index. Using values for typical 'BK7' low-fluorescence glass employed in optical microscopy of 1.52 and for water of 1.33 this indicates a critical angle of

~61°. However, a by-product is the generation of an evanescent field into the water-based medium whose intensity $I(z)$ decays exponentially as a function of vertical distance z from the glass–water interface, governed by a depth of penetration exponential factor d:

$$I(z) = I(0)\exp(-z/d) \tag{3.9}$$

$$d = \frac{\lambda}{4\pi\sqrt{n_{glass}^2 \sin^2\theta - n_{water}^2}} \tag{3.10}$$

Typical values of d are in the range 50–150 nm. Thus, from Equation 3.9 it is clear that after ~100 nanometres the field intensity is ~1/3 of the value at the slide surface, whereas at ~1000 nanometres from the slide (the typical width of a bacterial cell), the intensity has dropped to just a few thousandths of a per cent. This means that only fluorophores very close to the slide are excited, whereas those in the rest of the specimen, or any present in the surrounding solution, are not. This delimited excitation field in effect creates an optical slice of ~100 nm depth, which can increase the imaging contrast to such an extent that detection of single molecules in living cells becomes possible above the level of background noise.

There are two predominant ways to achieve TIRF in practice, either via a *prism method* involving the biological sample positioned at the top of a typical microscope flow-cell with fluorescence emissions collected through the bulk to the solution, or more commonly via an *objective lens method* in which the sample sits at the glass coverslip bottom of the flow-cell with the objective lens used both for generating the evanescent field and for collection of fluorescence emission photons. Arguably the prism method has slightly lower noise due to the absence of any excitation scatter effects within the objective lens, but this also requires the use of a special *water-immersion* objective lens to image through the bulk of the sample solution to avoid incurring spherical aberration effects, and these lenses have a marginally lower numerical aperture than those used in the objective lens method, and so the photon collection efficiency is lower.

The biggest factor in deciding which TIRF method to use is probably the ease of optical alignment, which marginally favours the objective lens method. Looking at Equation 3.10 the term $n_{glass}\sin\theta$ is identical to the NA of such an objective lens, and so clearly to achieve TIRF using the objective lens method the NA must be greater than the refractive index of water, ~1.33. In practice, objective lenses are used with an NA of more like 1.4–1.5 to achieve TIRF.

Although the lateral resolution of TIRF is still diffraction limited at 200–300 nm, the axial resolution is set by the extent of the evanescent excitation field (i.e. ~100 nm) and so may in general be less than the optical resolution limit. Thus TIRF microscopy is rightly regarded as a super-resolution method. TIRF was first utilized in studying single biological molecules in the form of ATP molecules tagged with the organic green dye Cy3 binding to and dissociating from surface-bound myosin molecules (Funatsu *et al.*, 1995), with its first application to live-cell work being to image so-called epidermal growth factor receptors tagged with the organic dye Cy3 (Sako *et al.*, 2000, and see Chapter 7).

3.9.4 Stochastic activation, switching and blinking of fluorescent dyes

The techniques of *photoactivated localization microscopy* (PALM) (Betzig *et al.*, 2006), essentially synonymous with *fluorescence photoactivated localization microscopy* or

FPALM, and *stochastic optical reconstruction microscopy* (STORM) (Rust *et al.*, 2006) use photoactivatable or photoswitchable fluorophores respectively to allow a high density of target molecules to be labelled and tracked. These techniques use typically *ultraviolet* light stochastically either to *activate* a fluorophore from a non-fluorescent into a fluorescent form, which can then be subsequently excited at longer visible light wavelengths, or to *switch* a fluorophore from one colour emission to another. They have been implemented both with organic dyes (most commonly Dronpa) and with DNA-encoded fluorescent proteins (for example photoactivatable GFP, paGFP, and photoswitchable FPs such as Eos).

Both techniques rely on the photoconversion process to the final detected fluorescent state being stochastic in nature, thus allowing only a small sub-population to be present in any given image frame, thereby reducing the effective local concentration to below the optical resolution threshold and permitting FIONA-type localization fitting to be performed on individual detected fluorescent spots. Over many repeated cycles, this allows a super-resolution image to be built up of, for example, a spatially extended sub-cellular structure.

Existing disadvantages include the slowness of the total acquisition time, meaning that relatively dynamic processes cannot be investigated. Recently, however, a significantly faster method of STORM was demonstrated that utilized brighter organic dyes conjugated via SNAP-Tags (see section 3.5) which permitted dual-colour three-dimensional dynamic live-cell STORM imaging up to about two frames per second (Jones *et al.*, 2011), which may in the near future be applied more generally. In general though, samples are typically not alive because of the chemical fixation required to minimize unwanted movement within the sample. Also, the use of ultraviolet light to activate/switch the fluorophores is potentially highly damaging to biological material, especially since multiple repetition cycles are required to build up a full image – this is undesirable not so much because it could kill a cell relatively quickly as the samples are generally dead anyway, but rather because the presence of high local concentrations of free-radicals due to the ultraviolet irradiation could potentially modify and ultimately degrade the sub-cellular structures that are being investigated, thus producing pronounced artifacts.

In a similar manner, the reversible *photoblinking* of fluorophores has future potential for being applied to map out sub-cellular structures with super-resolution precision. Here, the stochastic nature of the photoblinking is utilized by carefully selecting a suitable range of excitation light intensity over which the fluorophore in question will undergo potentially several reversible blinking cycles before finally succumbing to irreversible photobleaching. It this way it may be possible to reduce effectively the local density of photoactive fluorophores in any given image frame to permit subsequent FIONA-type localization fitting as for PALM/STORM. As a technique this is in its infancy, but as proof-of-principle has been utilized in living bacterial cells to map out DNA binding proteins using the photoblinking of YFP (Lee *et al.*, 2011), which can be modified by fusing YFP to a membrane-localized protein to generate super-resolution details of cell membrane features in a technique imaginatively described as SPRAIPAINT (Lew *et al.*, 2011). This dye blinking method also works on brighter organic dyes and was named by other researchers *blinking assisted localization microscopy* or BaLM (Burnette *et al.*, 2011). A potential advantage over PALM/STORM is that it can be performed on living samples, and over faster sampling time scales.

KEY POINT

There is an unfortunate tradition in modern fluorescence microscopy in particular for competing research groups to 'invent' different techniques and then assign them a novel acronym. The observant reader should note that many such techniques are often in reality just variants of a few key, underlying methods, and is advised to **get to grips with the underlying science**, as opposed to attempt to memorize the less meaningful jargon!

3.9.5 Shrinking the point spread function

The standard Abbe limit can also be broken using techniques which reduce the physical size of the point spread function. These include *4Pi microscopy* (Hell and Stelzer, 1992), in which the sample is illuminated with excitation light from above and below using matched high NA objective lenses. The effect is to improve the axial resolution from the normal value of 500–700 nm to more like 100–150 nm; this equates to an almost spherical focal spot with a volume typically ~6 times less than that using in conventional confocal imaging.

Another increasingly common approach to shrink the size of the point spread function is to use *stimulated-emission depletion* (STED) microscopy (Hell and Wichmann, 1994), with related similar techniques known as *ground-state depletion* (GSD) and *saturated structured-illumination microscopy* (SSIM). In STED, the effective spatial extent of the point spread function is limited by inhibiting its fluorescence emissions from the outer regions for cases in which the nearest-neighbour separation of fluorophores in the biological sample is less than the standard optical resolution limit. This can be achieved by having a second 'stimulated emission' laser beam in addition to an excitation beam, whose intensity maximum is offset by 200–300 nm from the centre of the excitation beam (originally this was achieved using two offset beams, but now more commonly a doughnut shaped beam profile is used, generated using phase modulation optics, Figure 3.5F).

This beam has a slightly longer wavelength than the laser excitation beam and will stimulate emission from the fluorescence excited state (Figure 3.2C). The effect is to deplete this excited state in the outer doughnut region but leave a non-depleted central region whose area in the sample plane is much smaller than the original point spread function. In this way, researchers have been able to reduce the effective size of the point spread function to widths of less than 100 nm. In theory, the lateral width w of this effective point spread function, assuming an objective lens of numerical aperture NA, is then given roughly by:

$$w = \frac{\lambda}{2\mathrm{NA}\sqrt{1 + I/I_s}} \tag{3.11}$$

Here λ is the wavelength of the STED depletion beam with a 'saturating' intensity of I_s (the intensity required for reducing the fluorescence by depletion of the excited state by a factor of two for the fluorophore in question) and I is the excitation intensity centred on the fluorophore. Hence, for reasonably large I there is a $\sim 1/\sqrt{I}$, dependence on w, so in principle the width could be vanishingly small for very large I. STED microscopy does not improve the spatial resolution by narrowing the laser beams used since these are still unavoidably diffraction limited in extent, but instead utilizes the inhibited fluorescence from the depletion of the excited fluorescence state in surrounding fluorophores which are a distance less than the Abbe limit away. The limiting factor here is generally irreversible photobleaching, which can be improved by designing tailored STED dyes, but even so current resolution limits are a few tens of nanometres for most applications, more than an order of magnitude below the Abbe limit.

STED potentially has advantages over PALM/STORM in performing better in deep-tissue imaging, especially if combined with Bessel beam illumination (see section 5.2). In addition, the maximum achievable image sampling rates are currently higher for STED, at ~28 Hz, than for PALM/STORM, exceptionally up to ~2 Hz (but more generally an order of magnitude slower than this), and also of structured illumination at ~11 Hz (see the section below).

3.9.6 Near-field approaches

Optical effects which occur over distances less than a few wavelengths are described as near-field, which means that the light does not encounter significant diffraction effects. This is utilized in *scanning near-field optical microscopy* referred to as SNOM or sometimes NSOM (Hecht *et al*., 2000). For visible light applications this usually involves scanning a very thin optical fibre across a fluorescently labelled sample, with the excitation and emission light conveyed via the same fibre, and the vertical distance from the sample to fibre tip being kept roughly constant at less than a typical wavelength of the emitted light (the scanning technology is very similar to that used for *atomic force microscopy*, or AFM, which is discussed in detail in Chapter 4).

Here, the lateral spatial resolution is limited by the diameter of the optical fibre itself (typically as low as ~20 nm) with the vertical resolution limited by the reliability of the scanning (~5 nm). The main drawbacks are that scanning is generally very slow (several seconds to acquire a typical image) and the fluorescence imaging is limited to topographically accessible features on the sample (so not a relevant technique for imaging inside cells).

Near-field fluorescence excitation fields may also be generated from so-called *photonic waveguides*. Here, narrow waveguides are typically manufactured out of etched silicon to generate channels of width ~100 nm. By directing a laser beam through the silicon an evanescent excitation field can be generated in much the same way as for TIRF microscopy. Solutions containing fluorescently labelled biological molecules can then be flowed though a given channel and excited by the evanescent near-field. Many such flow-channels can be manufactured in parallel, with surfaces pre-coated by antibodies which then recognize different biological molecules, and this therefore is a mechanism to allow bio-sensing.

Until recently the sensitivity of such *optical microcavities* did not permit single-molecule capability. However, recent improvements using the so-called *whispering gallery mode* (WGM), in which an optically guided wave is *recirculated* in silicon crystal of typically circular shape, enhance the sensitivity of detection in the evanescent near-field to the single-molecule level (Vollmer and Arnold, 2008).

3.9.7 Structured illumination

Structured illumination (SI), also known as *patterned illumination*, is a super-resolution method which utilizes the *Moiré interference fringes* made between the sample and a specifically patterned illumination (Gustafsson, 2000). These are in effect equivalent to a beat pattern, which when measured in the *Fourier transform* of the image, in so-called *reciprocal* or *frequency space*, allows higher spatial frequency (i.e. higher resolution) information to be accessed than would normally be obtainable through conventional diffraction-limited microscopy. The standard enhancement of resolution by this method relies on a linear increase in spatial frequency due to the addition of spatial frequencies from both the sample, k_0, and the pattern illumination, k_1; the latter is itself diffraction limited and so the maximum enhancement factor for spatial resolution is two.

However, if the rate of light emission can be made to depend in a non-linear fashion on the illumination intensity (for example, using very high intensities) then the effective illumination pattern may contain *harmonics* with spatial frequencies that are integer multiples of k_1, which can therefore generate greater enhancement in spatial resolution. This has led to related non-linear SI techniques of *saturated pattern excitation microscopy* (SPEM) and *saturated structured illumination microscopy* (SSIM), generating a practical spatial resolution as low as ~50 nm.

The principal disadvantages are that the intensities required are currently high and therefore long-term sample photodamage is an issue, and also the imaging speeds are relatively low at typically tens of frames per second. These SI techniques currently have limited application to single-molecule imaging, however they are mentioned here for their clear future potential.

3.9.8 Förster resonance energy transfer

The most useful and widely applied super-resolution technique for investigating single-molecule interactions is *Förster energy resonance transfer* (FRET). This technique utilizes an energy transfer between the electronic orbitals of neighbouring molecules which are sufficiently close that the orbitals can transfer energy via electronic resonance as opposed to photon emission/absorption, and so is described as being a *non-radiative* transition. In biophysical single-molecule applications this is normally utilized in conjunction with fluorescence detection, and so is sometimes referred to as 'fluorescence resonance energy transfer' (for a good practical guide see Roy *et al.*, 2008). This technique was first performed on single pairs of fluorescently tagged biological molecules (so-called *smFRET*) using a SNOM/NSOM imaging technique (Ha *et al.*, 1996), but is now applied more frequently to standard forms of far-field imaging.

Here, a *donor* fluorophore will emit at a lower peak wavelength than the *acceptor* fluorophore, but if there is significant *overlap* between the emission spectrum of the donor and the excitation (i.e. absorption) spectrum of the acceptor then FRET may occur. This is then manifest in a small *reduction* in the fluorescence emission output from the donor, and a small *increase* in the fluorescence emission of the acceptor (Figure 3.5G).

The majority of studies to date have used organic dyes, a common FRET-pair being variants of Cy3 for the (green) donor and variants of Cy5 for the (red) acceptor, but the technique has also been successfully applied to QD pairs and FP pairs (see Miyawaki *et al.*, 1997 for the first FRET study using FPs). Measurable energy transfer from the donor to the acceptor can occur via FRET provided the fluorophores are separated by roughly less than the so-called *Förster radius*, R_0. The efficiency E of this energy transfer as a function of the length separation R of the donor–acceptor FRET pair is characterized by:

$$E = \frac{k_{\mathrm{FRET}}}{\sum_{\mathrm{donor}} k_i} = \frac{k_{\mathrm{FRET}}}{k_{\mathrm{FRET}} + k_{\mathrm{radiative}}} + \sum_{\mathrm{other\ losses}} k_j = \frac{1}{1 + (R/R_0)^6} \tag{3.12}$$

Here, k_{FRET} is the rate of FRET energy transfer from donor to acceptor, the summed k_i parameters represent the rates from the donor of all energy transfer processes, which include radiative ($k_{\mathrm{radiative}}$), FRET (k_{FRET}) and any other non-FRET and non-radiative processes ($\sum k_j$). In the absence of any acceptor, the donor will transfer energy at a rate ($k_{radiative}+\sum k_j$) and so the mean lifetime τ_D is equal to $1/(k_{radiative}+\sum k_j)$. In the presence of an acceptor FRET can occur at the rate k_{FRET} such that the donor lifetime τ_{DA} is then equal to $(R_0/R)^6/k_{FRET}$. Rearranging Equation 3.6 implies that $E=1-\tau_{DA}/\tau_D$. This can also be written as $E=1-I_{DA}/I_D$ where I_{DA} and I_D are the total fluorescence intensities of the donor in the presence and the absence of an acceptor respectively. The full dependence of R_0 is given by:

$$R_0 = \left(\frac{9000 Q_Y . \ln 10 . \kappa^2 . \left[\int f_{donor}(\lambda) \varepsilon_{acceptor}(\lambda) \lambda^4 d\lambda \right]}{128 \pi^5 n^4 N_A} \right)^{1/6} \tag{3.13}$$

Here, Q_Y is the quantum yield of the donor fluorophore in the absence of the acceptor, κ^2 is the so-called *dipole orientation factor*, n is the refractive index of the surrounding medium (nominally water at 1.33), N_A is Avogadro's number and the term inside the square brackets is the spectral overlap integral such that f_{donor} is the normalized donor emission spectrum and ε_A is the molar extinction coefficient for the acceptor. Typical values of R_0 for common FRET pairs are in the range 3–6 nm. The R^6 functional dependence on E in Equation 3.12 results in a highly sensitive response of E with respect to R. In other words, this technique is potentially very good at distinguishing between putative molecular interaction (R is less than ca. 5 nm, and E is 0.5–1), and no molecular interaction (R is greater than ca. 5 nm and E is 0–0.5). Thus, although far-field imaging still results in the Abbe optical resolution limit for the separate donor and acceptor fluorophore images, a FRET signal is only measured for molecular separations of less than ~5 nm, which is therefore super-resolution.

The κ^2 factor can in principle vary over a range 0–4 but is often assumed to be exactly 2/3, which is based on the theoretical assumption that donor and acceptor molecules are freely rotating and can therefore transfer energy in an effectively *isotropic* fashion. This is only true if the rotational diffusion time scale for the fluorophore tags is significantly less than the sampling time scale used in a given experiment. Typical rotational time scales are $\sim 10^{-9}$ s, and so for fluorescence imaging experiments where the fastest sampling times are $\sim 10^{-3}$ s the assumption is valid. However, for some fast non-imaging methods such as confocal fluorescence detection performed over the microsecond time scale (essentially similar to the detection method used in fluorescence correlation spectroscopy discussed previously) for fluorophore tags whose rotation is potentially inhibited for example by steric effects if a large tag is employed, then the assumption may no longer be valid.

The largest disadvantage with FRET is its relatively limited application in practice for FP fusion systems when applied at the single-molecule level. Although certain paired combinations of FPs have reasonable spectral overlap (for example CFP/YFP type pairs for blue/yellow, and GFP/mCherry for green/red), R_0 values are typically relatively high at ~6 nm and also since the FPs themselves are ~3–4 nm large this means that only FRET efficiency values of ~50% or less can be measured since the FPs cannot get any closer to each other. This means that the technique is far less sensitive compared to using smaller, brighter organic dye pairs either for measuring small changes in separation of less than ~5 nm (for example, using FRET as a *molecular ruler* to monitor very small relative conformational changes of 1 nm or less), or for being able to tell definitively whether the tagged proteins of interest themselves are actually interacting or not.

Therefore, many practical applications of smFRET involve optimized organic dye pairs. However, this comes with an additional problem that the chemical binding efficiency of the dye to the biological molecule of interest is never as high as 100% and so there will always be a sub-population of unlabelled molecules present in the system. Similarly, even when both dyes in the pair have bound successfully there may be a further sub-population that is photo-inactive, for example due to free-radical damage. Such 'dark' molecules will obviously not generate a FRET response and so may falsely indicate that there is no molecular interaction in some cases.

Also, since the emission spectrum of dyes is a continuum, there is always unavoidable bleed-though of each dye signal into the other's respective detector channel, which can be difficult to distinguish from 'real' FRET effects in conventional FRET setups unless meticulous control experiments are performed. These issues are largely overcome by the application of *alternating laser excitation*, or ALEX (Kapanidis *et al.*, 2005). Here, the donor and acceptor fluorophores are excited alternately with their respective fluorescence emission detection synchronized to this excitation, which allows us to distinguish the real FRET population very accurately. This technique has been applied to measuring DNA transcription for example (see Chapter 9).

Often, these smFRET approaches are applied in controlled experiments in vitro as opposed to in the living cell, primarily because of the sensitivity detection issue as outlined above in using fluorescent proteins in the context of potentially high native autofluorescence from the cell contaminating the signal. One issue with solution-phase confocal smFRET, one of the most common methods applied, in which interacting components are essentially free in solution to diffuse into and out of a confocal excitation volume, is that the small approximately micrometre length scale of the extent of the confocal volume sets a very low upper limit on the time over which one can observe a molecular interaction by FRET since the interacting pair will diffuse across the lateral extent of the confocal volume in the focal plane over a time scale of typically a few tens of milliseconds.

One recent approach taken to increase this FRET measurement time is to confine the mobility of the interacting molecules in some way, either by tethering to a surface (for example see Ha *et al.*, 2002) or better still via confinement of interacting molecules in solution inside a lipid bilayer *nanovesicle* immobilized to a microscope slide (Benitez *et al.*, 2010) since this approach suffers from potentially less contamination of surface forces on the interacting molecules. These approaches permit continuous smFRET observations to be made over a time scale greater by ~3 orders of magnitude compared to solution-phase confocal smFRET.

A very recent application of live-cell smFRET has come in the form of *mechanical force detection* across the cell membrane (Stabley *et al.*, 2011). Here, a specially designed probe can be placed in the cell membrane such that a red Alexa647 dye molecule and a FRET acceptor molecule, which acts as a quencher to the donor at short distances, are separated by a short extendable linker made from the polymer *polyethenlene glycol* (PEG). Local mechanical deformation of the cell membrane results in extension of the PEG linker which therefore has a *de-quenching* effect. With suitable calibration, this phenomenon can be used to measure local mechanical forces across the cell membrane.

KEY POINT

Keen readers will realize by now that there are several super-resolution fluorescence imaging studies published that are described as 'live-cell' or 'in vivo', or something similar, suggesting that this technique gives super-resolution information on biological processes under rigorously physiological conditions. **Caution needs to be applied!** The laser excitation intensities used in these studies, including PALM, STORM, STED and recent photoblinking reconstructive methods, are all very high compared to standard widefield fluorescence imaging. Furthermore, the wavelengths often employed for some of these techniques can be in the 'long-uv' of ~400 nm. Also, techniques which rely on image reconstruction following iterative Gaussian fitting of individual point spread functions in general require very prolonged cumulative laser exposures to the biological sample, since they may need typically several thousand image frames to reconstruct a single super-resolution image. All three factors are well known to increase the **risk of photodamage** to the cell, often terminally, with the build up of free-radical chemicals that bind indiscriminately not only to vital molecules in the cell but also potentially directly to the substructures actually under observation (though note, there are some promising recent developments using novel organic and FP dyes which are excited by longer infrared wavelengths, resulting potentially in less sample damage.) A mindful research scientist might be sceptical as to whether a cell were truly functional/alive during and following such an imaging procedure even if it were so at the start, in the absence of any functional information concerning the same single cell being imaged, and might question whether the imaging procedure had resulted in free-radical related artifacts or not.

3.10 'Multi-dimensional' imaging

Increasingly commonly, visible light imaging experiments on single biological molecules involve measuring several different physical parameters simultaneously. These include measuring not only different colour emissions for differently tagged molecules, but also parameters such as emission polarization. In *fluorescence polarization microscopy*, typically a fluorescent sample is excited using a laser with one direction of linear polarization, with fluorescence emissions being separated into two orthogonal linear polarization components. Asymmetrical *anisotropy* differences between these components can give information on whether the tagged molecules are undergoing some type of rotational diffusion, either due to normal thermal relaxation processes or through energized molecular conformational changes (for example see Forkey *et al.*, 2003). This technique has also been used to gain single-molecule biological information concerning the local architecture of artificial lipid bilayers (Harms *et al.*, 1999), as well as native cell membranes (Sund *et al.*, 1999).

It is now feasible to measure both dual-polarization and dual-colour fluorescence intensity simultaneously at the single-molecule level. In addition, in certain experiments FRET efficiency can also be estimated, as can rotational and lateral diffusion coefficients, and even molecular stoichiometry and variations in the local chemical environment. Thus, the potential is there to expand the number of dimensions measured from the standard three spatial dimensions and one of time (four-dimensional imaging), up to in effect more than ten dimensions. The key here is to measure physical phenomena that are as independent from each other as possible using visible light. Ultimately, all such measurements come down to dividing up the finite light signal, which can only be done so many times before the detection noise limit is reached. However, with improvements to both dye and camera detector technologies exponentiating year-on-year, there is significant future scope for such multidimensional imaging.

THE GIST

- Fluorescence microscopy is a method of choice for investigating single biological molecules with minimal physiological perturbation.
- Several methods exist which increase optical contrast by delimiting the excitation volume, the most common of which for single-molecule imaging is TIRF.
- Several different types of fluorescent tag exist, but the most useful are the fluorescent proteins, despite their being relatively poor as dyes.
- Normal 'far-field' light imaging is limited by diffraction given by the Abbe optical resolution limit of 200–300 nm, but several 'super-resolution' methods exist which can overcome this barrier and potentially provide spatial information at the nanometre level.
- Molecular interactions can be measured at the single-molecule level using techniques such as FRET.

TAKE-HOME MESSAGE Visible light techniques can allow us to detect single biological molecules both in the test tube and in the living cell, as well as collections of different molecules in the same system, and to monitor spatial changes down to nanometre precision, in some cases at the millisecond level or faster.

References

GENERAL

- Axelrod, D., Burghardt, T. P. and Thompson, N. L. (1984). Total internal reflection fluorescence. *Annu. Rev. Biophys. Bioeng.* **13**: 247–268.
- *Chemical Society Reviews*, 2009, **38**: 2813: this issue contains several good recent reviews concerning fluorescent protein structure, photochemistry and application.
- Chiu, S.-W. and Leake, M. C. (2011). Functioning nanomachines seen in real time in living bacteria using single-molecule and super-resolution fluorescence imaging. *Int. J. Mol. Sci.* **12**: 2518–2542.
- Feynman, R. P. (1959). 'There's Plenty of Room at the Bottom,' lecture, transcript deposited at Caltech Engineering and Science, Volume **23**:5, February 1960, pp. 22–36 (quote used with kind permission), available at www.its.caltech.edu/~feynman/plenty.html
- PubSpectra. An excellent free-access online resource for downloading fluorescent dye, protein, filter and light source spectra: http://home.earthlink.net/~pubspectra/
- Roy, R., Hohng, S. and Ha, T. (2008). A practical guide to single-molecule FRET. *Nature Methods* **5**: 507–516.

ADVANCED

- Appelhans, T. *et al.* (2012). Nanoscale organization of mitochondrial microcompartments revealed by combining tracking and localization microscopy. *Nano Lett.* January 13. [Epub ahead of print]
- Badieirostami, M., Lew, M. D., Thompson, M. A. and Moerner, W. E. (2010). Three-dimensional localization precision of the double-helix point spread function versus astigmatism and biplane. *Appl. Phys. Lett.* **97**: 161103.
- Benitez, J. J., Keller, A. M. and Chen, P. (2010). Nanovesicle trapping for studying weak protein interactions by single-molecule FRET. *Methods Enzymol.* **472**: 41–60.

- Betzig, E. *et al.* (2006). Imaging intracellular fluorescent proteins at nanometer resolution. *Science* **313**: 1642–1645.
- Bostock, J. and Riley, H. T. (1855). Pliny the Elder, The Natural History. *Plin. Nat.* Book 32, Chapter 52. For an online resource see: www.perseus.tufts.edu/hopper/text?doc=Plin.+Nat.+toc – a little incongruous to cite a nineteenth century translation of a first century AD Latin text for a twenty-first century single-molecule biology textbook, however the range of natural history phenomena covered is fascinating in its breadth, and the addictive reverence with which 'Nature' is described may be helpful to the modern scientist whose experiment has just failed for the fourth time. And, although some of the descriptions are terse and largely anecdotal, there is much still to be learnt of the scientific importance of describing real case study examples.
- Burnette, D. T. *et al.* (2011). Bleaching/blinking assisted localization microscopy for superresolution imaging using standard fluorescent molecules. *Proc. Natl. Acad. Sci. USA* **108**: 21081–21086.
- Chandrasekhar, S. (1943). Stochastic problems in physics and astronomy. *Rev. Mod. Phys.* **15**: 1–89.
- Chiantia, S., Kahya, N. and Schwille, P. (2007). Raft domain reorganization driven by short- and long-chain ceramide: a combined AFM and FCS study. *Langmuir* **23**: 7659–7665.
- Chirico, G. *et al.* (2001). Single molecule studies by means of the two-photon fluorescence distribution. *Microsc. Res. Tech.* **55**: 359–364.
- Churchman, L. S. *et al.* (2005). Single molecule high-resolution colocalization of Cy3 and Cy5 attached to macromolecules measures intramolecular distances through time. *Proc. Natl. Acad. Sci. USA* **102**: 1419–1423.
- Forkey, J. N., Quinlan, M. E., Shaw, M. A., Corrie, J. E. T. and Goldman, Y. E. (2003). Three-dimensional structural dynamics of myosin V by single-molecule fluorescence polarization. *Nature* **422**: 399–404.
- Funatsu, T., Harada, Y., Tokunaga, M., Saito, K. and Yanagida, T. (1995). Imaging of single fluorescent molecules and individual ATP turnovers by single myosin molecules in aqueous solution. *Nature* **374**: 555–559.
- Gall, J. G. and Pardue, M. L. (1969). Formation and detection of RNA-DNA hybrid molecules in cytological preparations. *Proc. Natl. Acad. Sci. USA* **63**: 378–383.
- Gelles, J., Schapp, B. J. and Sheetz, M. P. (1988). Tracking kinesin-driven movements with nanometre-scale precision. *Nature* **331**: 450–453.
- Ghosh, I., Hamilton, A. D. and Regan, L. (2000). Antiparallel leucine zipper-directed protein reassembly: application to the green fluorescent protein. *J. Am. Chem. Soc.* **122**: 5658–5659.
- Gordon, M. P., Ha, T. and Selvin, P. R. (2004). Single-molecule high-resolution imaging with photobleaching. *Proc. Natl. Acad. Sci. USA* **101**: 6462–6465.
- Gustafsson, M. G. (2000). Surpassing the lateral resolution limit by a factor of two using structured illumination microscopy. *J. Microsc.* **198**: 82–87.
- Ha, T. *et al.* (1996). Probing the interaction between two single molecules: fluorescence resonance energy transfer between a single donor and a single acceptor. *Proc. Natl. Acad. Sci. USA* **93**: 6264–6268.
- Ha, T. *et al.* (2002). Initiation and re-initiation of DNA unwinding by the *Escherichia coli* Rep helicase. *Nature* **419**: 638–641.
- Harms, G. S., Sonneteer, M., Schütz, G. J., Gruber, H. J. and Schmidt, T. (1999). Single-molecule anisotropy imaging. *Biophys. J.* **77**: 2864–2870.
- Hecht, B. *et al.* (2000). Scanning near-field optical microscopy with aperture probes: fundamentals and applications. *J. Chem. Phys.* **112**: 7761–7774.
- Hell, S. and Stelzer, E. H. K. (1992). Fundamental improvement of resolution with a 4Pi-confocal fluorescence microscope using two-photon excitation. *Opt. Commun.* **93**: 277–282.
- Hell, S. W. and Wichmann, J. (1994). Breaking the diffraction resolution limit by stimulated emission: stimulated-emission-depletion fluorescence microscopy. *Opt. Lett.* **19**: 780–782.
- Huisken, J. *et al.* (2004). Optical sectioning deep inside live embryos by selective plane illumination microscopy. *Science* **305**: 1007–1009.
- Ji, N., Milkie, D. E. and Betzig E. (2010). Adaptive optics via pupil segmentation for high-resolution imaging in biological tissues. *Nature Methods* **7**: 141–147.
- Jones, S. A., Shim, S. H., He, J. and Zhuang, X. (2011). Fast, three-dimensional super-resolution imaging of live cells. *Nature Methods* **8**: 499–508.
- Kapanidis, A. N. *et al.* (2005). Alternating-laser excitation of single molecules. *Acc. Chem. Res.* **38**: 523–533.

- Keller, P. J., Schmidt, A. D., Wittbrodt, J. and Stelzer, E. H. K. (2008). Reconstruction of zebrafish early embryonic development by scanned light sheet microscopy. *Science* **322**: 1065–1069.
- Kitagawa, M. *et al.* (2005). Complete set of ORF clones of *Escherichia coli* ASKA library (a complete set of *E. coli* K-12 ORF archive): unique resources for biological research. *DNA Res.* **12**: 291–299.
- Leake, M. C. (2001). *Investigation of the extensile properties of the giant sarcomeric protein titin by single-molecule manipulation using a laser-tweezers technique*. PhD dissertation, London University, UK.
- Lee, S. F. *et al.* (2011). Super-resolution imaging of the nucleoid-associated protein HU in *Caulobacter crescentus*. *Biophys. J.* **100**: L31–L33.
- Lew, M. D. *et al.* (2011). Three-dimensional superresolution colocalization of intracellular protein superstructures and the cell surface in live *Caulobacter crescentus*. *Proc. Natl. Acad. Sci. USA* **108**: E1102–E1110.
- Los, G. V. *et al.* (2008). HaloTag: a novel protein labeling technology for cell imaging and protein analysis. *ACS Chem. Biol.* **3**: 373–382.
- Maestro, L. M. *et al.* (2010). CdSe quantum dots for two-photon fluorescence thermal imaging. *Nano Lett.* **10**: 5109–5115.
- Magde, D., Elson, E. L. and Webb, W. W. (1972). Thermodynamic fluctuations in a reacting system: measurement by fluorescence correlation spectroscopy. *Phys. Rev. Lett.* **29**: 705–708.
- Marriott, G. *et al.* (2008). Optical lock-in detection imaging microscopy for contrast-enhanced imaging in living cells. *Proc. Natl. Acad. Sci. USA* **105**: 17789–17794.
- Miseta, A. and Csutora, P. (2000). Relationship between the occurrence of cysteine in proteins and the complexity of organisms. *Mol. Biol. Evol.* **17**: 1232–1239.
- Miyawaki, A. *et al.* (1997). Fluorescent indicators for Ca^{2+} based on green fluorescent proteins and calmodulin. *Nature* **388**: 882–887.
- Nirmal, M. *et al.* (1996). Fluorescence intermittency in single cadmium selenide nanocrystals. *Nature* **383**: 802–804.
- Park, H., Toprak, E. and Selvin, P. R. (2007). Single-molecule fluorescence to study molecular motors. *Q. Rev. Biophys.* **40**: 87–111.
- Plank, M., Wadhams, G. H. and Leake, M. C. (2009). Millisecond timescale slimfield imaging and automated quantification of single fluorescent protein molecules for use in probing complex biological processes. *Integr. Biol.* **1**: 602–612.
- Qu, X., Wu, D., Mets, L. and Scherer, N. F. (2004). Nanometer-localized multiple single-molecule fluorescence microscopy. *Proc. Natl. Acad. Sci. USA* **101**: 11298–11303.
- Ritter, J. G. *et al.* (2010). Light sheet microscopy for single molecule tracking in living tissue. *PLoS One* **5**: e11639.
- Rust, M. J., Bates, M. and Zhuang, X. (2006). Sub-diffraction-limit imaging by stochastic optical reconstruction microscopy (STORM). *Nature Methods* **3**: 793–795.
- Sako, Y., Minoghchi, S. and Yanagida, T. (2000). Single-molecule imaging of EGFR signalling on the surface of living cells. *Nature Cell Biol.* **2**: 168–172.
- Schmidt, T. *et al.* (1996). Imaging of single molecule diffusion. *Proc. Natl. Acad. Sci. USA* **93**: 2926–2829.
- Shimomura, O., Johnson, F. H. and Saiga, Y. (1962). Extraction, purification and properties of aequorin, a bioluminescent protein from the luminous hydromedusan, Aequorea. *J. Cell. Comp. Physiol.* **59**: 223–239.
- Stabley, D. R., Jurchenko, C., Marshall, S. S. and Salaita, K. S. (2011). Visualizing mechanical tension across membrane receptors with a fluorescent sensor. *Nature Methods*. **9**: 64–67.
- Sund, S. E., Swanson, J. A. and Axelrod, D. (1999). Cell membrane orientation visualized by polarized total internal reflection fluorescence. *Biophys. J.* **77**: 2266–2283.
- Thompson, R. E., Larson, D. R. and Webb, W. (2002). Precise nanometer localization analysis for individual fluorescent probes. *Biophys. J.* **82**: 2775.
- Tinnefeld, P. *et al.* (2000). Confocal fluorescence lifetime imaging microscopy (FLIM) at the single molecule level. *Single Mol. 1* **3**: 215–223.
- Tsien, R. Y. (1998). The green fluorescent protein. *Annu. Rev. Biochem.* **67**: 509–544.
- Voie, A. H., Burns, D. H. and Spelman, F. A. (1993). Orthogonal-plane fluorescence optical sectioning: three-dimensional imaging of macroscopic biological specimens. *J. Microsc.* **170**: 229–236.

- Vollmer, F. and Arnold, S. (2008). Whispering-gallery-mode biosensing: label-free detection down to single molecules. *Nature Methods* **5**: 591–596.
- Warshaw, D. M. *et al.* (2005). Differential labeling of myosin V heads with quantum dots allows direct visualization of hand-over-hand processivity. *Biophys. J.* **88**: L30–L32.
- Waterman-Storer, C. M., Desai, A., Bulinski, J. C. and Salmon, E. D. (1998). Fluorescent speckle microscopy, a method to visualize the dynamics of protein assemblies in living cells. *Curr. Biol.* **8**: 1227–1230.
- Yang, W. and Musser, S. M. (2006). Visualizing single molecules interacting with nuclear pore complexes by narrow-field epifluorescence microscopy. *Methods* **39**: 316–328.
- Yuste, R. (2005). Fluorescence microscopy today. *Nature Methods* **2**: 902–904.

Questions

FOR THE LIFE SCIENTISTS

Q3.1. The numerical aperture of the human eye is estimated to vary between roughly 0.1 and 0.3 from dark to light conditions. Is there any correlation to be drawn between the theoretical optical resolution under these conditions and the appearance of the histological sections of the retina in Figure 3.1?

Q3.2. A single-molecule confocal microscopy experiment is performed where the focused laser volume had a lateral width measured at 270 nm and an axial width measured at 650 nm. An *E. coli* bacterial cell in the sample was measured as being 2.5 μm long and 0.9 μm wide, and had a shape approximated by a cylinder capped at either end by a hemisphere. (a) What proportion of the cell could the focused laser excite in principle? (b) If the focused laser beam is occupied by a single GFP-tagged protein in the cytoplasm of the cell for ~55% of the time, what is the molarity of the protein being examined? The laser beam was focused on the mid-point of the cell in the cytoplasm; from measuring the width of fluorescence pulses from a few fluorescently tagged molecules, the range of time taken to traverse the confocal spot in the sample plane was estimated at 1–2 ms. (c) Assuming that the diffusion coefficient D can be approximated from Stokes law as $6\pi\eta/r$ where the cytoplasm viscosity η of ~0.001 Pa s and r is the effective radius of the GFP-tagged molecule, estimate the effective diameter of the protein.

Q3.3. A protein is labelled with a donor and acceptor fluorophore to study a conformational change from state I to II using FRET. The FRET acceptor–donor pair has a known Förster radius of 5.5 nm, and the measured fluorescence lifetimes of the isolated donor and the acceptor fluorophores are 4.9 ns and 1.1 ns, respectively. (a) Show that $E=1-\tau_{DA}/\tau_D$ where τ_{DA} and τ_D are the fluorescence lifetimes of the donor in the presence and absence of acceptor respectively. (b) What is the distance between the donor and acceptor if the measured donor lifetime in conformation I is 50 ps? Previous structural data suggest that the fluorophore separation may change in a distinct step by ca. 0.9 nm when the protein makes a transition between conformations I and II. (c) What is the associated donor fluorescence lifetime of conformation II, and how small would the experimental error in measuring the FRET efficiency need to be in order to see this small step change? (d) Using the same donor and acceptor, what could be done to measure this step change more clearly, and what is the maximum possible change in FRET efficiency that could be measured here?

FOR THE PHYSICAL SCIENTISTS

Q3.4. An experiment measuring the ion-current from rod photoreceptor cells in the eye indicated that controlled flashes of visible light of peak emission wavelength ~500 nm generated a measured ion-current from a single rod cell of 34 pA which resulted from ~50 single-photon-induced transitions of the photoreceptor molecule rhodopsin. In the absence of light, the variance in the dark current was measured at 0.03 pA due to spontaneous thermal isomerization of rhodopsin. Stating clearly any assumptions you make, is it possible to 'see' a single molecule of green fluorescent protein (GFP) with the naked eye?

Q3.5. A STED microscope was built using a 'white-light' laser which emits laser light over a continuum of wavelengths from ca. 400–1400 nm of total power 4 W. The output was generated in measured pulse-bursts of 80 ps, which were repeated at a frequency of 1 MHz. (a) What is the average spectral power per nm from the laser, and how does this compare with the 'peak' power output? In an experiment, the beam was split by a dichroic mirror to generate first a 'green' excitation beam which was filtered using a narrow 'bandpass' filter that would transmit ca. 95% of light over a 10 nm range centred on 532 nm wavelength, and zero elsewhere, before being attenuated by a factor of 10 and focused by a high-power oil-immersion objective lens onto a sample over an area equivalent to a circle of diameter 600 nm. The 'red' part of the beam was passed through a 'monochromator' prism that would select just wavelengths over a range 640–660 nm, before being attenuated by a factor of 10^4, and then phase modified and subsequently focused by the same objective lens to generate a doughnut-shaped STED beam in the sample of area ~10^{-9} cm^2. The objective lens had an NA of 1.4 and a transmittance equivalent to 70%. A biological sample in the STED microscope consisted of fibres of the protein tubulin (used in the cytoskeleton, see Chapter 7) labelled with fluorescent antibody stuck to a microscope slide. Earlier electron microscopy (see Chapter 4) had indicated an average diameter of the labelled fibres of ca. 50 nm. (b) Explain, giving your reasons, whether or not this microscope could indicate whether originally separate fibres had bound together to form a bundle.

Q3.6. In a FLIM-FRET experiment performed at room temperature using CFP and YFP as the donor and acceptor FRET fluorophores respectively, the fluorescence lifetimes of both proteins were measured as being in the range 1–4 ns. Both proteins have a molecular weight of ca. 28 kDa and have an effective diameter of ca. 3 nm if the shape of each is approximated as a sphere. (a) Show using the equipartition theorem what the typical rotational time scale is for a freely rotating fluorescent protein, stating any assumptions you make. The experimenters hoped that any dynamic changes observed in the estimated FRET efficiency could be assigned as being due to relative changes in separation of the CFP and YFP as opposed to being due to relative changes in orientation between the two fluorescent proteins. (b) Show, with reasoning, whether this is valid or not. (c) In practice, why is this experiment not likely to succeed at the single-molecule level?

Q3.7. (a) Describe briefly the technique of total internal reflection fluorescence (TIRF) microscopy, and give an example of its application in modern single-molecule cellular biophysics. (b) Most TIRF live-cell applications tend to investigate membrane complexes, why is this? (c) Would TIRF still be effective if there was a high concentration of autofluorescent proteins in the cytoplasm? (d) Could TIRF be applied to monitor single molecules in the nucleus of cells? The wave equation has solutions of the form:

$$E = E_0\exp\{i[(k\sin\theta)x + (k\cos\theta)y - \omega t]\}$$

For real angles θ, this represents a travelling wave in the focal (x,y) plane of the sample. (e) Show that the same solution but with a complex θ describes the electric field in water near a water–glass interface in the case where plane wave illumination is totally internally reflected within the glass at the interface. (f) Obtain an expression in terms of the angle of incidence in glass, the angular frequency ω and the refractive indices of glass and water, for the electric field in water for the above case. (g) Describe using a sketch the wave that this represents. A cell contains protein molecules which are both integrated into the membrane at an area surface density of n_m and in the cytoplasm at a number concentration of n_c. The cell is excited by incident laser light. (h) How deep a vertical section of the cell layer illuminated by an equally intense plane wave at normal incidence (using 'epifluorescence'), would give the same fluorescence intensity as for TIRF above?

FOR THOSE WHO HAVE NOT MADE UP THEIR MIND

Q3.8. The minimum angle subtended by a single photoreceptor in the human eye (for example, a cone cell) is estimated to be around 0.3°, however under some conditions the angular resolution of the eye is much better, sometimes by a factor of ca. 10. Why is this?

Q3.9. A high-power oil-immersion objective lens has an NA of 1.49, and transmits ~75% of visible light. (a) What is the total photon capture efficiency of the lens? In a TIRF imaging experiment using the objective lens method with a laser excitation of wavelength 514 nm, a protein is labelled with YFP. (b) Using PubSpectra (see references) select a suitable dichroic mirror and emission filter for exciting and visualizing YFP molecules, and estimate what is the maximum proportion of emitted fluorescence photons which can reach a camera detector. (c) If the camera detector converts 95% of all incident visible light photons into a signal, has a pixel size a of 24 μm and a background noise level b equivalent to ~100 counts, what is the best spatial precision with which one can estimate the position of a single YFP molecule stuck to the surface of a glass coverslip? In an actual experiment, the best precision was measured to be about ten times worse than the theoretical estimate. (d) What is the most likely cause of this discrepancy?

Q3.10. Does TIRF microscopy require a laser source, or would a non-coherent white light source such as a mercury-arc lamp be acceptable?

4 Making the invisible visible: part 2 – without visible light

Seeing is Believing.

(MANFRED VON HEIMENDAHL IN *ELECTRON MICROSCOPY OF MATERIALS: AN INTRODUCTION*, 1981. © ELSEVIER)

GENERAL IDEA

Here we discuss the miscellaneous experimental techniques that allow us to monitor single biological molecules using physical approaches which do not rely primarily on visible light.

4.1 Introduction

There now exist several methods which permit measurement of the presence of single biological molecules using physical principles which do not rely primarily on the detection of visible light. These include a variety of *scanning probe microscopy* techniques, including atomic force microscopy, which are discussed in detail in the first section of this chapter. In addition, significant advances in our understanding of single-molecule biology have come from methods using electron microscopy, which is one of the pioneering techniques used for obtaining structural information on fixed single-molecule samples. Recent advances in the measurement of small ion currents through both solid-state and native physiological nanometre length scale pores have furthered our knowledge of many areas of single-molecule bioscience. Furthermore, *Raman spectroscopy* has now advanced to a level of sensitivity such that measurements of single biological molecules are feasible. And finally, there are several microscopy methods which allow us to deduce the position of single molecules using primarily infrared *optical tweezers*.

4.2 Scanning probe microscopy

Scanning probe microscopy (SPM) covers a range of techniques which in essence allow *topographical detail* from the surface of samples to be obtained by laterally scanning some form of physical probe across the surface. There are over 20 different types of SPM that have currently been developed which can measure a variety of different physical parameters as the probe is placed in proximity to a sample surface. One of these techniques is SNOM which was discussed in Chapter 3. The most useful SPM technique in terms of obtaining information on single-molecule biology, which does not utilize visible light detection directly, is *atomic force microscopy* (AFM), but there are other SPM techniques such as *scanning tunnelling microscopy* (STM) and surface conductance microscopy which both have application to single-molecule biological investigations.

4.2.1 Scanning tunnelling microscopy

Historically, STM (Binnig *et al.*, 1982) was the precursor to all SPM techniques, developed in the 1980s by Binnig and Rohrer (who subsequently received a Nobel Prize

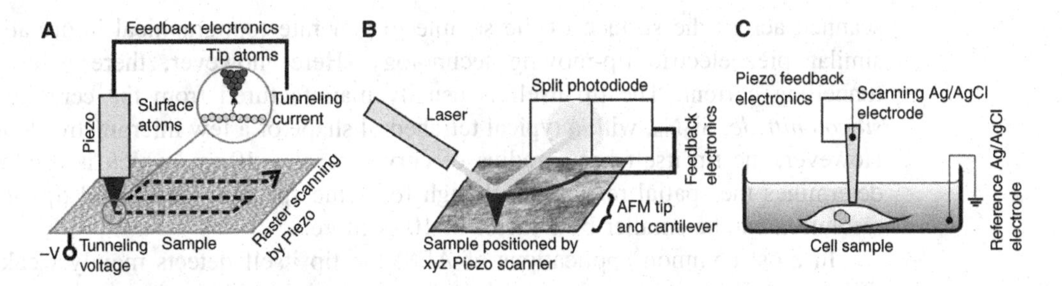

FIGURE 4.1 Scanning probe microscopy techniques. Schematics illustrating (**A**) scanning tunnelling, (**B**) atomic force and (**C**) scanning conductance microscopy.

for their efforts). It involves a solid-state scanning probe whose tip can conduct electrons away from the surface of a conducting sample to a suitable current measuring device. Tips are made from a highly conductive material; gold, platinum/iridium alloy and tungsten are popular choices. They can be manufactured either by chemical etching or by cutting a narrow wire, which when optimized can produce a tip 'point' which is in effect a single atom. Recently, non-metallic carbon nanotubes have also been utilized as tips – these have the advantages of being narrow but also highly conducting electrically with arguably a better manufacturing reproducibility and working shelf-life. In all cases, there is no physical contact between the tip and sample surface, and so there is a classically forbidden potential energy gap through which electrons must *quantum tunnel* for conduction to occur (Figure 4.1A).

The tunnelling current varies exponentially with tip–sample distance, and is very difficult to measure if the distance is greater than a few tenths of a nanometre from a weakly conducting surface such as a biological sample. STM is usually operated in a *constant current* imaging mode. Here, the tunnelling current is kept constant using feedback electronics to vary the height of the tip from the sample, typically using a highly sensitive low noise *piezoelectric device*, while the probe is laterally raster-scanned. The variation in measured sample height is thus a measure of the sample topography, which can then be converted into a two-dimensional contour plot of the sample surface.

A less common mode of operation is for the tip–sample distance to be kept constant so that the variation in tunnelling current itself can be converted into topographical information. This has the advantage of not requiring electronic feedback, which ultimately can permit faster imaging, though it does require samples to be atomically smooth and so is not of relevance to biological material.

The spatial resolution is limited primarily by the size and shape of the tip, as opposed to being diffraction limited, as is the case for the many far-field optical microscopy methods we encountered in Chapter 3, and is ~0.1 nm laterally and ~0.01 nm vertically. The lateral resolution is the same length scale as the diameter of a single hydrogen atom, so at best this method allows for true imaging at a sub-atomic spatial resolution. The main limitation for its use in the life sciences is that most biological material generates only a weak tunnelling current. However, STM has been used to generate images of single DNA molecules, large proteins called *macroglobulins* and single virus particles in vitro (Arkawa *et al.*, 1992).

4.2.2 Atomic force microscopy

The SPM technique applied most frequently for single biomolecule studies is AFM (Binnig *et al.*, 1986), sometimes known as *surface force microscopy* (SFM). Many of the principles of AFM are common to STM, in that a small, solid-state probe tip is scanned across the surface of the sample to generate topographical information, using similar piezoelectric tip-moving technology. Here, however, there is no quantum tunnelling current. The tip itself is usually manufactured from the ceramic insulator *silicon nitride*, Si_3N_4, with a typical tetrahedral shape of a few micrometres length scale. However, the tip itself has a radius of curvature of ~10 nm, which is primarily what determines the spatial resolution, though for some specially sharpened tips this spatial resolution can be smaller by a factor of 10 or more.

In most common applications of AFM the tip itself detects mainly weak *Van der Waals atomic forces* from the sample surface when it is within a few nanometres distance away (there are also lesser used variants which detect electrostatic, chemical and sometimes magnetic forces). The base of the tip probe is conjugated to a thin, flexible metallic *cantilever* strip, which is ~0.1 mm wide and a few tenths of a millimetre long, and as the tip approaches the surface the Van der Waals forces cause the metallic strip to bend upwards (Figure 4.1B). This deflection can be detected very accurately; usually this involves focusing a laser beam onto the back of the cantilever and imaging the scatter image onto a split *photodiode* detector, whose voltage response can thus be converted ultimately to a corresponding distance displacement of the tip, provided the cantilever stiffness k is known.

The most common way of deducing k is to measure the vertical mean-squared displacement $\langle z^2 \rangle$ far away from the sample surface. Thus, if the voltage output V from photodiode results in α volts per nanometre of cantilever vertical deflection, then the stiffness in pN nm^{-1} satisfies:

$$k = k_B T / \langle z^2 \rangle \approx 4.1/\alpha^2 \langle V^2 \rangle \quad (4.1)$$

Here, k_B is Boltzmann's constant and T is the absolute temperature, such that at a room temperature, ~20 °C, $k_B T$ is ~4.1 pN nm (see Chapter 1). The stiffness of the spring may also be calculated from measuring the resonant angular frequency ω_0 of the cantilever, since it obeys the simple harmonic oscillator formula of $\omega_0 = \sqrt{(k/m)}$, however this requires accurate knowledge of the effective mass m of the cantilever and tip combined, which is often difficult to measure. Typical cantilever stiffness values used in AFM are ~0.1 pN nm^{-1}.

PHYSICS-EXTRA

The vertical oscillations of the AFM cantilever result in two **independent quadratic terms** in the energy equation, one due to the cantilever's kinetic energy and the other due to the elastic potential energy, and so the **equipartition theorem** predicts that each equivalent molecule in the oscillator system is associated with an average energy of $k_B T/2$, with the mean elastic potential energy equal to $k\langle z^2 \rangle/2$.

In most cases the tip is operated in a *constant force mode*, similar to the constant current mode in STM, in that the sample–tip distance is adjusted continuously using feedback electronics to ensure a constant detected vertical force of kz. During imaging, the AFM is used normally in either a *contact/static* or *non-contact/dynamic* mode. In the contact mode, the cantilever deflection is kept constant throughout as the tip is scanned across the sample surface, such that the overall force detected by the tip from the sample is in the short range repulsive regime.

In non-contact (or *tapping*) mode the tip is oscillated slightly above its *resonant frequency* to generate distance amplitudes of a few nanometres. As the tip approaches the surface during its oscillation cycle the resonant frequency decreases a little, as does the amplitude of oscillation, with an associated change of phase between the forcing oscillation and the cantilever motion. Depending on the measurement system, either the change in frequency (*frequency modulation*) or the change in amplitude or phase (*amplitude modulation*) can be detected; the latter is relatively sensitive to the type of sample material being imaged. These detected signal changes can all be converted after suitable calibration into a distance measurement from the sample.

Non-contact mode imposes smaller forces on soft samples compared to contact mode and so is in general preferable when imaging typically compliant biological material. It also has a lower risk of actually scraping the sample as the tip is scanned across it and thereby contaminating the tip probe. However, if the sample is in physiological water-based pH buffer environment (as opposed to a high vacuum as would be the case to generate the lowest measurement noise), then non-contact mode will in general image not only the sample surface but also the first few shells of water molecules (see Chapter 2), which in effect reduces the imaging spatial resolution. In contact mode the tip can penetrate beyond the water layers bound to the surface to image the sample molecules directly. Similarly, the finite sharpness of the AFM tip itself means that some surface features of the sample will be inaccessible, with a resultant tip broadening convolution artifact (see Q4.12).

AFM imaging has been able to measure several types of single biological molecules (Arkawa *et al.*, 1992), including static snapshots of the motor protein myosin (see Chapter 9). A traditional weakness of AFM is the relatively slow imaging speeds due to slow scanning and feedback electronics. Recent improvements have been made in improving the imaging speed, so-called *high-speed AFM* (HS-AFM), using a combination of sharp tips with greater natural resonance, or by using *torsional oscillations* of the cantilever as opposed to vertical oscillations, both of which have resulted in video-rate imaging speeds (Hobbs *et al.*, 2006), which have been employed for measuring real time stepping motions of myosin molecules.

AFM has recently been applied, in combination with scanning tunnelling microscopy and *Kelvin probe* microscopy, in which a very cold probe tip at a temperature of ~5 K is used, to measure directly the actual distribution of electronic charge within a single molecule, in this case an organic molecule called *naphthalocyanine* which has been used in the context of developing *single-molecule logic switches* for bionanotechnological purposes (Mohn *et al.*, 2012).

4.2.3 Scanning ion conductance microscopy

In *scanning ion conductance microscopy* (SICM), the scanning probe consists of a very narrow glass pipette drawn out such that its end diameter is only ~20–30 nm (Hansma *et al.*, 1989). This technique combines the standard scanning approaches of SPM with the ion-flux measurement methods of *patch clamping* (see section 4.4.1). A voltage potential difference is applied between the end of the tip and the bulk solution which, in physiological ionic buffers, results in a measurable ion current. As the tip is moved to within less than a distance of roughly its own diameter then the ion flow is impeded.

Using feedback electronics similar to those described previously, this drop in ion current can be used to maintain a constant height of the tip above the sample, which can then generate topographical information as the tip is scanned laterally across the surface (Figure 4.1C). This offers a relatively poorer spatial resolution compared to STM or AFM of ~50 nm, but has the advantage of causing less sample damage. Recent improvements, primarily in narrowing the diameter of the pipette to ~10 nm, have resulted in non-contact imaging of collections of single protein molecular complexes on the outer surface membrane of live cells. This improved technique has also been used in conjugation with single-molecule folding kinetics studies of fluorescent proteins, using the same nanopipette to deliver a denaturant to unfold chemically, and hence photobleach, single fluorescent protein molecules, prior to their refolding and gaining photoactivity (Klenerman *et al.*, 2011).

KEY POINT

Surface probe microscopy offers great insight into **single-molecule topology** of biological molecules, however its use in live cells is currently limited to monitoring molecules on the cell membrane surface. Some developments are being made towards intra-cellular SPM imaging, for example using ultra-sharp tips such as **carbon nanotubes** as extensions to AFM probes that have been made hydrophobic to allow them to pass through the cell membrane barrier while minimizing mechanical distortion. As yet, however, practical in vivo SPM imaging has not been achieved.

4.3 Electron microscopy

Historically, the first true single-molecule imaging technique developed was electron microscopy (see Chapter 1). Here, electrons are accelerated via *thermionic emission* from a hot electrode source. The electron beam can be focused onto a thin sample via a series of electrostatic or electromagnetic lenses that function analogously to the glass lenses which focus visible light photons in an optical microscope. The primary difference is wavelength, λ; if an electron of mass m and charge e is accelerated via a potential difference of voltage V then its momentum p from its kinetic energy must be generated by the electrostatic potential energy, which can be approximated as:

$$eV \approx p^2/2m \tag{4.2}$$

The equivalent wavelength of such an electron is then given by the *de Broglie relation*, which states that momentum and wavelength are inversely dependent:

$$\lambda = h/p \approx h/\sqrt{2meV} \tag{4.3}$$

Here, h is *Plank's constant*. The potential electrostatic energy of a charged particle is equivalent to the product of the charge (in Coulombs) and the voltage difference. The kinetic energy for a particle of speed v and mass m is normally written as $\frac{1}{2}mv^2$, which is also equal to $p^2/2m$ where the momentum p is given by mv. Usually the potential difference is expressed in terms of kiloelectronvolts (keV, or 1000 times the potential energy of 1 V applied to a single electron, or $\sim 1.6 \times 10^{-16}$ J), and 10–100 keV is typical for most applications, which in principle generates electrons with equivalent matter-wave wavelengths of 10^{-12}–10^{-11} m.

If electrons are incident on a sample surface that has periodic features composed in effect of multiple planes, then it is possible for the back scattered electrons to diffract to form constructive interference patterns from the interference between the electrons scattered back (i.e. reflected) from the different planes, with the condition for such nth order intensity maxima being given by *Bragg's law* (as is also the case for x-ray crystallography):

$$n\lambda = 2d \sin\theta \tag{4.4}$$

Here, n is a positive integer, d is the characteristic spacing between parallel planes and θ is the angle incidence of the scattered electron beam normal to the planes.

Similarly, electrons can be transmitted through the sample (see section 4.3.1) and experience diffraction through a periodic lattice as they do, but here the condition for constructive interference is:

$$n\lambda = d \sin \theta \qquad (4.5)$$

Here, d is now the equivalent 'aperture' spacing in the plane of the sample (typically the characteristic spacing of a repeating biological substructure) and θ is the angle of diffraction normal to the plane. Although not a single-molecule technique per se, both Bragg reflection and electron diffraction can be used to gain insight into the physical separation of single molecules within a biological sample (see Q4.6).

PHYSICS-EXTRA

The speed of the accelerated electrons is typically ~60% of the speed of light, so a small **relativistic correction** to Equation 4.1 would reflect an associated increase in electron mass of ~10%, which means that the approximation of Equation 4.2 is still accurate to within ~5%.

4.3.1 Transmission electron microscopy (TEM)

Here, the interaction of electrons with the sample is predominantly due to scattering from atomic nuclei, which increases with atomic number. For biological matter the range of atomic number is relatively low and so *heavy-metal contrast reagents* must be applied, using so-called *negative* or *positive staining* methods.

For negative staining, biological cell or tissue samples must first be chemically fixed using a harsh multi-stage process involving typically a combination of *acetone* and *glutaraldehyde* that chemically crosslinks different molecular structures in the sample. The sample is then embedded in wax, sliced using a device called a *microtome* to generate thin (~50 nm) sections, and then a contrast reagent containing commonly either *uranyl acetate* or *osmium tetroxide* is applied.

This stain fills the spaces not occupied by biological matter and so ultimately generates a *negative* image of the sample. It has been used to excellent effect to generate differential contrast between heterogeneous biological tissues (Figure 4.2A), though in using such an involved multi-stage preparation it is sometimes difficult to avoid experimental artifacts. One method employed to minimize such problems has been to snap-freeze (*cryo-fix*) samples using liquid nitrogen instead of potentially damaging chemical fixation.

The other most common contrast reagent incorporation method involves positive staining via metallic shadowing, generally of evaporated platinum. A typical approach applied to visualizing single molecules with TEM is to spray a dilute purified solution of the molecules onto a thin sheet of evaporated carbon which is supported by a standard EM-grid sample holder. The buffer is dried off in a vacuum and then platinum is evaporated onto the sample from a low angle of typically 10° or less as the sample is rotated laterally. This creates a reasonably uniform metallic shadow of the topographical features of any single molecules stuck to the surface of the carbon, which are then suitably electron-dense to provide a high scatter signal from an electron beam whereas the supporting thin carbon sheet is relatively transparent to electrons. In effect this results in a *positive* image.

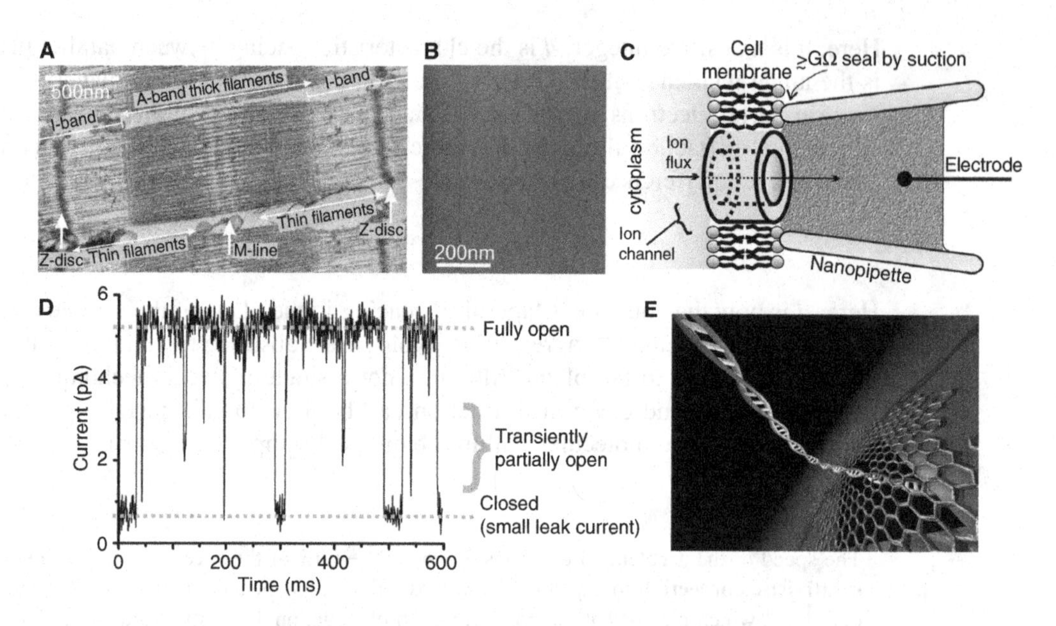

FIGURE 4.2 Electron micrographs using **(A)** negative staining, here on a thin section of muscle tissue illustrating a functional 'sarcomere' unit of an individual myofibril, with single-molecule filaments clearly visible, and **(B)** positive staining of single myosin molecules using platinum evaporative shadowing (author's own data, see Leake, 2001). **(C)** Schematic of single ion channel recording setup using cell membrane patch-clamping with **(D)** typical recording of a single 'gated' potassium ion channel (Kir4.1) whose fully open state corresponds to ~5 pA current, with evidence of transient partial open states (the small non-zero current for the closed state is due to the finite gigaseal resistance allowing some ion leakage between the membrane and micropipette (data courtesy of Lijun Shang, Oxford University)). **(E)** An instructive cartoon showing a single DNA molecule translocating through a graphene nanopore (reproduced with permission of Cees Dekkar, Delft University).

A modification of this approach is used in *freeze-fracture electron microscopy* (FFEM) or *freeze-etch* experiments. Here, the sample is *cryo-fixed*, typically using liquid nitrogen, but is then fractured using the tip of the microtome, revealing a random, fracture picture of the structural makeup immediately beneath the surface of the sample. This fracture pattern is then metal-shadowed from typically a relatively high angle of ~45° so as to reach recessed surface features, and a supporting layer of carbon is often evaporated onto that. The sample metal replica may then be viewed using TEM, with this approach being used to excellent effect to visualize the lipid bilayer architecture of cell membranes.

Non-scattered (i.e. *transmitted*) electrons are detected for both positive and negative staining methods typically using a combination of phosphor plate and high-efficiency camera. This has been used to generate structural information for several different types of biological molecule (Figure 4.2B). TEM generally uses relatively high potential differences, with ~100 keV being typical. The spatial resolution is generally 10 times poorer than the predicted electron wavelength due to spherical aberration, and so in reality values of $\sim 5 \times 10^{-11}$ are usual, or roughly half the diameter of a typical atom.

Single biological molecules often exist in multiple structural states (see Chapter 2) with resultant heterogeneity in the observed TEM images. By assigning separate TEM single-molecule images to different sub-classes, an average TEM image of each different structural state may be calculated. This has been used to excellent effect to characterize the different structural states of myosin (Walker *et al.*, 2000); this comes with a notable caveat in that great care should be taken to objectify the system of sub-classification to avoid imposing an apparent structural classification on such images which in reality may not exist.

4.3.2 Scanning electron microscopy (SEM)

Here a smaller potential difference is generally used compared to TEM, typically in the range 10–30 keV, with the lower resultant electron energies generating a greater proportion of elastically back scattered electrons from the sample as well as secondary electrons produced by inelastic scattering, which both form the detected signal. Metallic contrast reagents must still be applied as with TEM, but the primary advantage of SEM is that a larger scale sample area can be viewed. The larger associated electron energies in combination with spherical aberration effects limit the spatial resolution to typically 1–20 nm, with improved measurement stability upon imaging on a cold stage cooled by liquid nitrogen. The resolution is just at the limit for biological single-molecule detection, and SEM is arguably a more appropriate technique for investigating relatively large molecular aggregates or multi-molecular complexes than single subunit molecules themselves.

4.4 Ionic currents through nanopores

If a small hole is made in a sheet of electrical insulator surrounded by an ionic solution, and an electrostatic potential difference is applied across the sheet, then ions will flow through the hole. If the hole itself has a length scale of nanometres (a *nanopore*) then this simple principle can form the basis of several single-molecule detection techniques.

4.4.1 Single ion channel recording

At various locations, the lipid bilayer architecture of cell membranes is disrupted by natural nanopores of ion channels. These are protein structures which permit the controllable flow of ions into and out of the cell, generally involving exceptional specificity in terms of the ions that are allowed to pass through the pore and often employing additional highly sensitive *voltage-gating* mechanisms. The presence of these single nanopore molecular complexes can be detected and studied using the electrophysiological technique of patch clamping.

It is relatively easy to show that the typical resistance of an open ion channel is a few gigaohms (see Q4.3), provided one accounts not only for the resistance due to the pore but also for the resistance due to ion flow access effects into and out of the pore aperture (interested physical scientists should read the brief but elegant paper by Hall, 1975). Thus, any probe measuring ion flux through the channel needs to have a seal with the membrane which has a greater effective electrical resistance than the channel itself to stand any decent chance of measuring it. Typically a glass micro- or nano-pipette is pressed into suction contact to make a seal with very high electrical resistance generally at greater than the gigaohm level (hence the term *gigaseal*). Towards the tip of the micropipette is a single silver electrode, which can measure ion currents (Figure 4.2C).

Experiments may either be performed with the micro/nano-pipette remaining in contact with the cell, or with the associated patch of membrane excised. Electrical measurements may then be made either by keeping the current-flow clamped using feedback circuitry and measuring changes in voltage with time, or more commonly by clamping the voltage to a fixed value and measuring changes in current. This generally follows some form of physical or chemical intervention which is likely to affect whether the ion channel is open or closed, for example adding a ligand or drug inhibitor, or changing the fixed voltage level, typically of the order of ~100 mV (Figure 4.2D).

The physical basis of the equilibrium level of voltage across a *semi-permeable* barrier, such as the cell membrane with pores through which ions can diffuse selectively, is established when the osmotic force due to any differences in ion concentration either side of the membrane is balanced by the net electrostatic force due to the difference in electrochemical potential on the charged ion in the presence of the membrane voltage potential. The former force is in effect *entropic* in origin whereas the latter is *enthalpic*, and the combination of both gives rise to the well-known *Nernst equation*:

$$V_{\mathrm{m}} = \frac{nRT}{zF} \ln \frac{C_{\mathrm{outside}}}{C_{\mathrm{inside}}} \tag{4.6}$$

Here, V_{m} is the *equilibrium voltage* across the cell membrane with ions each of charge z having in general different concentrations C inside and outside the cell (R is the molar gas constant, T the absolute temperature and F the Faraday constant). With several different charged ions the equilibrium potential can be calculated from the fractional contribution of each (characterized by the more general *Goldman equation*). Typically for many cell types the size of V_{m} is in the range -50 to -200 mV, with the negative sign being generally due to energized net pumping out of Na^{+} ions compared with a smaller influx of K^{+} ions.

The size of the measured current when a channel is open is around 1 pA to a few tens of picoamps. This equates to 10^{6}–10^{8} ions per second, which even when sampled at the microsecond time scale still struggles to be single *ion* detection per se. However, the ion-flux is rather the detection signature for the presence of a single ion channel and of its state of opening or closure, or indeed somewhere in between as appears to be the case for some channels.

The random area of membrane encapsulated by the patch clamp may contain more than one ion channel, which impairs the ability to measure single-molecule properties (for example, investigating whether there are heterogeneous short-lived states between a channel being fully open or closed) since the measured current will be the sum of the currents through each channel. Since each channel may be open or closed in a stochastic manner this leads to difficulties in interpretation of the experimental ion-flux data. One way to minimize this problem is to genetically modify the cell to generate a lower surface density of ion channels, either by trying to inhibit their expression level, or by keeping the expression level fixed but growing larger cells. Neither of these techniques is ideal as they both affect the native physiological state of the cell.

A further improvement is to use smaller diameter pipettes; standard glass micropipettes may be heated and drawn out controllably to generate inner diameters down to a few tens of nanometres. Finally, ion-flux measurements may be performed in combination with fluorescence measurements; if a fluorescence marker can be placed on a component of the nanopore than it may be possible to count how many ion channels are present in the patch-clamp region, though controllably placing a fluorescent tag on a nanopore but avoiding impairment of the ion channel function is in general non-trivial.

BIO-EXTRA

In many cells the size of the equilibrium transmembrane voltage potential is **homeostatically controlled** very robustly, for example in bacteria V_m is normally very closely regulated to roughly −150 mV. In some cells, for example during nerve impulse conduction in axons, the sign of V_m can vary due to a wave of depolarization of voltage. In the resting state Na^+ ions are actively pumped out in exchange for K^+ ions which are pumped into the cell, energized by the hydrolysis of ATP. These sodium–potassium **ion exchange pumps** are essentially Na^+/K^+ selective ion channels which exchange with the ratio of three Na^+ to every two K^+, hence resulting in a net negative V_m, with the so-called resting potential typically of ca. −70 mV. During nerve impulse conduction the ion pumps open transiently to both Na^+ and K^+, causing **depolarization** of V_m, rising over a period of ~1 ms to between +40 to +100 mV depending on nerve cell type, with the resting potential re-established a few milliseconds later. The recovery time required before another action potential can be reached is typically ~10 ms, so the maximum nerve **firing rate** is ~100 Hz.

4.4.2 Solid-state nanopores

Using modern *nanofabrication* techniques it is now possible to manufacture nanopores reproducibly using a silicon-based substrate. One popular method involves transmission electron microscopy; a high intensity beam of electrons is focused onto a thin membrane consisting of silicon nitride which generates a hole through the membrane. By varying the power of the beam, the size of the nanopore can be tuned, with reproducible pore diameters as low as ~5 nm possible (van den Hout *et al.*, 2010). Such nanopores have have been applied successfully in the detection of single molecules of a variety of biological polymers including RNA and DNA (Rhee and Burns, 2006). The principle involves first applying a voltage across either side of the nanopore, which induces ion flow through the pore in the case of a typical physiological ionic solution.

Biological polymer molecules present in the solution will in general possess a net non-zero charge due to the presence of charged chemical subunits on the surface, with the effect that the whole molecule will migrate down the voltage potential gradient, but due to its relatively large size the typical drift speed will be significantly slower than the ion-flux through the pore. As a given biopolymer molecule approaches the nanopore itself, the flow of ions is impeded, most significantly as the molecule eventually passes through the pore. As it does so the drop in ion current may be measurable experimentally if the drift speed through the pore is slow enough, and the specific shape of the drop in current with time is in effect a signature for that specific type of molecule, and so can be used as a method of single-molecule detection.

With greater precision the hope is to be able to measure consistently the different nucleotide bases of nucleic acids as, for example, a single molecule of DNA migrates through the pore, hence leading to the ability to sequence single DNA molecules rapidly. The main problem with this is that the speed of migration, even for the lowest controllable voltage gradient, is still relatively high for unconstrained DNA molecules leading to unreliability in experimental measurements of the ion-flux signature. One attempt to slow down molecules as they migrate through a nanopore involves controllably pulling on the molecule from the opposite direction to the electrostatic force using optical tweezers (Keyser *et al.*, 2006), discussed in greater detail in Chapter 5.

A further problem with this sequencing method is the finite length of the solid-state nanopore. Typically, the minimum width of a silicon nitride membrane required to produce a structurally stable nanopore reproducibly is ~20 nm; this is equivalent to roughly 50 nucleotide base pairs of DNA (see Chapter 2) assuming the molecule is stretched parallel to the pore axis. Recent attempts to circumvent this problem have involved reducing the thickness of the substrate by using just a monolayer of *graphene* (Schneider *et al.*, 2010). Graphene is a two-dimensional single atomic layer of carbon atoms which are packed into a honeycomb shape with a thickness of only 0.3 nm, comparable to just a single nucleotuide base pair, which despite such minimal thickness is a structurally stable membrane (Figure 4.2E).

That being said, recent simulation studies for the translocation of single biopolymers through a nanopore which incorporate some degree of realistic flexibility of the pore wall actually suggest that allowing the pore some level of compliant wiggling can improve the consistency of biopolymer translocation (interested physical scientists should see Cohen *et al.*, 2011), in which case pores composed of a less stiff material than graphene or silicon nitride might be an advantage (section 4.4).

4.4.3 Engineered protein nanopores

Several natural pore-forming proteins exist which can self-assemble to create a nanopore within a lipid bilayer, and are indeed more compliant than the artificial pores seen in the previous section. The best studied of these is the protein *α-haemolysin*. This is a poison naturally secreted by the *Staphylococcus aureus* bacterium which binds to cell membranes of other competing bacteria and spontaneously punches a hole in the lipid bilayer, significantly impairing the cell's viability by disrupting the protonmotive force across the membrane by allowing protons to leak uncontrollably through the hole (see Chapter 2). It does this by the self-assembly of a nanopore consisting of seven α-haemolysin subunits (Figure 4.3A). These nanopores can be incorporated in a controlled environment in an

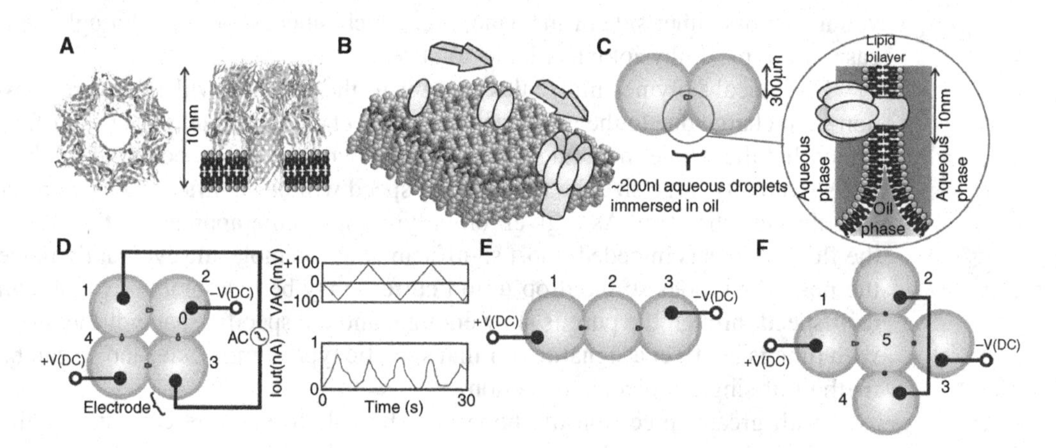

FIGURE 4.3 **(A)** Structure of α-haemolysin heptamer (plan view left panel, PDB ID:7AHL), with cartoon of it integrated into a lipid bilayer (side view right panel). **(B)** A possible assembly model for the nanopore complex derived from rapid, cooperative self-assembly of monomer subunits into the fully functional nanopore. **(C)** Two nanodroplets sharing a common bilayer interface in which a heptamer α-haemolysin nanopore is integrated. **(D)** Four nanodroplet network resulting in full-wave rectification of AC input (via droplets 1 and 3) to generate rectified DC output (droplets 2 and 4) utilizing modified α-haemolysin nanopores which are voltage gated (left panel) with schematic depiction of typical input AC voltage and rectified current output from one of the heptamer nanopores shown (right panel). **(E)** Three nanodroplet 'bio-battery' network in which a combination of positive and negative ion nanopore selectivity for the two interfaces of droplet 2 leads to a small DC voltage potential across droplets 1 and 3. **(F)** The five droplet 'nano-eye' in which droplets 2, 3 and 4 share several bacteriorhodopsin molecules with the lipid bilayer interface with droplet 5, and droplets 1 and 5 share an unmodified α-haemolysin nanopore.

artificial lipid bilayer and utilized in a similar manner to solid-state nanopores to study the translocation of a variety of different single biological molecules through the nanopore by measuring the molecular signature of the ion current as the molecule passes through (for a good review see Bayley, 2009).

These naturally derived protein nanopores confer some advantages over solid-state nanopores. Firstly, their size is highly consistent and not prone to manufacturing artifacts. Secondly, these pores can be engineered to operate both with additional *adapter molecules* such as *cyclodextrin*, which allows greater ion current measuring sensitivity for translocating molecules such as DNA. In addition the amino acid residues which make up the inside surface of the pore can be modified, for example to alter their electrostatic charge, which can be utilized to provide additional selectivity as to what biological molecules are permitted to pass through the pore. This nanopore technology is a prime candidate to achieve the goal of reliable and consistent single-molecule bio-sensing and single-molecule sequencing of important biological polymers such as DNA in the near future.

Such nanopores can also be established in complex *nanodroplet network systems*. Here, ~200 nl droplets have an internal aqueous phase separated by an artificial lipid monolayer which remains structurally stable owing to the centrally acting hydrophobic force imposed from the external oil phase (Figure 4.3C). Such droplets can be positioned directly on the tip of an agarose-coated Ag/AgCl 100 μm diameter electrode by surface tension from the aqueous phase which in turn is connected to a micromanipulator. Multiple droplets may then be positioned adjacent to each other relatively easily in a two-dimensional array, with droplets sharing common lipid bilayer interfaces and joined by one or more α-haemolysin nanopores integrated into the bilayer (Figure 4.3C, expanded section).

By modifying the amino acid residues pointing towards the lumen of the nanopore itself to give them all a positive charge, it was found that these nanopores would be open in the presence of a positive voltage potential, but closed in the presence of a negative potential, presumably due to some induced conformational change blocking the pore lumen (Maglia *et al.*, 2009). Thus, this modified nanopore is in effect voltage-gated and so acts as an electrical diode – as a proof-of-principle it was possible to join four such nanodroplets to form a *full-wave AC-to-DC rectification unit* (Figure 4.3D).

Other arrangements of such nanodroplets have led to a tiny *bio-battery* (Figure 4.3E). In its simplest form this is made from a linear arrangement of three nanodroplets in which the central droplet is connected to the others via either a positive or a negative ion selective nanopore, resulting in a small but measurable current flow between the connected electrode termini of ~50 pA between the outer two nanodroplets.

A *nano-eye*, which can detect photons of light has also been developed (Holden *et al.*, 2007). This bio-mimetic system consists of five nanodroplets (Figure 4.3F) with integrated nanopores consisting of either α-haemolysin or the photo-sensitive protein *bacteriorhodopsin*. Bacteriorhodopsin is a natural membrane nanopore which utilizes the absorbed energy from a single ~green wavelength photon of light to pump a single proton across a lipid bilayer. This therefore constitutes a small current, which the nanodroplet arrangement can detect. Although it is possible to implement such a system controllably with only 1–2 α-haemolysin complexes in each common lipid bilayer interface, the number of bacteriorhodopsin molecules required to generate a measurement current is of the order of thousands, but as a proof-of-principle this shows great promise.

Currently such nanodroplet arrangements are two-dimensional, but there is significant scope to implement far more complex nanoscale bio-synthetic systems in three dimensions. Although nanodroplet systems are clearly not 'natively cellular' per se, they represent a *synthetic biological system* which is moving in the direction of an exceptionally *cell-like* physiological behaviour.

PHYSICS-EXTRA

Modelling the **translocation** of a polymer through a nanopore is non-trivial; the theoretical approaches which agree best with experimental data involve difficult molecular dynamics simulations (for example, see Matysiak *et al.*, 2006), and the manner in which these simulations are constructed illustrates well the general need for exceptionally fine precision not only in experimental measurement but also in theoretical analysis at this level of single-molecule biophysics research. Here, the polymer is typically modelled as interlinked, jointed segments (a **freely jointed chain**, see Chapter 5), and then a non-trivial potential energy function U is constructed corresponding to the summed effects from each segment, including contributions from a bond-link potential, chain bending, excluded volume effects due to the physical presence of the polymer itself not permitting certain spatial conformations, electrostatic contributions in assuming an imposed E-field gradient to bring about the translocation, and Van der Waals effects between the polymer and the nanopore wall. Then, the **Langevin equation** is applied which equates the force experienced by any given segment to the sum of the negative gradients of U with an additional contribution due to thermal bath coupling in the presence of random thermal fluctuations. By solving this equation for each chain segment in very small time steps and imposing pseudo-random noise for the thermal fluctuations of the surrounding water solvent molecules, predictions can be made as to the position of each given segment as a function of time, hence allowing us to simulate the translocation process through the nanopore.

4.5 Raman spectroscopy

As alluded to in Chapter 3, the Raman effect occurs when a photon of light is scattered inelastically by a molecule, resulting in the scattered photon having either a greater energy (anti-Stokes scattering) or, more commonly, a lower energy (Stokes scattering), principally due to transitions in the molecule's vibrational and rotational energy states. Usually the shift in scattered energy is measured in terms of the change in wavenumber (the inverse of the photon wavelength) generally quoted in units of cm^{-1}, with a range of 500–2000 cm^{-1} being typical for biological molecules. A near infrared laser source (wavelength $\sim$1 μm) is often employed to generate the incident photons. In principle, the Raman spectrum generated, depicting scattering peaks of different amplitudes at different wavenumbers, is a signature for the specific type of single molecule in the sample, and so forms the basis for a biomolecule sensing technique (Figure 4.4A).

4.5.1 Surface enhanced Raman spectroscopy (SERS)

However, to detect the presence of single molecules in a sample using Raman spectroscopy requires significant enhancement to the standard method used to acquire a scattering spectrum from a bulk, homogeneous sample. To date, the most effective means of achieving this is to utilize surface enhancement methods, the most popular of which is SERS (Kneipp *et al.*, 1997). Here, the sample is placed in a colloidal substrate of gold or silver nanoparticles typically tens of nanometres in diameter (a variation of this is tip-enhanced Raman spectroscopy, or TERS, where the end of a metalized surface probe tip performs a similar enhancement function to a metal nanoparticle). Photons from the laser

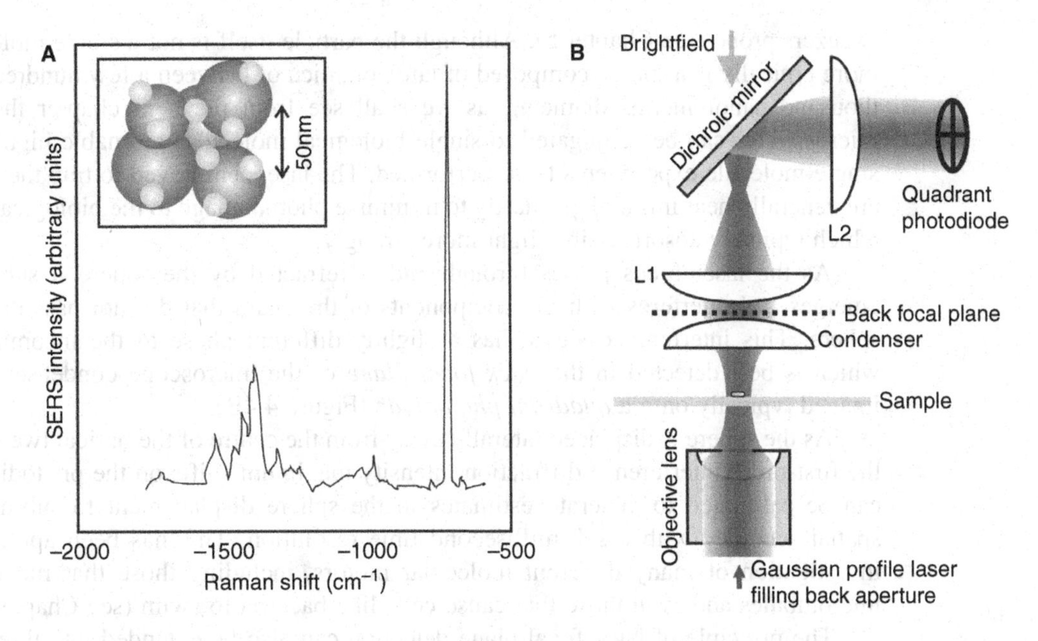

FIGURE 4.4 **(A)** Schematic depiction of a typical SERS spectrum taken from a nanomolar sample of the nucleotide adenine, with inset illustrating adenine molecules (light grey) bound to silver or gold nanoparticle aggregates (dark grey). **(B)** Schematic optical diagram for back focal plane detection, lenses L1 and L2 forming a telescope which images the back focal plane of the condenser (dotted line) onto a quadrant photodiode via a dichroic mirror which reflects the laser light focused on the sample (typically near infrared of ca. 1 μm wavelength) but transmits visible light from a brightfield lamp source to illuminate the sample.

will induce surface plasmons in the metallic particles, in much the same way as they do in surface plasmon resonance techniques (see Chapter 1).

In the vicinity of the surface the local electric field E associated with the photons is enhanced by a factor E^4. This enhancement effect depends critically on the size and shape of the nanoparticles, but typically can generate an enhancement in measurement sensitivity of up to $\sim 10^{14}$, and is particularly effective if the molecule itself is conjugated to the nanoparticle surface. This enhancement allows sample volumes of a few tens of picolitres to be probed at dilutions as low as $\sim 10^{-14}$ M, sufficient to detect single molecules (for a recent review see Kneipp *et al.*, 2010).

In terms of detection of single biological molecules this technique is still in its infancy, however it has been applied successfully at, or very near to, the single-molecule level to detect nucleotide bases relevant to DNA/RNA sequencing, amino acids and large protein complexes such as haemoglobin, in some cases pushing the sample detection volume down to ~ 100 fl (Kneipp *et al.*, 1999). This technique is beginning to be applied to bio-sensing in live cells that can engulf suitable nanoparticles via the process of *endocytosis*, which is the process by which eukaryotic cells engulf large external particles (typically either food particles or as part of the immune response) via a mechanism of cell membrane invagination to produce an internalized liposome (see Chapter 7).

4.6 Interference-based detection

Back focal plane detection (Gittes and Schmidt, 1998) is an *optical interference technique* which allows the spatial displacement to be measured to sub-nanometre precision for a single particle which passes through the laser focus at the centre of an optical tweezers probe (see Chapter 5). Although the particle itself is not a single molecule, but more typically is a sphere composed of latex or silica of between a few hundred to a few thousand nanommetres diameter, as we shall see from the next chapter these nano/microspheres can be conjugated to single biological molecules to enable highly precise single-molecule experiments to be performed. The lasers employed to trap these spheres are generally near infrared, primarily to minimize photodamage to the biological material which typically absorb visible light more strongly.

As the laser focus passes through and is refracted by the sphere it subsequently emerges and interferes with the components of the beam that did not pass through the sphere. This interference signal has a slightly different phase to the incoming beam, which is best detected in the *back focal plane* of the microscope condenser and then imaged typically onto a *quadrant photodiode* (Figure 4.4B).

As the sphere is displaced laterally away from the centre of the optical tweezers trap, the first-order interference diffraction intensity maximum shifts on the photodiode. This can be calibrated to generate estimates in the sphere displacement to sub-nanometre spatial accuracy with a sub-millisecond time resolution. This has been applied to the investigation of many different molecular motors, including those that run on DNA, microtubules and even those that cause cells like bacteria to swim (see Chapter 8).

The principle of back focal plane detection can also be extended into three dimensions. For example, by using this interference method to map out the random motions over time of a microsphere which is tethered by a suitably long single molecule, the surface topography of the environment immediately in the vicinity of the sphere can be mapped out, such as the surface of a cell. Also, this technique can be modified to act as a *holographic microscope* to map out the positions of multiple gold-coated nanoparticles in colloidal suspension, which can then be used to monitor phenomena such as fluid-flow patterns around the surface of live cells.

THE GIST

- Scanning probe microscopy, in particular AFM, can be used to image the topography of single biological molecules, with recent technical advances allowing certain molecules to be monitored in real time at 'video' rate.
- Although artifacts are more prevalent in transmission electron microscopy compared to many other single-molecule techniques, as a result of the involved nature of the sample preparation stages, it can still generate very detailed structural information for single biomolecules at a resolution of less than 0.1 nm.
- Measuring picoamp level ion currents through either natural or engineered nanopores can give us information not only about the molecular nature of the pore itself, but also about biopolymer molecules translocating through the pores, which can be utilized for bio-sensing and molecular sequencing.
- SERS has a sensitivity 10^{14} greater than normal Raman spectroscopy applicable to volumes comparable to small single cells, and is starting to be utilized for single biological molecule detection, both in vitro and in live cells.
- Back focal plane detection is a powerful interference detection method which allows sub-nanometre displacement measurements of laser-trapped particles.

TAKE-HOME MESSAGE Single-molecule detection techniques not primarily using visible light are very diverse, and permit molecular structures to be imaged, important molecules such as DNA to be sequenced and very precise optical tweezers measurements to be made.

References

GENERAL

- Arkawa, H., Umemura, K. and Ikai, A. (1992). Protein images obtained by STM, AFM and TEM. *Nature* **358**: 171–173.
- Bayley, H. (2009). Piercing insights, *Nature* **459**: 651–652.
- Binnig, G., Quate, C. F. and Gerber, C. (1986). Atomic force microscope. *Phys. Rev. Lett.* **56**: 930–933.
- Kneipp, K. *et al.* (1997). Single molecule detection using surface-enhanced Raman scattering (SERS). *Phys. Rev. Lett.* **56**: 1667–1670.
- Kneipp, K. *et al.* (1999). Ultrasensitive chemical analysis by Raman spectroscopy. *Chem. Rev.* **99**: 2957–2976.
- Kneipp, J. *et al.* (2010). Novel optical nanosensors for probing and imaging live cells. *Nanomed. Nanotechnol. Biol. Med.* **6**: 214–226.
- Rhee, M. and Burns, M. A. (2006). Nanopore sequencing technology: research trends and applications. *Trends Biotechnol.* **24**: 580–586.

ADVANCED

- Binnig, G. *et al.* (1982). Tunneling through a controllable vacuum gap. *Appl. Phys. Lett.* **40**: 178–180.
- Cohen, J. A., Chaudhuri, A. and Golestanian, R. (2011). Active polymer translocation through flickering pores. *Phys. Rev. Lett.* **107**: 238102.
- Gittes, F. and Schmidt, C. F. (1998). Interference model for back-focal-plane displacement detection in optical tweezers. *Opt. Lett.* **23**: 7–9.
- Hall, J. E. (1975). Access resistance of a small circular pore. *J. Gen. Physiol.* **66**: 531–532.
- Hansma, P. K. *et al.* (1989). The scanning ion-conductance microscope. *Science* **243**: 641–643.
- Hobbs, J. K., Vasilev, C. and Humphris, A. D. L. (2006). Video AFM – A new tool for high speed surface analysis. *Analyst* **131**: 251–256.
- Holden, M. A., Needham, D. and Bayley, H. (2007). Functional bionetworks from nanoliter water droplets. *J. Am. Chem. Soc.* **129**: 8650–8655.
- Keyser, U. F., van der Does, J., Dekker, C. and Dekker, N. H. (2006). Optical tweezers for force measurements on DNA in nanopores. *Rev. Sci. Instrum.* **77**: 105105.
- Klenerman, D., Korchev, Y. E. and Davis, S. J. (2011). Imaging and characterization of the surface of live cells. *Curr. Opin. Chem. Biol.* **15**: 696–703.
- Leake, M. C. (2001). *Investigation of the extensile properties of the giant sarcomeric protein titin by single-molecule manipulation using a laser-tweezers technique*. PhD dissertation, London University, UK.
- Maglia, G. *et al.* (2009). Droplet networks with incorporated protein diodes show collective properties. *Nature Nanotechnol.* **4**: 437–440.
- Matysiak, S. *et al.* (2006). Dynamics of polymer translocation through nanopores: theory meets experiment. *Phys. Rev. Lett.* **96**: 118103.
- Mohn, F., Gross, L., Moll, N. and Meyer, G. (2012). Imaging the charge distribution within a single molecule. *Nature Nanotechnol.* [Epub ahead of print]
- Schneider, G. F. *et al.* (2010). DNA translocation through graphene nanopores. *Nano Lett.* **10**: 3163–3167.
- van den Hout, M. *et al.* (2010). Controlling nanopore size, shape, and stability. *Nanotechnology* **21**: 115304.
- Walker, M. L. *et al.* (2000). Two-headed binding of a processive myosin to F-actin. *Nature* **405**: 804–807.

Questions

FOR THE LIFE SCIENTISTS

Q4.1. A membrane protein complex was imaged using AFM in contact mode. In one experiment the protein complex was purified and inserted into an artificial lipid bilayer on a flat mica surface. This indicated protein topographic features ca. 0.5 nm extruding above the bilayer. When the AFM tip was pushed into these features and then retracted it was found that the tip experienced a measurable pulling force. When the same experiment was performed using a living cell, similar topographic features could be imaged but when the tip was pushed into the sample with the same force limit set as before and then retracted, no such pulling force could be measured. Comment on these observations.

Q4.2. A solid-state nanopore of ca. 5 nm diameter was used to detect the translocation of a filamentous protein in phosphate buffered saline solution following the application of a 100 mV voltage across the nanopore. Secondary structure prediction suggested that the protein consisted mainly of 5 α-helices containing 10–20 amino acids each connected by a random coil region of between 5–10 amino acids. On the basis of the mix of amino acids the protein had a marginally net positive charge and it was found that there were just two cysteine residues separated by 20 amino acids. When the ion flux through the nanopore was measured as a function of time it indicated that for most of the time the current had a reasonably stable value of ca. 50 pA, but also there were much shorter-lived currents of 40, 42, 44 and 46 pA. However, when 5 mM DTT was added to the solution the short-lived current values were measured at 40, 42, 44, 46 and 48 pA. Comment on these results.

Q4.3. Physiological 'Ringer' solution has a resistivity ρ of ~80 Ω cm. Assume that the electrical resistance through a nanopore of cross-sectional area A and length l is given by $\rho\, l/A$, and that there is an additional 'access resistance' due to either ion entry or ion exit to/from a circular aperture radius a of $\rho/4a$. What is the total electrical resistance measured across a typical open sodium ion channel of length ~5 nm and pore diameter ~0.6 nm?

Q4.4. Using a narrow pipette, scanning conductance microscope images of purified F1-ATPase enzymes could just be discerned when the enzymes were stuck to a coverslip surface. However, when potassium channels were over expressed in a cell and a membrane patch excised and imaged on a flat surface, no clear images of the channels could be obtained. Comment on these observations.

Q4.5. A cell is placed in a physiological solution of 100 mM NaCl, 20 mM KCl at room temperature. The cell membrane has several open chloride ion channels. Using single-molecule fluorescence imaging the internal concentration of Cl^- ions was measured at 22 mM, while that of K^+ was estimated to be 30 mM. (a) Show how the Nernst equation can predict the transmembrane voltage – explain why you must use the Cl^- ion concentrations for this and not the K^+. (b) It is found that you need an energy input to pump K^+ ions out of the cell – why does this not occur spontaneously? (c) A chemical is applied to open up all Na^+ and K^+ ion channels – would you expect the K^+ ion concentration inside and outside the cell to be equal?

FOR THE PHYSICAL SCIENTISTS

Q4.6. A dried, unstained cell membrane was placed in a 150 keV electron microscope, which resulted in some of the transmitted electrons being diffracted, with a first-order diffraction deflection of 10 milliradians, and some being scattered back with much weaker interference maxima at a deviation of only ca. 0.5 milliradians from the normal to the plane of the membrane. If the cell membrane was composed primarily of tightly packed lipids, comment on the angular deflections and intensity of the scattered and diffracted electron beam observed.

Q4.7. In an AFM imaging experiment the maximum vertical displacement of the cantilever was limited by the height of the silicon nitride tip of roughly 10 μm, giving a full-scale deflection of the photodiode output of 10 V. At room temperature with the tip far away from the sample the root mean square photodioide output was 5.5 mV with the laser reflecting onto the back of the cantilever switch on, and 1.9 mV when the laser was switched off. The machine was used to image single myosin molecules on a flat surface, whose head regions were found to generate forces of ~5 pN each when performing a 'power stroke'. When the tip is just in contact with a head region what offset voltage should be applied to just cause a power stroke to stall?

Q4.8. Fick's First law of diffusion states that the vector particle flux $\mathbf{J} = -D\nabla(n)$ where D is the diffusion coefficient and m is the number of particles per unit volume. (a) Modelling an ion channel as a one-dimensional cylinder of radius a, find an expression for the channel current due solely to diffusion of univalent ions of molar concentration C, stating any assumptions you make. In a patch clamp experiment an extracted region of cell membrane contained an estimated ~10 sodium ion channels each of diameter ~1 nm. When a voltage of −100 mV was applied across the membrane patch in a solution of 150 mM NaCl the measured current was found to fluctuate with time from a range of zero up to a maximum at which the observed resistance of the patch was measured as $3 \times 10^9\ \Omega$. (b) Estimate the current through a single open channel and the hypothetical sampling frequency required to monitor the passage of a single ion. (c) How significant is diffusion to the ion-flux through a single channel?

FOR THOSE WHO HAVE NOT MADE UP THEIR MIND

Q4.9. Graphene, as well as being a very thin yet strong structure, is also electrically conducting. Is this an advantage or disadvantage to using it as the nanopore substrate for sequencing single DNA molecules?

Q4.10. An experiment was devised using back focal plane detection to monitor the lateral position of a latex microscope attached to bacterial flagellar motor (see Chapter 8) via a stiff filament stub of a live bacterial cell, which was free to rotate in a circle a short distance above the cell which itself was stuck firmly to a microscope coverslip. The motor is expected to rotate at typical speeds of 100 Hz, and is made up of around 20 individual subunits arranged in a circle, each of which is thought to generate torque independently to push the filament around in a ratchet mechanism. (a) What in principle is the minimum sampling bandwidth of the quadrant photodiode used in the back focal plane detection in order to see the microsphere ratchet around all of the torque-generating units? In practice it is difficult to make a completely stiff filament; in a separate experiment using a completely unstiffened filament attached to a 0.5 μm diameter bead it was found that the filament compliance resulted in a relaxation-drag delay to bead movement following each ratchet of a few tenths of a millissecond, whereas a 1.0 μm diameter bead had an equivalent response time ca. 10 times slower. (b) Which bead is the best choice to try to monitor the ratchet mechanism? In principle it is possible to make some of the 20 ratchet subunits dysfunctional without affecting the functionality of the others. (c) How many subunits need to be made dysfunctional in order to detect individual ratchets clearly? (d) New evidence suggested that there may be some level of co-operativity between the subunits – how does this affect your answers above?

Q4.11. A population of motor proteins was prepared for TEM imaging on a 100 keV machine using an evaporative platinum rotary-shadow technique. Each platinum atom is about 0.5 nm in diameter but it was found that the smallest observable metal particles were composed of ca. 5 atoms. (a) What is the real practical spatial resolution in this experiment? The motor protein was known to exist in two different structural states: state I for ca. 20% of the time, and state II for the remaining 80%. It was hoped that each single-molecule TEM image could be assigned as being in one of the two states and that an 'average image' from each class could then be constructed. Other experiments suggested that in changing from state I to state II one of the structural features in the molecule moves by ca. 3 nm. (b) How many single molecules in total should we expect to have to analyse on the TEM images to be able to measure this 3 nm change to greater than a 10% accuracy?

Q4.12. An AFM image was obtained for a hard spherical nanoparticle acting as a firmly bound surface marker between live cells stuck to the surface. However, the image obtained for the nanoparticle did not indicate a sphere but rather a hump-like shape whose width was ~150 nm, larger than separate TEM estimates had predicted for the nanoparticle diameter. (a) Why is this? (b) Assuming the AFM tip is a tetrahedron of base edge length 1 μm and base-tip height 10 μm, what is the diameter of the nanoparticle?

5 Measuring forces and manipulating single molecules

If everything seems under control, you're just not going fast enough.

(ATTRIBUTED TO MARIO GABRIELLE ANDRETTI, BORN 1940, FORMER WORLD CHAMPION RACING CAR DRIVER)

GENERAL IDEA

In this chapter we encounter the biophysical methods which can be used to measure forces exerted by single biological molecules, and also techniques which can allow us to manipulate single molecules controllably.

5.1 Introduction

There now exist several methods which permit highly controlled measurement and manipulation of the forces experienced by single biological molecules. These varied tools all come under the banner of *force transduction devices*, since they convert mechanical molecular forces into some form of amplified, measurable signal. They share other common features, for example, in general, the single molecules are not manipulated directly but are in effect physically conjugated, usually via one or more chemical links, to some form of adapter which is the the real force transduction element in the system. The principal forces which are used to manipulate the relevant adapter include optical, magnetic, electrical and mechanical forces, and in general all these forces are implemented in an environment of complex feedback electronics and stable, noise-minimizing microscope stages, for the purposes of both *measurement* and *manipulation*.

5.2 Optical tweezers

The ability to trap particles using laser radiation pressure was reported by Arthur Ashkin, the forefather of optical (or laser) tweezers, as early as 1970 (see Ashkin, 1970). The modern form in which we know 'standard' optical tweezers today (formerly described as *single-beam gradient force traps* which devotees often refer to simply as *optical traps*), which result in a net optical force on refractile/dielectric particles of higher refractive index than the surrounding medium roughly towards the intensity maximum of a focused laser, was developed in the early 1980s by Ashkin and co-workers (Ashkin *et al.*, 1986). These optical force transduction devices have since been used for very diverse applications in the study of single-molecule biology (for good reviews see Svoboda and Block, 1994 for an early but very accessible explanation of the physics, and Moffitt *et al.*, 2008 for a more recent compilation of the biological applications).

5.2.1 The single-beam gradient force optical trap

(I) TRAPPING THEORY

Photons of light carry momentum as do particles of matter, but here the momentum p is given by:

$$p = E/c = h\nu/c = h/\lambda \tag{5.1}$$

Here, E is the energy per photon of frequency ν, wavelength λ and speed c, with h Plank's constant. This embodies the famous de Broglie relation encountered previously in Chapter 4. This momentum can be imparted to an object as radiation pressure, for example when a beam is incident on a particle, resulting in scattering of the photons, or at the point of emerging from an optically transparent particle if a photon is transmitted through it (simply on the basis of the Third law of motion, namely that every action has an equal and opposite reaction), as well as during other optical processes such as absorption and emission. Standard optical tweezers utilize a *gradient force*. Here the material in the particle through which photons are passing can be considered to be composed of multiple tiny electrical dipoles on the same length scale as individual atoms and molecules.

An electric dipole occurs when positive and negative charges are physically separated by some distance, in its simplest form modelled as two equal and opposite charges with the dipole moment given by the size of the positive charge multiplied by the distance vector; hence the dipole has intrinsic directionality. The atomic/molecular length scale is 3–4 orders of magnitude smaller than the wavelength of visible light, which is called the *Rayleigh regime* (i.e. the length scale is much *smaller* than the wavelength of light). Conversely, when the length scale is much *greater* than the wavelength of light (at least an order of magnitude) this is described as the *Mie regime*. In the Mie regime, wave-like effects are comparatively small when averaged over the much larger size of the particle and so beams of photons can be modelled as rays (so-called *ray optics regime*).

A photon particle can also be considered as an electromagnetic wave, and so as it passes through *optical media* it will induce a small force on the electric dipoles which averages out over time to point in the direction of the gradient of the intensity of the photon beam. A full derivation of the magnitude and direction of the forces involved requires a solution of Maxwell's four electromagnetic equations (interested physical scientists should see Rohrbach, 2005), however we can gain significant qualitative insight into the phenomenon of optical trapping by considering a ray-optic depiction of the passage of light through a particle (Figure 5.1A, B).

Typically, a single mode laser beam (so-called TEM00 mode, which is the lowest order or fundamental transverse mode of the laser resonator head and has the same cross-sectional profile as a Gaussian beam) is focused using a high NA objective lens onto a *refractile, dielectric* particle. An example of this is a latex or silica microsphere whose refractive index n can be typically in the range 1.4–1.6, higher than that of the surrounding water solution ($n = 1.33$). The focused laser forms a confocal intensity volume in the vicinity of the particle (see Chapter 3). Although optical trapping does not require symmetrical particles, most often they will be spheres.

A stably trapped particle will sit roughly at the centre of the laser focus (though slightly displaced by the forward scatter momentum from the laser beam), such that the spatial gradient of the intensity of the focused laser light is zero. If the particle is displaced from the focus centre then the refraction of the higher intensity light fraction through the particle close to the focus causes an equal and opposite force on the particle

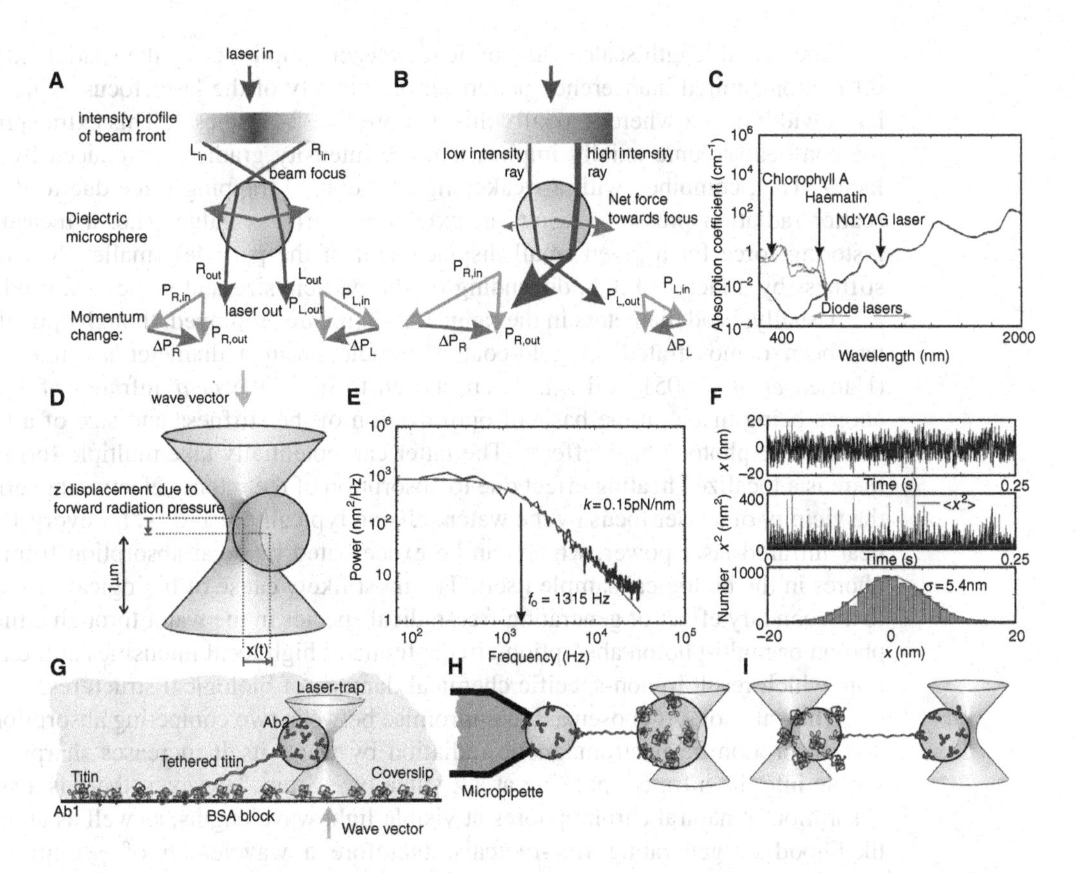

FIGURE 5.1 **(A)** A ray-optic depiction of the gradient force for an optically trapped particle. A parallel Gaussian-profile laser beam is focused and refracted by the trapped particle, such that the equal and opposite changes in momentum on either side of the particle cancel out, resulting in net zero force when the particle is roughly at the laser focus. **(B)** When the particle is laterally displaced from the laser focus the net change in momentum experienced by the particle due to the reaction forces when refracted beams of light emerge from the particle are directed back towards the centre of the laser focus, as illustrated by the momentum vector diagrams. **(C)** Absorption spectrum for water (dark line) compared with two typical biological chromophores. **(D)** Displacement of a 1 μm diameter bead in optical tweezers. **(E)** Power spectrum of the bead of **(D)** with Lorentzian fit, compared with **(F)** mean square displacement. **(G)** Single trap stretch of a titin molecule tethered to a microscope coverslip via antibodies Ab1 and Ab2 that bind to opposite ends of the titin molecule. **(H)** Similar titin stretches using a micropipette combined with a single optical trap and **(I)** two optical tweezers. (Original data, author's own work.).

which is greater than that experienced in the opposite direction due to refraction of the lower intensity portion of the laser beam (Figure 5.1B). The particle therefore experiences a net restoring force back to the focus centre – hence, it is *trapped*, providing any external force perturbations on the particle do not displace it beyond the physical extent of the optical tweezers.

(II) THE LENGTH SCALE OF OPTICAL TRAPS

In practice, stable optical tweezers require a tight, diffraction-limited focus. This means that many photons will enter the focal waist at steep angles relative to the optical axis, which results in high intensity gradients across the trap profile, and so contribute the most to the optical restoring force. To achieve this therefore requires using high NA objective lenses, typically in the range 1.2–1.5 (see Chapter 3), as well as marginally over-filling the back aperture of the objective lens with incident laser light.

The actual length scale of the optical tweezers trap is set by the spatial extent of the diffraction-limited interference pattern in the vicinity of the laser focus, which laterally has a width of $\sim\lambda$ whereas axially this is more like 2–3 times λ. This extra spreading of the confocal volume axially implies that the intensity gradient is reduced by the same factor. This, combined with a weakening of the axial trapping force due to the forward scatter radiation pressure, results in axial trap stiffness values (i.e. a measure of the restoring force for a given small displacement of the particle) smaller than the lateral stiffness by a factor of 3–8, depending on the particle size and value of λ used.

Usually, bead diameters in the range 0.2–2 μm are employed, though optical trapping has been demonstrated on gold-coated particles with a diameter as small as 18 nm (Hansen *et al.*, 2005), and λ is often chosen to be in the *near* infrared of $\sim$1 μm, the choice being made on the basis of optimization of the stiffness and size of a trap while minimizing photodamage effects. The latter can potentially take multiple forms. Firstly, there is a localized heating effect due to absorption of the intense electromagnetic field in the vicinity of a laser focus by the water solvent, typically of 1–2 °C for every 100 mW of near infrared laser power, which can be exacerbated by local absorption from chromophores in the biological sample itself. The most likely cause of biological damage is due to a secondary effect of generating free-radical species in the water through either single-photon or multi-photon absorption effects found at high local intensities at the focus of a trap, which result in non-specific chemical damage of biological structures.

The value of λ is chosen as a compromise between two competing absorption factors. The absorption of electromagnetic radiation by water itself increases sharply from the visible into the infrared, peaking at a λ value of $\sim$3 μm. However, there is a substantial absorption of natural chromophores at visible light wavelengths, as well as an increasing likelihood of generating free-radicals, therefore a wavelength of $\sim$1 μm is a good compromise between these competing absorption-related biological damage effects. At wavelengths of 1000–1200 nm there is also a local dip in the water absorption spectrum, which makes Nd:YAG ($\lambda = 1064$nm) and Nd:YLF ($\lambda = 1047$nm) excellent choices as commercially available crystal laser sources (Figure 5.1C).

(III) TRAP STIFFNESS

The stiffness k of an optical trap can be estimated by measuring the motion of a trapped particle and modelling this with the Langevin equation. This equation takes into account the restoring optical force on a trapped particle along with its frictional drag coefficient γ due to the viscosity of the surrounding water-based solution, as well as random thermal fluctuations in force – the *Langevin force* denoted below as a random functional of time $F(t)$ due to the trapped particle being immersed in a thermal bath (see Chapter 1):

$$\gamma v + kx \approx F(t) \tag{5.1}$$

Here x is the lateral displacement of an optically trapped bead relative to the centre of the trap and v is its speed (Figure 5.1D). The *inertial* term which might normally be expected in this equation (mass times acceleration) is substantially smaller than the other two drag and optical spring force terms owing to the relatively small mass, and so can be neglected.

The motion regime in which such optically trapped particles operate can be characterized by a very small *Reynolds number*, which in effect is the measure of the ratio of inertial to drag forces (drag coefficient times speed) of $\sim10^{-5}$. This is roughly the same ratio seen for the motility of small cells such as bacteria (see Chapter 8), and it means that there is no significant 'gliding' motion as such, rather once an external force is no longer applied then the particle, barring random thermal fluctuations from the surrounding water, comes to a halt.

Equation 5.1 is essentially motion in a so-called *harmonic or parabolic energy potential* which is characterized by having a power-spectrum $P(f)$ as a function of frequency f of a *Lorentzian* shape (eager physical scientists should see Wang and Uhlenbeck, 1945) given by:

$$P(f) = \frac{k_B T}{2\pi^3 (f^2 + f_0^2)} \tag{5.2}$$

Here, f_0 is the so-called *corner frequency* given by $k/2\pi\gamma$. Typically, the corner frequency will be of the order of ~1 kHz, and so provided frequencies are acquired roughly an order of magnitude greater than this, namely ~10 kHz, a reasonable fit can be obtained (Figure 5.1E) to estimate the stiffness k (typically in the range 0.1–0.2 pN nm^{-1}). Similarly, using the equipartition theorem (see Chapter 1), the *mean square displacement* of the motion of an optically trapped bead should satisfy:

$$k\langle x^2 \rangle / 2 = k_B T / 2 \tag{5.3}$$

Thus, by estimating the mean square displacement of the trapped bead the stiffness may also be estimated (Figure 5.1F). The Lorentzian method has a marginal advantage in that it does not require specific knowledge of the bead displacement calibration factor for the positional detector used for the bead, typically a quadrant photodiode using *back focal plane detection* (see Chapter 4), simply a reasonable assumption that the response of the detector is approximately linear with small bead displacements. However, strictly speaking, both methods only generate estimates for the trap stiffness at the centre of the trap. For many applications at low force this is acceptable since the trap stiffness is reasonably constant.

However, some single-molecule stretch experiments require access to relatively high forces of 100 pN or more, such that the bead is very close to the physical edge of the trap, and in this regime there can be significant deviations from a linear dependence of trapping force with displacement. To characterize this the position of an optical trap can be oscillated using a square wave at typically 100 Hz of amplitude ~1 μm; the effect at each square wave alternation is to displace the trap focus instantaneously (or at least within a response time of less than a microsecond) such that the bead is then at the very edge of the trap. Then, the speed v of movement of the bead back towards the trap centre can be used to calculate the drag force. Using Equation 5.1 and averaging over many cycles such that the mean of the Langevin force is zero implies the average drag force should equal the trap restoring force at each different value of x, and thus the trap stiffness can be characterized for the full lateral extent of the trap.

PHYSICS-EXTRA

The drag coefficient γ can be calculated from **Stokes' law** for spherical particles as $6\pi\eta a$ where η is the viscosity of the water and a is the radius of the sphere. It is possible to determine the value of the drag coefficient analytically for other well-characterized particle shapes, for example ellipsoids, rods and discs, but in general an alternative method involves calculating the value experimentally from measurement of the diffusion coefficient D of the particle; then generally $\gamma = k_B T/D$, where k_B is the Boltzmann constant with absolute temperature T, known as the

Einstein–Stokes relation. The **Reynolds number** Re is $v\rho a/\eta$ for a spherical particle of density ρ; if the macroscopic length scale world of a swimming human were associated with the same small value of Re of $\sim 10^{-5}$ found for optically trapped particles and swimming bacteria then this would imply a viscosity equivalent to that of treacle, which gives some idea of what is meant by an absence of gliding.

(IV) OPTICAL TRAPPING IN PRACTICE

Optical tweezers have proved to be an exceptionally powerful tool for manipulating single biological molecules and characterizing many aspects of their force-dependent features. Although single molecules themselves cannot be optically trapped with any great efficiency (some early optical tweezers experiments toyed with rather imprecise manipulation of chromosomes and other organelles), they can be manipulated via the approximately micrometre sized optically trapped bead. Latex microspheres can be commercially engineered to have a coating of a variety of different chemical groups, most importantly *carboxyl*, *amino* and *aldehyde*. Using standard bulk conjugation chemistry these groups can either be bound directly to biological molecules or more commonly linked to an adapter molecule such as a specific antibody or a biotin group (see Chapter 2) which will then bind to a specific region of a biological molecule of interest. By chemically functionalizing the beads in this way, single molecules can be attached to the surface of an optically trapped bead and tethered to a fixed surface such as a microscope coverslip (Figure 5.1G).

Many of the first experiments of this nature involved the large muscle protein titin (Tskhovrebova *et al.*, 1997) which allowed researchers to characterize the mechanical elasticity of single titin molecules as a function of its molecular extension by laterally displacing the microscope stage to stretch the molecule relative to the trapped bead. This approach was further modified to tether a single titin molecule between an optically trapped bead and a suction micropipette (Figure 5.1H), which offered some improvement in fixing the tether axis to be parallel to the lateral plane of movement of the trap, thus making the most out of the lateral trapping force available (Kellermeyer *et al.*, 1997). This method was also employed to measure the mechanical properties of single DNA molecules (Smith *et al.*, 1996) as well as the force dependence of unfolding and refolding of model folded molecular motifs such as the RNA hairpin (Liphardt *et al.*, 2001, and see Chapter 2).

These experiments showed directly that the mechanism for refolding of a molecule is not a simple reversal of the unfolding mechanism (for a good discussion see Sali *et al.*, 1994 and Q5.4). A view of molecular folding which has emerged recently is a representation of the process of refolding as a *funnel* shape on a *free energy diagram*, with the most stable state being at the very bottom of the funnel. However, recent developments now indicate that a better way to view the kinetics of molecular folding is as a series of interconnected *network hubs*, such that each hub represents a *metastable* transition state which may be reached via several other alternative hub states en route to the final, stable folded conformation (Bowman and Pande, 2010).

Similarly, improvements were made by tethering titin and related molecules between two independent trapped beads (Figure 5.1I); this offered further advantages of allowing fast feedback experiments to clamp both the molecular force and position while monitoring the behaviour of two separate beads at the same time (Leake *et al.*, 2004, and the section below on multiple laser traps). By monitoring the displacement of a trapped bead relative to the centre of the optical trap, the molecular force can be determined from a

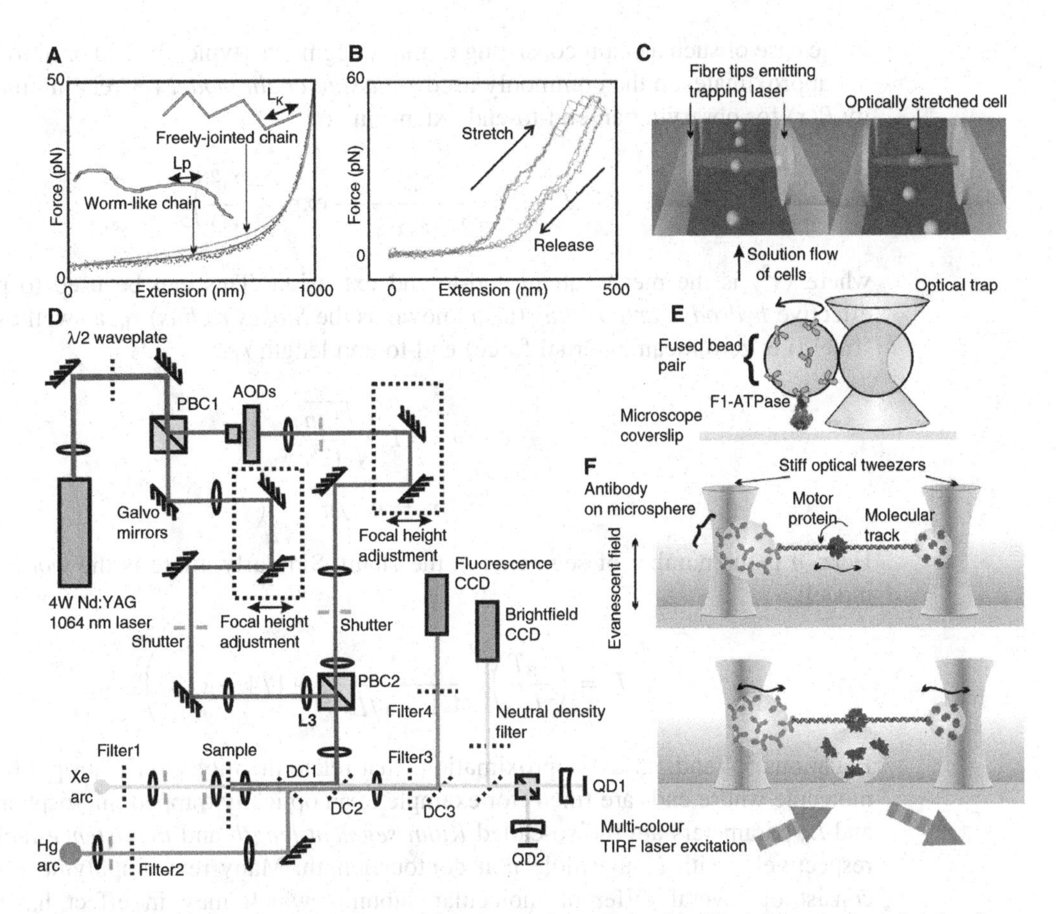

FIGURE 5.2 **(A)** Experimentally measured force–extension relationship (dots) for a single molecule of full length muscle protein titin using the mean values from several stretch–release cycles for two-bead optical tweezer stretching, fits to freely jointed and worm-like chain indicated (L_K and Lp are the corresponding segment and persistence lengths), and **(B)** a single stretch–release cycle for a related molecule kettin indicating unfolding of molecular domains (original data, author's own work, for full details see Leake, 2001). **(C)** Schematic of a microfluidic flow-cell of an optical stretcher (upper panel) with a captured cell being optically stretched (lower panel) (modified with permission from Gück *et al.*, 2005). **(D)** Optical diagram of a typical dual optical trap setup including fluorescence imaging (AODs acousto-optic deflectors, PBC polarizing beam cube, DC dichroic mirror, QD quadrant photodiode). **(E)** Optical spanner generated from a fused latex bead pair attached to surface-bound rotary motor protein F1-ATPase with opposing bead optically trapped. **(F)** Schematic of transverse magnetic tweezers, here used to apply torque to a single DNA molecule with the trapped bead monitored using back focal plane detection (see Chapter 4) and multi-colour TIRF used to monitor the location of bound protein complexes on the DNA.

knowledge of the trap stiffness. The relationship between this force and the end-to-end extension of the molecule can then be measured experimentally. The principal contribution to this force is entropic in origin and can be modelled using a variety of formulations to determine parameters such as equivalent chain segment lengths in the molecule (Figure 5.2A, B). One popular model is the so-called *freely jointed chain*:

$$x = L_c\left(\coth\left(\frac{FL_k}{k_BT}\right) - \frac{k_BT}{FL_k}\right) \quad (5.4)$$

In the case of such a chain consisting of many segments (typically 100 or more) Equation 5.4 approximates to the commonly used *Gaussian chain model* for relating the probability $P(x)$ for observing an end-to-end extension x:

$$P(x) = \frac{4\pi x^2}{(2/3\pi\langle x^2\rangle)^{3/2}} \exp\left(\frac{-3x^2}{2\langle x^2\rangle}\right) \tag{5.5}$$

where $\langle x^2\rangle$ is the mean square end-to-end extension. This can be used to predict the effective *hydrodynamic radius* (also known as the *Stokes radius*) r_h, as well as the mean 'free' (i.e. zero mean external force) end-to-end length r_f:

$$r_h = L_k\sqrt{\frac{3\pi n}{128}} \tag{5.6}$$

$$r_f = L_k\sqrt{n} \tag{5.7}$$

Here, n is the number of segments in the chain. Similarly, there is the *worm-like chain* model:

$$F = \left(\frac{k_BT}{L_p}\right)\left(\frac{1}{4(1-x/L_c)^2} - 1/4 + x/L_c\right) \tag{5.8}$$

Equations 5.4 and 5.5 are approximations that relate the molecular entropic force F of a molecule whose ends are fixed (for example to an optically trapped microsphere). The L_k and L_p parameters are the so-called *Kuhn segment length* and *persistence length* values respectively, with L_c the molecular contour length. Many real biopolymers in practice consist of several different molecular subunits which may in effect have different stiffness values. To model such behaviour a common approach is to sum up the effects of N such different components (for example, see Leake *et al.*, 2004), such that for the freely jointed chain:

$$x = \sum_{i=1}^{N} x_i = \sum_{i=1}^{N} L_{ci}\left(\coth\left(\frac{FL_{ki}}{k_BT}\right) - \frac{k_BT}{FL_{ki}}\right) \tag{5.9}$$

For the worm-like chain, the relation is:

$$F = F_i = \left(\frac{k_BT}{L_{pi}}\right)\left(\frac{1}{4(1-x_i/L_{ci})^2} - 1/4 + x_i/L_{ci}\right) \tag{5.10}$$

Here again, the assumption is that the separate end-to-end values x_i for each component add up to the total end-to-end length x for the molecule, with the constraint for both models that the sum of the separate L_{ci} contour lengths sum up to the total contour length of the whole molecule.

Many elegant test tube single-molecule optical tweezers experiments have been performed at relatively low forces of just a few piconewtons, relevant to the physiological forces experienced in living cells for a variety of different motor proteins (see Chapter 2). These include seminal studies on the muscle protein *myosin* interacting with *actin* (Finer *et al.*, 1994), the *kinesin* protein involved in cell division (Svoboda *et al.*, 1993), and a variety of proteins that use DNA as a molecular track. The current state-of-the-art involves replacing the air between the optical components of a bespoke optical tweezers setup with *helium*, using sealed gas-tight tubes, to minimize noise effects due to the temperature-dependent refraction of lasers through gases. This has allowed the transcription of single nucleotide base-pairs on a single molecule of DNA by a single molecule of the *ribonucleic acid polymerase* (RNAP) motor protein enzyme complex to be monitored directly (Abbondanzieri *et al.*, 2005).

The biggest drawback with optical tweezers with regard to single-molecule cellular biophysics has been their limited application for live-cell studies. Barring some exceptional cases, for which an optically trapped microsphere probe has been attached to a molecular machine which is integrated in the cell membrane (such as the bacterial flagellar rotary motor, see section 8.3) and so is technically outside the cell itself, there have until recently been only relatively crude attempts to utilize optical tweezers to monitor cellular processes in vivo. This is likely to change dramatically in the near future however. Researchers recently have been able to use optical trapping of microspheres that have been internalized by specialized cells called macrophages, which function in the immune response in vertebrate animals, to monitor molecular cargo trafficking inside cells (see section 9.2). Since the internal environment of the cell has local variations in viscosity and refractive index, different to those of the water based solutions of test tube level assays, this necessitates careful internal calibration of the trap stiffness in the actual local vicinity of each internalized microsphere.

PHYSICS-EXTRA

When a molecule is stretched with its ends fixed, there are fewer molecular conformations that it can adopt compared to a less stretched molecule. Thus, the number of accessible **microstates** for a given **macrostate** energy level is smaller, implying a difference in absolute entropy between more and less stretched molecules. Thus, in an isolated system, work has to be put into the system to go from a less stretched to a more stretched state, which is manifest as the integral of the extension with an opposing molecular entropic force. The two most common physical models for this force are represented by the **freely jointed chain** and the **worm-like chain**. In the former the polymer is modelled as a series of universally jointed segments in a chain, with each segment the same length, and the end-to-end distance can be derived by modelling the motion of the chain segments as being essentially a cumulative random walk in space from each contributing connected segment. If the number of segments is very high then the probability for a given end-to-end extension is approximated by the **Gaussian chain model**. The worm-like chain considers the polymer as a semi-flexible tube which has a characteristic persistence length, and in the regime of low force at least both models converge (see Q5.6). The sizes of the persistence or segment lengths that emerge from fits to experimental data can be related back to real biological sub-structures in the stretched polymer. Improvements to these models can be made by applying several different equivalent chains with different elastic properties in series, by incorporating elements of steric hindrance effects both at segment joints and where parts of the chain potentially collide with other chain regions as well as introducing additional non-entropic enthalpic contributions in the model, for example due to electrostatic effects.

5.2.2 Bessel beam, fibre-based and evanescent field optical traps

Bessel beams represent a class of optical illumination which is in principle *diffraction free* (Durnin *et al.*, 1987), in that they have a Gaussian-like central peak intensity of width approximately one wavelength, as with the case of a single-beam gradient force optical trap. However, in theory they have zero divergence parallel to the optic axis. In practice, due to the finite sizes of optical components used, there is some small divergence, but this still results in minimal spreading of the intensity pattern over length scales of ~1 m. The primary advantage of optical trapping with such beams is that since there is minimal divergence of the intensity profile of the trap with depth into the sample then this could be applied to generating optical traps far deeper into a sample than is permitted with standard gradient force traps. This intensity profile is also relatively unaffected by the presence of small obstacles in the beam path. Unlike in standard Gaussian-based force gradient traps in which the beam is distorted after encountering a particle, a Bessel beam can reconstruct itself around an object provided some proportion of the light waves are able to move past the obstacle, and so these beams can be used to generate multiple optical traps (see the section below) which are separated by as much as several millimetres.

These Bessel beam techniques have not yet emerged into mainstream single biomolecule use, however there is significant potential for use in larger scale biological samples such as large cells or tissues. One additional application is to use a Bessel beam not for trapping but as a source of scanning illumination for exciting fluorophores in the sample, since then the information obtained is not limited by the axial extent of a standard confocal volume of typically a few micrometres or less (Planchon *et al.*, 2011).

Optical tweezer effects can also be generated by using *optical fibres*. The numerical aperture of a single-mode fibre is relatively low (~0.1), generating a slightly divergent beam from its tip. Optical trapping can be achieved using two juxtaposed fibres separated by a gap of a few tens of micrometres (Figure 5.2C). If a refractile particle is placed in the gap then the combination of balanced forward scattering forces and lateral forces from refraction of the two beams results in an optical trap, 10–100 times weaker in stiffness compared to typical single-beam gradient force traps for comparable input laser powers. However, such an arrangement can be used to trap relatively large single cells. The refractive index of the inside of a cell is in general highly heterogeneous, and marginally higher than that of the water based solution of the external environment (see Chapter 3). This, combined with the fact that cells have a defined compliance, results in an *optical stretching* effect in these fibre traps, which has been used to probe subtle mechanical differences between normal and marginally stiffer cancerous cells (Gück *et al.*, 2005), as well as allowing whole cells to be rotated controllably.

A potential weakness with the method is that the laser power required to produce measurable probing of cell stiffnesses also produces relatively large rises in local temperature at the near infrared wavelengths nominally employed – a few tens of degrees above room temperature being not atypical, which is dangerously close to denaturation temperatures of polypeptides/proteins (i.e. the temperature above which the subtle native secondary structures of proteins is broken and incorrect secondary structures are formed in effect irreversibly).

It is also possible to generate two-dimensional optical force effects using an *evanescent field* (see Chapter 3), however to trap a particle stably in such a geometry requires an opposing, fixed structure against the direction of the force vector, which is typically the solid surface from which the evanescent field emanates. This has been applied in the cases of nanofabricated photonic waveguides and at the surface of optical fibres. To date, there has been no significant application of either optical stretching/rotating or evanescent field trapping to single-molecule biology studies, however there is scope to develop these techniques, especially the latter since there is a huge potential advantage for being applied in a multiple array format of many such optical traps, which thus could have use in high-throughput biomolecular screening assays.

BIO-EXTRA

Protein secondary, tertiary and quaternary structures (see Chapter 2) are held together primarily by relatively weak hydrogen bonding in combination with weak Van der Waals type dispersion forces. At temperatures typically above 50–60 °C these bonds will break, resulting in destruction of the three-dimensional molecular conformation, or **denaturation**. Denaturation can similarly occur at extremes of pH. There are some well-known protein exceptions to this rule, which occur naturally in cells that experience extreme environmental conditions. One such exception is the so-called **thermophilic bacteria** such as *Thermus aquaticus* which can survive in hot water pools, for example in the vicinity of lava flow, to ~80 °C. An enzyme called **Taq polymerase** which is naturally used by these bacteria in the process of DNA replication is now routinely used in **polymerase chain reactions** (PCR) to amplify small samples of DNA. This is used in biomedical screening and forensic science, as well as being routinely used for biological research. A key step in PCR involves cycles of heating replicated (i.e. amplified) DNA to ~90 °C to denature the two helical strands from each DNA molecule, each of which then acts as a template for the subsequent round of amplification, and Taq polymerase will facilitate this replication at a rate in excess of 100 nucleotide base-pairs per second. The advantage of the Taq polymerase is that, unlike DNA polymerases from non-thermophilic organisms, it can withstand such high heating without significant impairment, and in fact even at 98 °C, near the boiling temperature of water, it has a stability half-life of ~10 minutes.

5.2.3 Multiple laser traps

The laser beam used to generate a standard gradient force optical trap can be split before reaching the sample, either by a *space-dividing* optical component such as a glass splitter cube, or by in effect *time-sharing* the beam along different optical paths in the microscope setup, to generate more than one optical trap (Figure 5.2D). Time-sharing is most popularly obtained by passing the initial beam through a device called an *acousto-optic deflector* (AOD). AODs are optical crystals through which a *standing-wave* can be established by applying radio-frequency oscillations to either end of the crystal. The standing-wave results in periodic changes in refractive index in the crystal which in effect serves as a diffraction grating for the incident light beam of wavelength λ. The effective grating aperture size d is a function of the radio-frequency input, the result being to diffract the light at an angle θ given by Bragg's law mentioned previously in Chapter 4:

$$\sin\theta = n\lambda/d \tag{5.11}$$

Here, n is the order of the diffraction intensity maximum. Normally the devices are configured to generate most power through the first order diffraction maximum, with a maximum efficiency of ~80%. An AOD has a frequency response typically of 10^7 Hz or more, and so the angle of deflection can be alternated rapidly between typically 5–10° on the sub-microsecond time scale, in effect resulting in two beams separated by a small angle, that can then each be manipulated separately to generate a separate optical trap.

Often two orthogonally crossed AODs are employed to allow not only time-sharing but also independent full two-dimensional control of each trap in the lateral focal plane of the microscope. This results in a reduced power efficiency of a further ~80%, but with the significant advantage of being able to apply changes in trap displacements over a time scale faster than the actual bead response time due to frictional drag. This therefore permits feedback-type experiments to be carried out, for example, if there are fluctuations to the molecular force of a tethered single molecule.

In principle the same methodology can be employed to generate more than two optical traps, however, in the case of several such traps, independent control of each is facilitated by use of a *spatial light modulator* (SLM), which works by generating controllable changes in both phase and deflection to generate the traps using a diffractive optical array in the light path. The primary disadvantages of SLM devices is that they have relatively low refresh bandwidths of tens of hertz which limits their utility to relatively slow biological processes, in addition to their being prohibitively expensive. They have an advantage, however, in being able to generate truly three-dimensional optical tweezers, so-called *holographic optical traps* (Dufresne and Grier, 1998).

This principle can be extended to using Bessel beams (see section 5.2) to permit multiple traps separated by relatively large distances of up to three orders of magnitude larger than the trap diameter itself in three spatial dimensions. But, splitting light into a number N traps comes with the obvious caveat that the stiffness of each trap is less by the same factor N. There are potentially several biological questions that can still be addressed with low stiffness traps, however the spatial fluctuations of trapped beads in these cases are very high, resulting in detection noise far in excess of $\sim$10 nm, which can often swamp the molecular signal that one is trying to detect. In addition, there is an ultimate upper limit to the number of usable traps allowed based on the lowest level of trap stiffness which will just be sufficient to prevent random thermal fluctuations pushing a bead out of the physical extent of the trap (see Q5.5). Arguably, the most useful multiple trap arrangement for single-molecule biological investigations involves two standard Gaussian-based force gradient traps, between which a single molecule is tethered.

5.2.4 Optical spanners

A rotationally symmetrical particle such as a microsphere trapped by standard single-beam gradient force optical tweezers experiences zero net angular momentum, and so it is not possible to manipulate torque on the particle controllably (that being said, a typical trapped bead does not undergo 'free' rotation in optical tweezers since the rotational drag forces associated with a micrometre diameter bead far exceed typical Langevin forces from impacting water molecules). There are, however, two practical ways which have been developed to achieve this, which can both lay claim to being in effect *optical spanners*.

The first requires introducing asymmetry into the trapped particle system to generate a lever. For example, it is possible to fuse two microspheres controllably such that one of the beads is chemically bound to a biological molecule of interest to be manipulated with torque, while the other is trapped using standard optical tweezers and its position is rotated controllably in a circle centred on the first bead (Figure 5.2E), thereby providing a wrench-like effect, which has been put to useful purpose for the study of the enzyme F1-ATPase (Pilizota *et al.*, 2007). This is a rotary molecular motor which functions to generate molecules of the universal biological fuel ATP (see Chapter 2).

A related technique has utilized diffractive spatial light modulation (see the previous section) to generate tens of relatively weak traps whose position can be programmed to create an *optical vortex* effect, though to date this has had no significant application for investigating single biological molecules.

The second optical spanner approach has involved using either Bessel traps, as discussed previously, or so-called *Laguerre–Gaussian* (LG) beams generated from higher order laser modes either by optimizing for higher order lasing oscillation modes from the laser head itself, or by applying phase modulation optics in the beam path. Combining such asymmetrical laser profiles (Simpson *et al.*, 1996) or Bessel beams with the use of helically polarized light on multiple particles (Volke-Sepulveda *et al.*, 2002) or single *birefringent* particles which have *differential optical polarizations* relative to different spatial axes such as certain crystal structures, for example *calcite* (La Porta and Wang, 2004), has provided recent promise for single-molecule applications, having been used to apply optical torque controllably for single-molecule studies of DNA Holliday junctions (Forth *et al.*, 2011, and see Chapter 2).

5.2.5 Combining optical-trapping with other single-molecule methods

Optical tweezers can be combined with other single-molecule techniques, the most practicable of which involves various forms of single-molecule fluorescence microscopy. To implement laser trapping with simultaneous fluorescence imaging is in principle relatively easy, in that a near infrared laser trapping beam can be combined along a visible light excitation optical path by using a suitable dichroic mirror (see Chapter 3) which, for example, will transmit any visible light excitation laser beams but reflect near infrared, thus allowing the laser trapping beam to be coupled in the main excitation path of a fluorescence microscope (Figure 5.2D). This has been used for example to combine optical tweezers with TIRF to study the unzipping of the two helices of single DNA molecules (Lang *et al.*, 2003) as well as to image extended segments of DNA (Gross *et al.*, 2010) amongst several other applications.

Dual optical trap arrangements can also be implemented to study motor proteins, including DNA motor proteins, by stretching a molecular track between two optically trapped microspheres while simultaneously monitoring using fluorescence imaging (Figure 5.2F, upper panel). Such an arrangement has been used in similar assays which use fluid flow to stretch out single tethered polymer molecules of biological importance, most importantly applied to segments of DNA to form in effect *DNA curtains* (Finkelstein *et al.*, 2010). However, the advantage with using two optical traps is that both motor protein motion and molecular force can be monitored simultaneously by observing the displacement fluctuations of the independently trapped microspheres. Lifting the molecular track from the microscope coverslip eradicates potential surface effects that could impede the motion of the motor protein.

A similar technique is the *dumbbell assay* (Figure 5.2F, lower panel), originally designed to study motor protein interactions between the muscle proteins myosin and actin (Finer *et al.*, 1994), but utilized since to study several different motor proteins including kinesin and DNA motor complexes. Here, the molecular track is again tethered between two optically trapped microspheres but is lowered onto a third surface-bound microsphere coated in motor protein molecules, which thus results in stochastic *power stroke* interactions which can be measured by monitoring the displacement fluctuations of the trapped microspheres. By combining this approach with fluorescence imaging such as TIRF, the position of the molecular track can be measured at the same time, resulting in a potentially very information-rich assay.

Another less widely applied combinatorial approach has involved using optical tweezers to provide a restoring force to electro-translocation experiments of single biopolymers (see Chapter 4) to slow down the biopolymer controllably as it translocates down an electric potential gradient through a nanopore, in order to improve the effective spatial resolution of ion flux measurements, for example to determine the base sequence in DNA molecule constructs (Schneider *et al.*, 2010). There have also been some generally abortive attempts at combining optical tweezers with AFM imaging, for example to stretch a single-molecule tether between two optically trapped beads while simultaneously imaging the tether using AFM. However, the vertical fluctuations in stretched molecules due to the relatively low vertical trap stiffness have to date been high enough to limit the practical application of such approaches.

5.3 Magnetic tweezers

Magnetic probes in the length scale range of hundreds to thousands of nanometres may be controlled directly via the manipulation of the magnetic *B*-field in the local vicinity of the probe. This can result in controllable deflections in such probes over the nanometre length scale, with the resulting force transduction device commonly referred to as *magnetic tweezers*. The most common arrangement consists of a ca. micrometre sized paramagnetic bead. This is typically manufactured using a latex bead in which iron oxide nanoscale particulates, or other similar paramagnetic materials, are embedded, with the whole resultant bead then magnetized in a strong uniform external *B*-field to induce a well-defined magnetic moment. Such beads can be chemically functionalized on their surface to permit conjugation to a variety of different biological molecules.

Magnetic tweezers have been used to great success with molecules such as DNA (Manosas *et al.*, 2010). The external *B*-field in the magnetic tweezers setup is usually built as a module on an inverted optical microscope, with either fixed magnets or electromagnetic coils placed around the magnetic probe (Figure 5.3A). By moving the microscope stage, a candidate paramagnetic bead can be captured in between the magnets. Usually, the magnetic bead probe will be conjugated to a single biological molecule of interest which in turn is tethered via its opposite end to a microscope slide or coverslip. By moving the microscope stage vertically relative to the magnets/coils, for example by changing the focus, the molecule's end-to-end extension can be adjusted

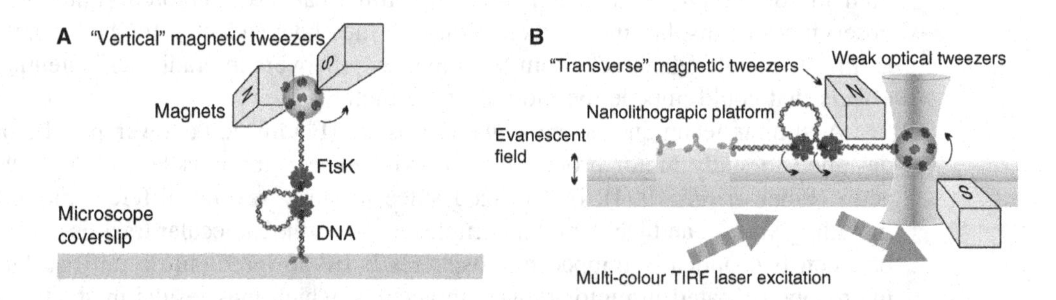

FIGURE 5.3 (A) Standard vertical magnetic tweezers, here stretching and twisting a single DNA molecule to which is bound an important motor protein complex called FtsK which is used in DNA segregation in bacterial cells. In such an experiment, shortening of the DNA can be observed upon addition of FtsK. However, since in this geometry it is not possible to visualize either the FtsK or DNA tether separately, it is largely guess-work as to whether this process involves some form of DNA 'loop-extrusion' as shown, or another mechanism such as wrapping up around the paramagnetic bead for example. To address such issues researchers are developing (B) transverse magnetic tweezer systems that should permit simultaneous twisting/stretching of the DNA with single-molecule fluorescence imaging.

controllably. Thus, the mechanical properties of individual molecules can be probed using this approach in much the same way as for optical tweezers.

Furthermore, since the magnetic probe will try to align with the external magnetic field, any relative rotation of that field, for example either by physically rotating the magnets around a vertical axis or by applying an alternating phase current to four electromagnetic coils arranged around a magnetic probe, will rotate the probe itself. This therefore permits a controllable torque to be applied to a single biological molecule, in arguably a more direct and technically simpler way than can be achieved for optical tweezers which would need either to utilize an extended optical handle or to use the rotating polarization of a non-Gaussian mode laser (see section 5.2). Such torque control has been used on DNA–protein complexes, for example to study DNA replication. For DNA to replicate it must first unwind, which is usually done controllably using enzymes called *helicases* – these are a vitally important class of molecules, as suggested by the fact that ~1% of the expressed content of the eukaryotic genetic code (i.e. the genome) encodes for helicases of some type. Thus, molecular rotation is intrinsically important and the ability to measure it gives potentially enormous insight into molecular mechanisms of DNA coiling.

BIO-EXTRA

There are several natural molecular machines which act on DNA to provide some form of torque. The DNA molecule is most stably a super-coiled structure (see Chapter 2), with the coiling facilitated by enzymes called **topoisomerases**. However, to undergo replication or repair, or to express peptides from the genes, this super-coiled structure needs first to relax into an uncoiled conformation, which is brought about by enzymes called **gyrases**. In order to access the individual strands of the double helix, this helical structure itself must be unwound, which in turn is made possible by enzymes called **helicases**. It is likely that many of these torque-generating nanomachines work in a highly coordinated fashion.

The disadvantages of magnetic over optical tweezers are that in general they are slower by a factor of at least 10^3 since they do not utilize fast acousto-optical deflector components, like optical tweezers can, and traditionally they require relatively large micrometre-sized beads to have a sufficiently large magnetic moment but with the caveat of a relatively large frictional drag which ultimately limits how fast they can respond to changes in external B-field (a typical frequency bandwidth for magnetic tweezers is ~1 kHz, so they are limited to monitoring changes over times greater than a millisecond). Also, traditionally, it has not been possible to visualize a stretched molecule at the same time as monitoring its extension and force, for example using fluorescence microscopy if the biomolecule in question can be tagged with a suitable dye. This is because the geometry of conventional magnetic tweezers is such that the stretched molecule lies parallel to the optical axis of the microscope and so cannot be visualized extended in the lateral focal plane.

To solve this problem some research groups are developing *transverse* magnetic tweezers systems (Figure 5.3B). The main technical problem with this is that there is normally very limited space in the microscope stage region around a sample to position physically the magnets or coils in essentially the same lateral plane as the microscope slide. To combat this some researchers are developing very small electromagnetic coils, potentially microfabricated into a bespoke flow-cell. Other recent improvements have involved using magnetic probes with a much higher magnetic moment which potentially may allow for reductions in the size of the probe with consequent improvements to

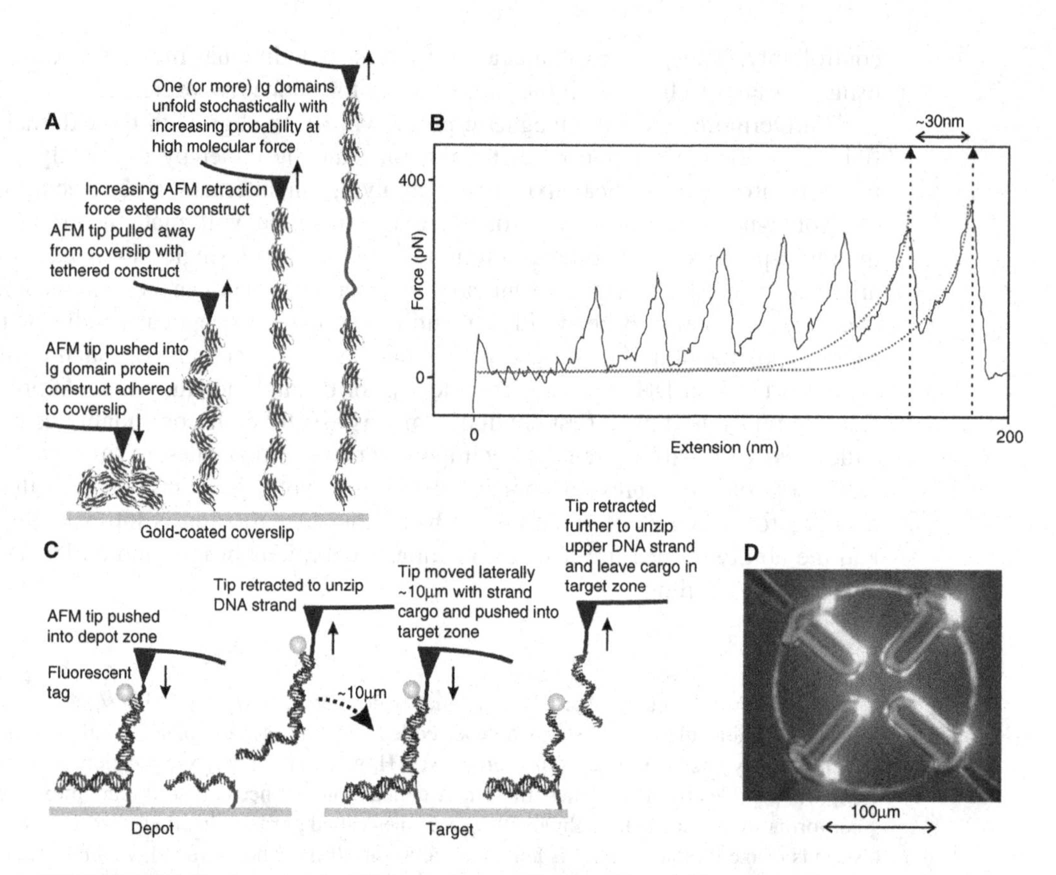

FIGURE 5.4 **(A)** Single-molecule AFM spectroscopy illustrated here with a schematic of a concatamer of eight Ig domain motifs, each of which can undergo forced unfolding to generate **(B)** (schematic depiction of) a characteristic 'saw-tooth' molecular signature with two worm-like chain fits shown (dotted lines), here indicating clearly six unfolding peaks with a characteristic spacing of fitted contour length values of ca. 30 nm (dotted lines). **(C)** AFM 'cut-and-paste of a DNA construct from a depot to a target zone involving a reproducible lateral displacement of a single molecule of ca. 10 µm. **(D)** Four microfabricated electrodes in quadrature with a ca. 40 µm gap used for electrorotation (image used by permission of Hywel Morgan, Southampton University).

maximum sampling speeds. One such probe uses a *permalloy* of nickel/chromium manufactured into a disc (Kim *et al.*, 2009). Such discs could potentially be manufactured to a consistency of ~100 nm diameter and still have a sufficiently high magnetic moment to permit molecular torque experiments on twisted molecules such as DNA, which would experience a significantly smaller viscous drag from the surrounding water-based solvent thus permitting much faster molecular rotation experiments to be performed.

5.4 Atomic force spectroscopy

Using basically the same design of AFM probe as used in AFM imaging (Chapter 4), single biomolecule mechanical stretch experiments can be performed (Figure 5.4A). The first generation of such AFM spectroscopy experiments involved non-specific conjugation of the molecule in question to a *gold-coated coverslip* followed by essentially a *fishing expedition* with an AFM probe in which the tip is pushed into the surface of the coverslip and then retracted back in the hope of tethering non-specifically another part of the molecule to the silicon nitride of the tip (possibly at least in part by a hydrophobic force mediated linkage).

The first such AFM devices were relatively crude in having only one axis of controllable movement (the vertical axis controlled by a piezo actuator to move the AFM tip relative to the sample), and in effect relied on stage *xy* drift to move to a different region of the sample! Surprisingly, such devices actually worked quite well. The first wave of single-molecule AFM experiments on such devices concerned modular proteins and protein constructs derived from the large muscle protein titin, which paralleled similar developments in force spectroscopy using optical tweezers. In one of the best examples of such experiments, single-molecule constructs consisting of concatamers (i.e. connected repeating units) of up to eight repeats of the same 'Ig' type module were used (Rief *et al.*, 1997).

BIO-EXTRA

Titin is an enormous molecule whose molecular weight lies in the MDa range. It consists of ~30,000 individual amino acid residues and is expressed as part of a filamentous system in muscle tissue acting as a passive spring to centralize the **Z-disc** component of **sarcomere** subunits of muscle **myofibrils**. Most of the molecule is composed of repeating units of **β-barrel** modules of ~100 residues each, which have sequence homology to either **fibronectin** (Fn) or **immunoglobin** (Ig) type domains, with a combined total in excess of 370 domains. The increased likelihood of unfolding of the β-barrel structure of Fn or Ig domains as molecular force is increased on the titin possibly confers a shock-absorber affect which ensures that the myofibril can maintain structural integrity even in the presence of high, pathophysiological forces. Titin is expressed as a variety of different **isoforms** of different molecular weights depending on the type of muscle tissue, and there is good evidence to indicate that this allows the titin molecular stiffness to cater for the range of force experienced in a given muscle type.

In fishing for such surface-bound constructs a variable number of Ig modules in the range 1–8 can be tethered between the gold surface and the AFM tip depending on the essentially random position of the non-specific binding to both. These domains will unfold in the same manner as those described for titin stretch experiments using optical tweezers (section 5.2), with a consequent sudden drop in entropic force from the molecule and increase in molecular extension of ca. 30 nm at each single domain unfolding event. Thus, the resultant force–extension traces have a characteristic *saw-tooth pattern*, with the number of *teeth* corresponding to the number of Ig domains unfolded in the stretch, and therefore varying in the range 1–8 in this case (Figure 5.4B).

Such patterns were very important: they indicated quite clearly that there was a single-molecule tether being stretched, as opposed to multiple tethers which might be anticipated if the surface density of molecules were sufficiently high. Aficionados of the technique (most notably Julio Fernandez, Columbia University) denoted such a pattern as a *molecular signature*.

This phrase 'molecular signature' has its due place in the history of single-molecule biophysics since even in the next generation of single-molecule experiments in which we currently reside, being able to determine definitively whether a single molecule is really being observed, or whether we are seeing the effects of several single molecules, is still technically non-trivial. Experiments are still performed in exceptionally low signal-to-noise ratio regimes of sometimes only barely above ~1, and as such it is all the more important to ensure that an appropriate molecular signature of some type can be observed irrespective of the single-molecule technique (see Q5.11).

Improvements have now been made in AFM spectroscopy by applying greater specificity of binding to a molecular construct by chemically functionalizing both AFM tip and the gold (or platinum) surface. The addition of lateral control allows reproducible movements to different regions of a sample, and the careful application of *force-feedback* in effect to clamp the molecular force to a set value allows the definitive observation of force-dependent kinetics of unfolding and subsequent refolding.

5.4.1 AFM 'cut-and-paste'

The spatial reproducibility of some AFM systems is now so high that single molecules can be pulled clear from a surface, moved laterally by anything from a few nanometres to several micrometres, and then repositioned controllably by pressing the AFM tip back into the surface (Figure 5.4C), thus in effect providing a molecular 'cut-and-paste' facility (Kufer *et al.*, 2008). By combining this approach with the specificity of DNA base-pairing (Chapter 2) it has been possible to use a complementary DNA strand conjugated to an AFM tip specifically to capture surface-bound DNA constructs from a 'depot' area on the sample, and then controllably reposition them elsewhere on the sample surface, offering future potential for designer nanotechnology applications.

5.4.2 Combining AFM spectroscopy with fluorescence microscopy

Modern single biomolecule experiments are increasingly characterized by a combinatorial approach to single-molecule techniques – essentially combining simultaneous measurements on the same single molecule using different single-molecule methods. One such combines AFM spectroscopy with fluorescence microscopy (Sarkar *et al.*, 2004). For example, it is possible to engineer similar Ig domain modular constructs as described in this section, but have a single Ig domain bounded by fluorescent protein FRET pairs – in the unfolded Ig conformation the separation of the FRET pairs is less than 5 nm and thus generates a measurable FRET efficiency (see Chapter 3). If the construct is then stretched using AFM spectroscopy then the unfolding of the Ig domain results in an increase in FRET separation of 20–30 nm, thus resulting in no measurable FRET efficiency between the acceptor and donor pair.

This in effect contributes a 'double' molecular signature which gives a significant increase in confidence that you are really observing a single-molecule event (analogous to having confidence in a person's identity if they have not only a signature which agrees reasonably well with that on the back of their credit card, but also know their correct PIN for that card). Single-molecule fluorescence microscopy has also been utilized as a quality control for the AFM cut-and-paste method (section 5.4.1) to confirm the correct placement of DNA hybrid constructs.

KEY POINT

Both optical and magnetic tweezers, and AFM spectroscopy, suffer similar issues with regard to being applicable for measurements inside living cells, as opposed to being more commonly applied to features which are accessible from the cell surface. New developments in special **transmembrane adapter molecules**, which can integrate into a cell membrane but bind both to internal cellular substructures/molecular complexes and to external molecules on the outside of the cell, may in the near future function as suitable **handles** for trapped microspheres and/or AFM tip probes to allow access to monitoring **internal cellular force dependent processes**.

5.5 Using force spectroscopy to explore non-equilibrium processes

5.5.1 Detailed balance

Kinetic systems in which each process is equilibrated with its reverse process, are said to be in *detailed balance*. The implicit assumption is one of *microscopic reversibility*. In the case of single biological molecules undergoing some reaction and/or molecular conformational change, this means that the transition would go in reverse if the time coordinate of the reaction were inverted. A useful way to think about kinetic transitions from one state (1) to another state (2) is to consider what happens to the free energy of the system as a function of some *reaction coordinate* (for example, the time since the start of the reaction/transition was observed). If the rate constant for 1→2 is k_{12} and that for the reverse transition 2→1 is k_{21} then detailed balance predicts that the ratio of these rate constants is given by the *Boltzmann factor*, namely that $k_{12}/k_{21}= \exp(-\Delta G/k_{B}T)$ where ΔG is the free energy difference between states 1 and 2 (note this expression is still generally true whether the system is in equilibrium or not).

Historically, rate constants were derived from bulk ensemble average measurements of chemical flux from each reaction and the ratio k_{12}/k_{21} is the same as the ratio of the bulk ensemble average concentrations of molecules in each state at equilibrium. However, using the ergodic hypothesis (see Chapter 1), in a single-molecule experiment each rate constant can be interpreted as the reciprocal of the average dwell time the molecule spends in a given state. By sampling the distribution of the lifetime of a given single-molecule state experimentally, for example the lifetime of a folded or unfolded state of a domain, these rate constants can in principle be determined and thus we have a single-molecule method to map out the shape of the free energy landscape using single-molecule data. In principle this is incredibly powerful since it tells us quantitatively how stable each molecular state is.

Single-molecule force spectroscopy in the form of optical tweezers or AFM can probe the mechanical properties of individual molecules. If these molecular stretch experiments are performed relatively slowly in *quasi* thermal equilibrium (for example, a sudden molecular stretch is imposed and then measurements are taken over at least several seconds before changing the molecular force again) then there is in effect microscopic reversibility at each force, meaning that the mechanical work done on stretching the molecule from one state to another is equal to the total free energy change in the system, and thus the detailed balance analysis above can be applied to explore the molecular kinetic process using slow-stretch molecular data.

PHYSICS-EXTRA

Explicitly, the principle of detailed balance, or microscopic reversibility, states that if a system is in thermal equilibrium then each of its degrees of freedom is separately in thermal equilibrium. For the kinetic transition between two stable states 1 and 2 the forward and reverse rate constants can be predicted using **Eyring theory**. Assuming the transition goes from a stable state 1 with free energy G_1 to a metastable transition state M with a higher free energy value G_M, back down to another stable state 2 with lower free energy G_2, then $k_{12} = \nu \exp[-(G_M - G_1)/k_BT]$ and $k_{21}=\nu \exp[-(G_M - G_2)/k_BT]$ where ν is the universal rate constant for a transition state, also given by k_BT/h from the Eyring theory of **quantum mechanical oscillators**, and has a value of $\sim 6\times 10^{12}\ s^{-1}$ at room temperature, and h is Plank's constant. These rate constant equations also equate to the **Arrhenius equation**, well-known to undergraduate biochemistry students.

5.5.2 Non-equilibrium processes, Kramers theory and the Jarzynski equality

Often, however, molecular stretches using force spectroscopy are relatively rapid, resulting in significant *hysteresis* on force–extension traces due to the time-dependent perturbations forcing rare molecular events (such as molecular domain unfolding) which drives the system away from equilibrium (in the case of domains unfolding they do not have enough time to refold again before the next molecular stretch cycle and so the transition is, in effect, irreversible, resulting in a *stiffer* stretch half-cycle compared to the relaxation half-cycle, see Figure 5.2B for example). In this instance, *Kramers theory*, which models systems in non-equilibrium states such as diffusing out of a potential energy well (interested physical scientists should read the seminal description in Kramers, 1940) can be extended to the specific case of investigating an unfolding type transition in a single-molecule mechanical stretch experiment (Evans and Ritchie, 1997) which indicates that:

$$F_{\mathrm{d}} = \frac{k_{\mathrm{B}}T}{x_{\mathrm{w}}} \log_{\mathrm{e}}\left(\frac{r x_{\mathrm{w}}}{k_{\mathrm{d}}(0) k_{\mathrm{B}} T}\right) \tag{5.12}$$

Here, F_{d} is the expected domain unfolding force, $k_{\mathrm{d}}(0)$ the spontaneous rate of domain unfolding at zero force due to a uniform rate of molecular stretch r, whilst x_{u} is the so-called *width of energy potential* for the unfolding transition.

Thus, by plotting the experimentally determined F_{d} values, measured using either optical tweezers or, more commonly, AFM spectroscopy, against the log of different imposed values of r, the value for x_{u} can be estimated. A similar analysis can also be performed for the refolding process, and for typical domain motifs which undergo such unfolding/refolding transitions a width of 0.2–0.3 nm is typical for the unfolding process, compared with an order of magnitude or more greater for the refolding process. This is consistent with molecular dynamics simulations suggesting that the unfolding is due to the unzipping of a few key hydrogen bonds which are of similar length to the estimated widths of energy potential (Chapter 2), whereas the refolding process is far more complex requiring cooperative effects over a much longer length scale of the whole molecule.

The *Jarzynski equality* is an important equation of statistical mechanics which relates the work done W on a thermodynamic system of a non-equilibrium process to the free energy difference ΔG between the states for the equivalent equilibrium process, through $\exp(-\Delta G/k_{\mathrm{B}}T) = \langle \exp(-W/k_{\mathrm{B}}T) \rangle$, where the angle brackets indicate a time average (unlike many theorems in statistical mechanics, this was actually derived relatively recently; interested physical scientists may wish to read Jarzynski, 1997). This is relevant to AFM spectroscopy of single biopolymers, since the equivalent mean work in unfolding a molecular domain is given by the work done in moving through a small distance x_{w} against a force F_{d}, $F_{\mathrm{d}} x_{\mathrm{w}}$, and so can be used as an estimate for the free energy difference which in turn can be related to real rate constants for unfolding.

5.6 Electric dipole induction in polarizable particles for torque and trapping

5.6.1 Electrorotation

An external electrical field can be generated to induce an electric dipole on an electrically polarizable particle between microelectrodes in a bespoke microscope flow-cell (Figure 5.4D),

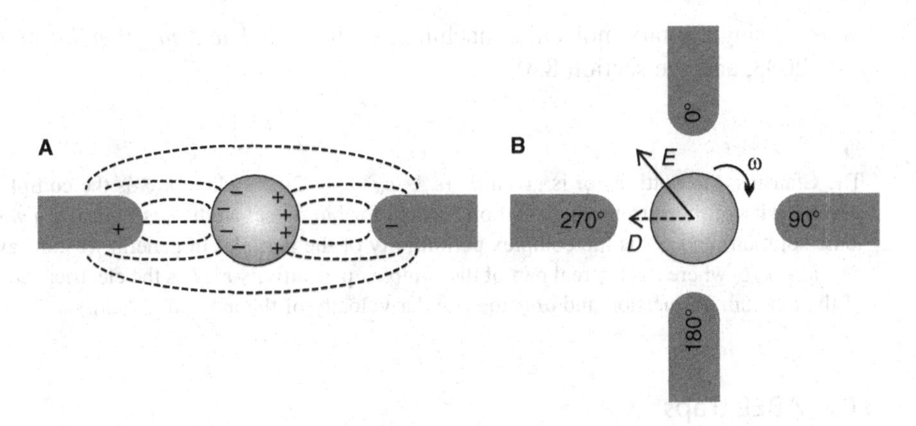

FIGURE 5.5 Schematics for the electrical induction of a dipole on a polarizable particle. In the case of (A) a DC *E*-field results in a dielectrophoretic force (dashed lines are field lines for constant electric potential), and in (B) an AC *E*-field results in a phase lag between the induced dipole axis *D* and the *E*-field vector *E*, resulting in torque with consequent bead rotation at angular velocity ω (the relative angle phase lag on the applied *E*-field is indicated on each electrode).

for example using a latex microsphere functionalized with charged chemical motifs, such as carboxyl anions or protonated amino groups (see Chapter 2), which allow ions to leap-frog over the bead surface. If the electric field is not uniform, i.e. there is a non-zero *E*-field gradient, this results in particle movement parallel to the *E*-field, termed *dielectrophoresis* (Figure 5.5A).

If the *E*-field is uniform, the dipole moment of the particle aligns along the field direction. If the *E*-field is oscillated so as to rotate it around the polarizable bead, and if the angular velocity of the field is sufficiently high, the time taken for the induced dipole to form, i.e. the *relaxation time* of the dipole, will result in this dipole *lagging* behind the *E*-field. This results in an angle between the *E* vector and the dipole at any given time, and the induced dipole will try to realign with the field, in other words the particle experiences a torque (Figure 5.5B). This causes the particle to *rotate* out of phase with the field, either with or against the *E*-field rotation, depending on whether the phase lag is less or more than half an *E*-field oscillation cycle. Therefore, in a rotating *E*-field of sufficiently high frequency, the particle continuously tries to align with the field, leading to *electrorotation*.

The ultimate speed of rotation is a function of charge surface density, bead radius a and electric field E and frequency f, with the torque G given by:

$$G = -4\pi\varepsilon_{\mathrm{w}} a^3 \mathrm{Im}[K(f)]E^2 \tag{5.13}$$

Here, ε_{w} is the permittivity of the surrounding water-based pH buffer which embodies the charge qualities of the solution, and $\mathrm{Im}[K(f)]$ is a factor which embodies the charge qualities of the bead (for interested physical scientists, this is the imaginary component of the *Clausius–Mossotti factor*).

A typical experimental arrangement consists of four microelectrodes phased in quadrature that produce a reasonably uniform AC *E*-field over an area of a few square micrometres at the geometrical centre of the microelectrodes (Figure 5.4D). For typical micrometre sized charged beads an *E*-field oscillation frequency of 1–10 MHz with a ~50 V amplitude potential across a microelectrode gap of ~40 μm can produce microsphere rotations of 100–1000 Hz. Such electrorotation experiments have been applied to studies of single rotary molecular machines, such as the *bacterial flagellar motor* (Rowe *et al*., 2003, and see section 8.3).

PHYSICS-EXTRA

The Clausius–Mossotti factor is given by ($\varepsilon_p^* - \varepsilon_w^*/\varepsilon_p^* + 2\varepsilon_w^*$) where ε_w^* is the **complex permittivity** of the surrounding solution (which, for biological studies, is typically a water-based ionic solution) and ε_p^* is the complex permittivity of the particle. In general we can say that $\varepsilon^* = \varepsilon - ik/\omega$ where ε is the real part of the complex permittivity ε^*, k is the electrical conductivity of the medium in question and ω is the angular velocity of the applied E-field.

5.6.2 ABEL traps

A similar arrangement of four microelectrodes in quadrature as for electrorotation can be used in DC mode to generate a controllably variable electric field. In doing so, a charged particle in the centre of the electrodes can be moved controllably by dielectrophoresis to compensate for random fluctuations in its position manifest as Brownian diffusion. The size of this dielectrophoretic force F is given by:

$$F = 2\pi a^3 \varepsilon_w \mathrm{Re}[K(f)]\, grad\, E^2 \tag{5.14}$$

Here, Re[$K(f)$] is a charge quality of the polarizable particle that relates to the DC electrical force (for physical scientists, this is now the real component of the Clausius–Mossotti factor). Thus, provided the electric field can be adjusted in feedback over a fast enough time scale, a particle in a suitable microscope sample cell can in effect be confined at least in two dimensions to a region of space covering an area of a few square micrometres. Such an arrangement is called an *anti-Brownian electrophoretic/electrokinetic*, or ABEL, trap. This works in principle on any object that can be imaged optically and that acquires an electric charge in water, and was first demonstrated on fluorescently labelled single microspheres on a device whose effective stiffness was $\sim 10^4$ smaller than that of a typical single-beam gradient force optical trap (Cohen and Moerner, 2005).

The potential electrical energy on a particle of total charge q in a field of magnitude E moving through a distance d parallel to the field is qEd, and by the equipartition theorem this implies that the mean distance fluctuations at the centre of the trap will be $\sim 2k_BT/qE$ (see Q5.9). Further refinements include elements of real time feedback to the vertical position of the sample (i.e. automated refocusing) to ensure that the particle lies roughly in the same lateral plane as the microelectrodes. The utility of such traps is that they permit significantly longer continuous observation of molecular machines in solution which otherwise would diffuse relatively quickly (~milliseconds) away from the detector field of view.

Earlier similar approaches for confining the Brownian motion of a single biological molecule directly (i.e. without using a relatively large adapter particle such as a micrometre sized bead) have utilized surface binding either via surface tethering of molecules or via surface binding of lipid vesicles containing a small number of molecules (see Chapter 3). However, the potential advantage of the ABEL trap is that there are no unpredictable surface forces present that could possibly interfere with the observed molecular properties. ABEL trapping has now been applied at the single-molecule level to provide ~ 1 nm precise trapping, which has been used to monitor differences in electrokinetic mobility of single fluorescently labelled DNA molecule constructs in the presence or absence of a DNA-binding protein called RecA over periods of several seconds (Fields and Cohen, 2011). This level of precision opens the possibility for measuring conformational transitions in single biological molecules in solution as they occur, provided they happen over a time scale between the diffusion time of milliseconds and the trapping dwell time of seconds.

THE GIST

- Optical tweezers most commonly function as single-beam gradient force traps on refractile micrometre sized particles allowing nanometre level displacements and piconewton precise forces to be measured.
- Optical tweezers can be modified to exert angular torque, and can also be multiplexed to allow for manipulation of single tethered biomolecules to investigate molecular mechanical properties.
- Magnetic tweezers are powerful, easily implemented tools for controling and monitoring molecular torque.
- AFM spectroscopy allows single biomolecules to be mechanically stretched and probed, and permits controllable surface placement of single molecules.
- AFM spectroscopy, magnetic and optical tweezers can all be combined with other single-molecule techniques to allow simultaneous monitoring of multiple single-molecule properties of the same molecule.
- Electrical induction can also be used to exert torque in AC mode and control the displacements of diffusing particles in DC mode.

TAKE-HOME MESSAGE Force transduction devices exist in a variety of formats utilizing several different physical forces exerted typically on some type of approximately micrometre-sized 'adapter', more rarely on the single molecules themselves, which can be moved controllably over a nanometre length scale to exert piconewton level forces on conjugated biological molecules, either to probe their force dependence or to manipulate them directly.

References

GENERAL

- Ashkin, A., Dziedzic, J. M., Bjorkholm, J. E. and Chu, S. (1986). Observation of a single-beam gradient force optical trap for dielectric particles. *Opt. Lett.* **11**: 288–290.
- Manosas, M. *et al.* (2010). Magnetic tweezers for the study of DNA tracking motors. *Methods Enzymol.* **475**: 297–320.
- Moffitt, J. R., Chemla, Y. R., Smith, S. B. and Bustamante, C. (2008). Recent advances in optical tweezers. *Annu. Rev. Biochem.* **77**: 205–28.
- Svoboda, K. and Block, S. M. (1994). Biological applications of optical forces. *Annu. Rev. Biophys. Biomol. Stuct.* **23**: 247–285.

ADVANCED

- Abbondanzieri, E. A. *et al.* (2005). Direct observation of base-pair stepping by RNA polymerase. *Nature* **438**: 460–465.
- Ashkin, A. (1970). Acceleration and trapping of particles by radiation pressure. *Phys. Rev. Lett.* **24**: 156–159.
- Bowman, G. R. and Pande, V. S. (2010). Protein folded states are kinetic hubs. *Proc. Natl. Acad. Sci. USA* **107**: 10890–10895.

- Cohen, A. E. and Moerner, W. E. (2005). Method for trapping and manipulating nanoscale objects in solution. *Appl. Phys. Lett.* **86**: 093109.
- Dufresne, E. R. and Grier, D. G. (1998). Optical tweezer arrays and optical substrates created with diffractive optical elements. *Rev. Sci. Instrum.* **69**: 1974–1977.
- Durnin, J., Miceli, J. J. and Erberly, J. H. (1987). Diffraction-free beams. *Phys. Rev. Lett.* **58**: 1499–1501.
- Evans, E. and Ritchie, K. (1997). Dynamic strength of molecular adhesion bonds. *Biophys. J.* **72**:1541–1555.
- Fields, A. P. and Cohen, A. E. (2011). Electrokinetic trapping at the one nanometer limit. *Proc. Natl. Acad. Sci. USA* **108**: 8937–8942.
- Finer, J. T., Simmons, R. M. and Spudich, J. A. (1994). Single myosin molecule mechanics: piconewton forces and nanometre steps. *Nature* **368**: 113–119.
- Finkelstein, I. J., Visnapuu, M. L. and Greene, E. C. (2010). Single-molecule imaging reveals mechanisms of protein disruption by a DNA translocase. *Nature* **468**: 983–987.
- Forth, S., Deufel, C., Patel, S. S. and Wang, M. D. (2011). A biological nano-torque wrench: torque-induced migration of a Holliday junction. *Biophys. J.* **101**: L05–L07.
- Gross, P., Farge, G., Peterman, E. J. G. and Wuite, G. J. L. (2010). Combining optical tweezers, single-molecule fluorescence microscopy, and microfluidics for studies of DNA–protein interactions. *Methods Enzymol.* **475**: 427–453.
- Gück, J. *et al.* (2005). Optical deformability as an inherent cell marker for testing malignant transformation and metastatic competence. *Biophys. J.* **88**: 3689–3698.
- Hansen, P. M., Bhatia, V. K., Harrit, N. and Oddershede, L. (2005). Expanding the optical trapping range of gold nanoparticles. *Nano Lett* **5**: 1937–1942.
- Jarzynski, C. (1997). Nonequilibrium equality for free energy differences. *Phys. Rev. Lett.* **78**: 2690–2693.
- Kellermayer, M. S., Smith, S. B., Granzier, H. L. and Bustamante, C. (1997). Folding-unfolding transitions in single titin molecules characterized with laser-tweezers. *Science* **276**: 1112–1116.
- Kim, D.-H. *et al.* (2009). Biofunctionalized magnetic-vortex microdiscs for targeted cancer-cell destruction. *Nature Methods* **9**: 165–171.
- Kramers, H. A. (1940). Brownian motion in a field of force and the diffusion model of chemical reactions. *Physica* **7**: 284–304.
- Kufer, S. K., Puchner, E. M., Gumpp, H., Liedl, T. and Gaub, H. E. (2008). Single-molecule cut-and-paste surface assembly. *Science* **319**: 594–596.
- Lang, M. J., Fordyce, P. M. and Block, S. M. (2003). Combined optical trapping and single-molecule fluorescence. *J. Biol.* **2**: 6.
- La Porta, A. and Wang, M. D. (2004). Optical torque wrench: angular trapping, rotation, and torque detection of quartz microparticles. *Phys. Rev. Lett.* **92**: 190801.
- Leake, M. C. (2001). *Investigation of the extensile properties of the giant sarcomeric protein titin by single-molecule manipulation using a laser-tweezers technique*. PhD dissertation, London University, UK.
- Leake, M. C., Wilson, D., Gautel, M. and Simmons, R. M. (2004). The elasticity of single titin molecules using a two-bead optical tweezers assay. *Biophys. J.* **87**: 1112–1135.
- Liphardt, J. *et al.* (2001). Reversible unfolding of single RNA molecules by mechanical force. *Science* **292**: 733–737.
- Pilizota, T., Bilyard, T., Bai, F., Futai, M., Hosokawa, H. and Berry, R. M. (2007). A programmable optical angle clamp for rotary molecular motors. *Biophys. J.* **93**: 264–275.
- Planchon, T. A. *et al.* (2011). Rapid three-dimensional isotropic imaging of living cells using Bessel beam plane illumination. *Nature Methods.* **8**: 417–423.
- Rief, M., Gautel, M., Oesterhelt, F., Fernandez, J. M. and Gaub, H. E. (1997). Reversible unfolding of individual titin immunoglobulin domains by AFM. *Science* **276**: 1109–1112.
- Rohrbach, A. (2005). Stiffness of optical traps: quantitative agreement between experiment and electromagnetic theory. *Phys. Rev. Lett.* **95**: 168102.
- Rowe, A., Leake, M. C., Morgan, H. and Berry, R. M. (2003). Rapid rotation of micron and sub-micron dielectric particles measured using optical tweezers. *J. Mod. Opt.* **50**: 1539–1555.
- Sali, A., Shakhnovich, E. and Karplus, M. (1994). How does a protein fold? *Nature* **369**: 248–251.
- Sarkar, A., Robertson, R. B. and Fernandez, J. M. (2004). Simultaneous atomic force microscope and fluorescence measurements of protein unfolding using a calibrated evanescent wave. *Proc. Natl. Acad. Sci. USA* **101**: 12882–12886.

- Schneider, G. F. *et al.* (2010). DNA translocation through graphene nanopores. *Nano Lett.* **10**: 3163–3167.
- Simpson, N. B., Allen, L. and Padgett, M. J. (1996) Optical tweezers and optical spanners with Laguerre–Gaussian modes. *J. Mod. Opt.* **43**: 2485–2491.
- Smith, S. B., Cui, Y. and Bustamante, C. (1996) Overstretching B-DNA: the elastic response of individual double-stranded and single-stranded DNA molecules. *Science* **271**: 795–799.
- Svoboda, K., Schmidt, C. F., Schnapp, B. J. and Block, S. M. (1993). Direct observation of kinesin stepping by optical trapping interferometry. *Nature* **365**: 721–727.
- Tskhovrebova, L., Trinick, J., Sleep, J. A. and Simmons, R. M. (1997). Elasticity and unfolding of single molecules of the giant muscle protein titin. *Nature* **387**: 308–312.
- Volke-Sepulveda, K. *et al.* (2002). Orbital angular momentum of a high-order Bessel light beam. *J. Opt. B* **4**: S82–S89.
- Wang, M. C. and Uhlenbeck, G. E. (1945). On the theory of the Brownian Motion II. *Rev. Mod. Phys.* **17**: 323–341.

Questions

FOR THE LIFE SCIENTISTS

Q5.1. In a two-bead optical tweezers 'tapping' molecular stretch experiment on DNA performed on 2000 separate bead pairs at constant tapping frequency using a triangle wave profile, there seemed to be roughly three populations of different groups of molecules characterized by different estimated values of persistence length based on fits applied to the force–extension data using a worm-like chain model: ca. 1950 molecules seemed to have a persistence length close to 50 nm, ca. 45 molecules had a persistence length of close to 100 nm, and the remainder had a persistence length of at least 150 nm. For the group of molecules with persistence length close to 50 nm the stiffness of DNA was observed to decrease dramatically at values of molecular force above ca. 65 pN. Explain all of these observations. (Hint: See Q5.2.)

Q5.2. In a typical molecular stretch experiment involving a single molecule tethered between two optically trapped microspheres which are tapped together and then pulled apart over several cycles at a frequency of a few hertz there is a non-zero probability that the number of molecules tethered between the two beads is actually greater than one (and hence it no longer becomes a 'single-molecule' experiment) since the standard incubation process for the microspheres results in a surface covering of molecules as opposed to just a single molecule per microsphere, and so subsequent tethers may form at each new tap. If the probability of a given tether forming is independent of the time since any previous tether forming event then this process can be modelled as a Poisson distribution, such that probability $P(n)$ for forming n tethers is given by $\langle n \rangle n \exp(-\langle n \rangle)/n!$, where $\langle n \rangle$ is the average number of observed tethers formed between two microspheres. If one tether binding event is observed on average once in every n_{tap} tap cycles, what is the probability of not binding? By equating this to $P(0)$ derive an expression for $\langle n \rangle$ in terms of n_{tap}. We can write the fraction f of 'multiple tether' binding events out of all binding events as $P(>1)/(P(1) + P(>1))$. (a) Use this to derive an expression for f in terms of $\langle n \rangle$. (b) If a microsphere pair are tapped against each other at a frequency of 1 Hz and the incubation conditions have been adjusted to ensure a low molecular surface density on the microspheres such that no more than 0.1% of binding events are due to multiple tethers (and so only a maximum of ca. 1 in 10^4 measured events do not involve single molecules), how long on average would we have to wait before observing our first tether formed between two tapping microspheres? (This question is good at illustrating how tedious single-molecule experiments can sometimes be!)

Q5.3. In a vertical magnetic tweezers experiment performed on a single molecule of DNA tethered between a microscope coverslip and a paramagnetic bead, when the molecular motor FtsK (which uses DNA as its track) was added to the flow-cell the length of the distance between the coverslip and the paramagnetic bead was observed to decrease. (a) Some researchers have used this as evidence that there might be two FtsK molecular motors acting together on the DNA, why is this? (b) What other explanations could there be?

Q5.4. Consider a typical single-molecule refolding experiment of a short biopolymer (for example, Liphardt *et al.*, 2001). A short polypeptide has a molecular weight of ca. 14 *k*Da. (a) Roughly how many peptide bonds does it have? Each peptide has two independent bond angles called ϕ and ψ, and each of these bond angles can be in one of about three stable conformations based on Ramachandran diagrams (see section 2.4). (b) Estimate roughly how many different conformations the protein can adopt. (c) If the unfolded protein refolds by exploring each of these conformations very rapidly in ca. 10^{-12} s and then subsequently exploring the next conformation if this was not the true stable folded state, estimate the average length of time taken before it finds the correct stable folded conformation. In practice unfolded cellular proteins will refold over a time scale of milliseconds up to several seconds, depending on the protein. (d) Explain why proteins in cells do not refold in the exploration manner described above, and what do they do instead? (Hint: The best current estimate for the age of the universe is ca. 14 billion years. This problem is well known to structural biologists and is called Levinthal's paradox, see Sali *et al*., 1994.)

FOR THE PHYSICAL SCIENTISTS

Q5.5. A near infrared single-beam gradient force optical trap using a focused laser of wavelength 1047 nm exerts a lateral force of 30 pN on a latex microsphere of diameter 1 μm suspended in water at room temperature when the bead is displaced 200 nm from the trap centre. (a) If the trapping laser power passing through the microsphere is 200 mW, estimate the average angle of deviation of laser photons, assuming that the lateral force arises principally from photons travelling close to the optical axis. (b) Make a labelled sketch of the frequency power spectral density of the microsphere's positional fluctuations. (c) At what frequency in this instance is the power spectral density half of its maximum value? The incident laser beam is then divided up using a time-share approach with an AOD of efficiency 80% into several beams of equal power to generate several independent optical traps. (d) If each trap is required to exert a continuous high force of 30 pN estimate the maximum number of traps that can be used. (e) For experiments not requiring a continuous high force, estimate the maximum theoretical number of optical traps that can be generated by this method. (Hint: A stable trap is such that mean fluctuations of the trapped particle position do not extend beyond the physical size of the trap.) (f) How many such trapped microspheres would be required to push on a single myosin molecule to prevent it from undergoing a 'power stroke' (see Chapter 4) of force ~5 pN?

Q5.6. An ideal freely jointed chain consists of n rigid links in the chain each of length L_K, which are freely hinged where they join up. (a) By treating each link as a position vector, and neglecting possible consequences of interference between different parts of the chain, derive an expression for $\langle x^2 \rangle$, the mean square end-to-end distance. The equivalent result for a worm-like chain, which models a polymer as a continuous filament with a non-zero bending modulus, is:

$$\langle x^2 \rangle = 2L_p^2\left(\exp(-L_c/L_p) - 1 + L_c/L_p\right)$$

where L_c is the contour length and L_p the persistence length. (b) Evaluate this expression in the limits $L_c \gg L_p$ and $L_c \ll L_p$ and comment on both results.

Q5.7. A single protein molecule exists in just two stable conformations, 1 and 2, and can undergo reversible transitions between them with rate constants k_{12} and k_{21} respectively, where the free energy difference between the two states is ΔG. (a) Use the principle of detailed balance to estimate the probability that the system will be in state 1 as a function of time, assuming that at zero time the molecule is in state 1. A different protein molecule undergoes a series of two irreversible conformational changes, $1 \rightarrow 2$ and $2 \rightarrow 3$ with rate constants k_{12} and k_{23} respectively. (b) What is the likelihood that a given molecule starting out in state 1 at zero time will be in state 3 at time t?

Q5.8. Consider a certain linear protein molecule as ideal chain with $N = 500$ chain segments, each segment length equal to the average α-carbon spacing for a single amino acid (see Chapter 2) carrying a charge of $\pm e$ at either end of each amino acid subunit where e is the unitary electron charge. What is the protein's relative elongation (its end-to-end length relative to its root mean square unperturbed length) parallel to an E-field, of 10,000 V per cm?

Q5.9. An ABEL trap was used to constrain a 30 nm latex nanosphere suspended in water for which the total charge was known to be ~4000e, where e is the unitary electron charge. A mean electric field strength of 8000 V m^{-1} was applied across the microelectrodes. Assuming the room temperature viscosity of water is 0.001 N sm^{-2} how fast would a camera have to image the nanoparticle to ensure that the expected distance diffused by Brownian motion of the particle in each image frame is less than the positional fluctuations due to the finite ABEL trap stiffness?

Q5.10. By assuming solutions of the form $A(\omega)\exp(i\omega t)$ for the Langevin force at angular frequency ω derive an expression for the displacement $x_w(t)$ at time t in an optical trap. (a) If the power spectral density $G(\omega)d\omega$ is defined as $|A(\omega)^2|$ derive an expression for the mean square displacement of displacements in terms of G. (b) Use the fact that for 'white-noise' G is a constant and that the equipartition theorem predicts that the mean square displacement at each separate frequency is associated with a mean energy of $k_BT/2$ to show that the power spectral density should indeed be a Lorentzian.

FOR THOSE WHO HAVE NOT MADE UP THEIR MIND

Q5.11. (a) Why is it so important to be able to confirm a 'molecular signature' for single-molecule biology experiments in general? (b) The characteristic sawtooth pattern of a force–extension trace as obtained for example from AFM spectroscopy on certain molecules is one such, but can you think of other molecular signatures for the techniques described in Chapters 3 and 4?

Q5.12. Currently, single-molecule force spectroscopy is performed either on purified molecules or on the surface of cells. Why is this? Can you think of ways in which force spectroscopy might be performed inside living cells?

Q5.13. Optical tweezers and AFM spectroscopy have both been used to investigate molecular elasticity by mechanically stretching single biological molecules as well as observing molecular domain unfolding and refolding. Which technique is better, and why?

Q5.14. At a biophysical awards dinner a helium balloon escaped and became loosely trapped on the ceiling. Assuming no significant lateral friction since the balloon is bobbing up and down rapidly by random motion in the air, how long would it take a typical red laser pointer of 1 mW power output to push the balloon 10 m across the length of the dinner hall ceiling using forward photon pressure alone? Would it make a significant difference to encourage all the other ca. 500 people attending the dinner to assist in getting out their laser pointers and performing this in parallel? (Hint: This is what the author attempted in the not too distant past, assisted by perhaps 'too much of the good stuff' as P. G. Wodehouse might have said. It shows if anything the great merit in doing theoretical calculations in advance of the experiment...)

6 Single-molecule biophysics: the case studies that piece together the hidden machinery of the cell

Gallia est omnis divisa in partes tres (All Gaul is divided into three parts)

(JULIUS CAESAR, DE BELLO GALLICO, 51 BC)

GENERAL IDEA

Here we introduce the seminal biophysics investigations which have transformed our understanding of biology at the single-molecule level, and lay the foundations for describing single-biomolecule experimentation on functioning live cells

6.1 Introduction

An instructive exercise for those learning about single-molecule biophysics is to compile a list of one's own *top ten* research papers of all time. The choice is obviously subject to a great deal of personal bias, and is a dynamic structure which may change with time, and some of the early, seminal papers on the list might later be superseded by subsequent incarnations with more of the original unresolved questions resolved (and maybe even with the 'right' answers, as opposed to what were perhaps novel but slightly incorrect 'best guesses' at the time!). Even so, as a process for understanding how, and why, single-molecule biophysics has evolved the way it has, and how it is likely to progress into the arena of far greater physiological relevance in the near future, the reader might find the exercise suprisingly fulfilling. In this chapter we discuss some of the strong candidates for this list of seminal papers, and lay the foundations for the remaining chapters in this book which describe real single-biomolecule experiments performed either on living cells or in an environment which has substantial physiological relevance.

6.2 What makes a 'seminal' single-molecule biophysics study?

There are at least five important elements to making an *outstanding* single-molecule biophysics study which is likely to stand the test of time.

(i) The fundamental importance of the biological system under study.
(ii) The fact that certain vital questions concerning the biological process(es) investigated could not be addressed using alternative multi-molecule techniques.
(iii) The novelty of the single-molecule technique(s) used.
(iv) The extent of objective automation in measurement and analysis.
(v) A clear indication that the experiments being performed are truly on single molecules, and not a collection of molecules.

It is not essential that all outstanding single-molecule biophysics papers possess all five ingredients – especially so for the novelty of the single-molecule technique, since in some ways it could be argued that the further application of an already established single-molecule technique is as important as the technique's novelty. However, great publications in this area of research will certainly have a good combination of many of these five elements. Point (iv) is very important in that, working in a noise-dominated measurement regime of single-molecule biophysics experiments, there need to be objective ways to extract and analyse all types of single-molecule data. Point (v) is the most essential feature of any single-molecule study but unfortunately it is, paradoxically, the one which is often the most glossed-over by many contemporary research authors.

KEY POINT

All biophysics experiments performed at the single-molecule level are technically challenging, despite there being a multitude of papers now published in this area. There should be no room for complacency here, and all researchers in this area need to go to strong, certain efforts to establish the presence of a categorical **molecular signature** in each of these studies (see Chapter 5) and to be suitably cynical if they cannot provide one.

Below is a list of *some* of the seminal single-molecule biophysics papers, both as examples of the efforts and attainment that new researchers in this field need to aspire to, but also to illustrate an historical trend in moving towards more complex, physiologically relevant experiments. There is then a brief discussion of some of the objective methods employed to analyse single-molecule effects.

6.2.1 The (highly personal and potentially biased) 'top ten' single-molecule biophysics papers of all time

In compiling my own top ten list, more as a platform from which readers can assemble their own list, I gave a significant bias to the number of times each paper had been cited by another original research article to date. The prime advantage of this is that these cumulative citations are a rough manifestation of how important the expert single biomolecule research community deems any given investigation, as opposed to being more a matter of personal taste. This obviously biases somewhat against relatively new but potentially very important papers, and so to remedy this I have added a few studies that have accrued fewer total citations than some others which have not made my cut of ten, but which I predict will in the near future be recognized by the expert community as being of the most fundamental importance.

Figure 6.1 gives a visual compilation of the essence of each of these investigations, with Table 6.1 offering a brief description of the work and why it is important, plus my own subjective ranking of each paper. As with earlier chapters there is a suggested reading list at the end of this one, however here readers are particularly encouraged to go back to the original research papers that appear on the top ten list and try to understand them critically in the light of the previous chapters.

The complexity of single-molecule biophysical experiments has increased dramatically over the past three decades, extending from the first generation of ingenious experiments on single, purified biomolecules studied in effect in *isolation*, to increases in biological relevance. This has been achieved by applying multiple single-molecule techniques in combination for test tube level assays, moving towards *maximum information* single-molecule experiments, and by increasing the complexity of multiple components

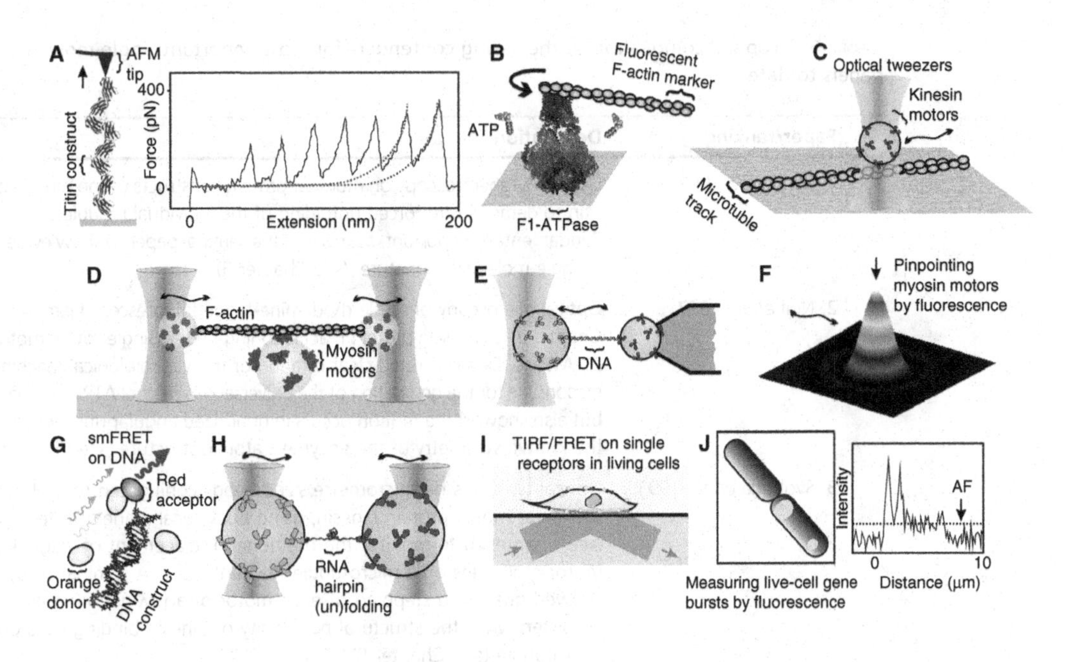

FIGURE 6.1 A suggested personal top ten set of the most important single-molecule biophysics papers of all time. Schematics depicting: **(A)** Rief *et al.* (1997), **(B)** Noji *et al.* (1997), **(C)** Svoboda *et al.* (1993), **(D)** Finer *et al.* (1994), **(E)** Smith *et al.* (1996), **(F)** Yildiz *et al.* (2003), **(G)** Ha *et al.* (1996), **(H)** Liphardt *et al.* (2001), **(I)** Sako *et al.* (2000), and **(J)** schematic depiction of results of Yu *et al.* (2006), AF indicates autofluorescence level.

used in *bottom-up* style in vitro approaches. In addition, the ability to perform single-molecule measurements in an environment which is very close, if not practically identical, to the native physiological context, namely in live cells, has increased (Leake, 2010).

KEY POINT

The majority of the seminal single-molecule biological papers in biophysics relate to studies performed in the test tube. However, it is clear from some of the more recent studies that we have entered the **next generation** of single-molecule biophysics in which a far greater **physiological relevance** is demanded from new studies.

6.2.2 Analytical methods employed to objectify single-molecule experiments

Single-molecule biophysics experiments are rife with noise, with the size of the detected signals being typically within the range of 1–10 times that of the noise amplitude. Therefore, the ability to detect the true signal and not erroneously detect noise represents a substantial analytical challenge. Most typically, molecular scale events are manifest as some form of transient *step* signal in a noisy time-series acquisition trace, for example a motor molecule might move via stepping actions along a given molecular track. Thus, the challenge becomes one of robust *step-detection* in a noisy data stream.

Edge-preserving filtration of the raw, noisy data is often the first tool employed. Standard filters, such as those that perform mean fits to running data windows or similar splinal/polynomial fits are not good in this regard, since they result in a blurring effect of the distinct edge event (Figure 6.2A). *Median filters*, or better still the *Chung–Kennedy algorithm* which consists of two adjacent running windows whose output is the mean from

TABLE 6.1 Top ten compilation of the strong contenders for most important single-molecule biophysics papers to date

Paper/ranking	Description
1. Rief *et al.*, 1997	Used AFM spectroscopy on modular protein constructs of the muscle protein titin to demonstrate forced unfolding of the individual modules. Fundamentally important, as this was the seminal paper to show evidence for a single-molecule 'signature' (see Chapter 5).
2. Noji *et al.*, 1997	Optical microscopy of single rhodamine-tagged fluorescent filaments (see Chapter 3) of muscle protein F-actin conjugated to single rotary motors of F1-ATPase. Demonstrated clear rotation of this vital biological machine responsible for the generation of the universal cellular fuel ATP (see Chapter 2), but also showed the motion occurs in quantized angular units which mirror the known symmetry of the enzyme's atomic structure (see Chapter 8).
3. Svoboda *et al.*, 1993	Optical tweezers on micrometre sized beads conjugated to molecules of the linear motor protein kinesin, using back focal plane interferometry (see Chapter 4) to monitor the positional displacement of single kinesin motor molecules on a microtubule filament track. A seminal study which showed quantized stepping of each motor of a few nanometres consistent with the structural periodicity of kinesin binding sites on the microtubule (see Chapter 9).
4. Finer *et al.*, 1994	Utilized dual optical tweezers on micrometre sized beads to tether a single filament of F-actin and lower it onto a third, surface-immobilized, bead functionalized with myosin motor molecules (see Chapter 5). This was the first study to measure clearly both the quantized nature of displacement and the force of a single molecular motor down to nanometre and piconewton level precision.
5. Smith *et al.*, 1996	Optical tweezers of a micrometre sized bead conjugated to a single DNA molecule tethered to a second micropipette-immobilized bead (see Chapter 5). The 'classic' single-molecule stretch study, seeding many more on other important biopolymers, was able to characterize single molecular force versus extension as well as observing individual 'over-stretching'.
6. Yildiz *et al.*, 2003	Applied single fluorescent dye imaging to tagged myosin motor molecules to show that they moved along F-actin tracks 'hand-over-hand'. This was the first study to show unconstrained walking of a single molecular motor, using nanometre precise localization in the form of FIONA (see Chapter 3).
7. Ha *et al.*, 1996	Tagged separate strands of a DNA construct with single dyes to measure single-molecule assembly of the double helix. This was the first study to use single-molecule pair FRET (smFRET) on a biomolecule (see Chapter 3).
8. Liphardt *et al.*, 2001	Optical tweezers on micrometre sized beads tethered to a single molecule of mRNA (see Chapter 2). This was the first study to demonstrate a clear example of detailed balance at the single-molecule level, between forced unfolded and reversibly refolded conformational states of RNA (see Chapter 5).
9. Sako *et al.*, 2000	Used single dye tags monitored by TIRF and FRET (see Chapter 3) on an 'EGF' ligand to monitor the dynamic behaviour of single receptor complexes in the cell membrane. This was the first study to use single-molecule detection on a functioning, live cell (see Chapter 7).
10. Yu *et al.*, 2006	Applied fluorescence imaging to protein products to monitor single-molecule gene expression bursts. This was the first single-molecule investigation on living cells to monitor the behaviour of single-molecule gene machinery in the cytoplasm as opposed to the cell membrane (see Chapter 9).

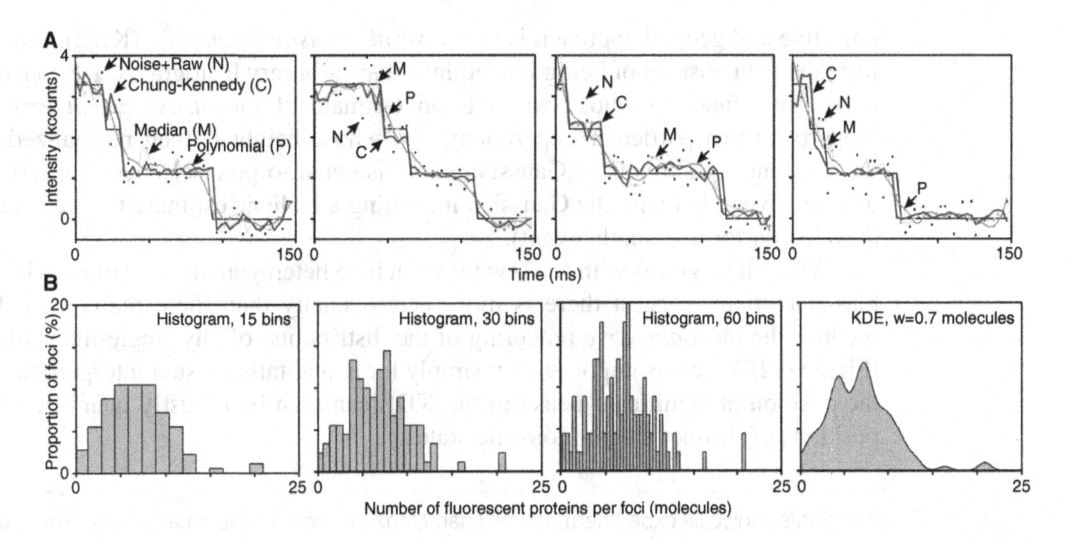

FIGURE 6.2 **(A)** Four examples of simulated single-molecule step-like events, here shown with step-wise photobeaching (see Chapter 8) of a hypothetical molecular complex that initially contains three photoactive fluorescent proteins, each of which is stochastically bleached during continuous laser excitation. The noise-free simulated data are shown as the straight line staircase in black, over which are shown the effects of the addition of realistic noise (N), which is then filtered either using a Chung–Kennedy (C), median (M) or polynomial filter (P). The polynomial filter clearly generates blurring artifacts at the step edges, whereas the Chung–Kennedy and median filters perform much better. **(B)** Histograms containing 15, 30 and 60 bins, compared against a kernel density estimation, here using real experimental data for the measurement of the number of molecules contained within detected fluorescent foci of a protein complex called MukBEF which acts as a DNA condensing unit in *Escherichia coli* bacteria (original data, author's own work).

the window possessing the smallest variance, are suitable. A filtered step event is then detected as being true on the basis of the change in the mean and variance (the *t-statistic*) between the two adjacent running windows being above a pre-defined threshold.

There are many such variants of methods which detect step events from a noisy time series, whose performance may be dependent upon the specific biophysical system and are thus *model dependent* such that the probability of observing any given step may be dependent upon the previous history of the time-series. These are so-called *Markovian* processes. A good review of several popular step-detection methods from time-series data is given in Carter *et al.* (2008).

The biggest issue, however, with all such *time-domain* detection methods is that they are potentially very sensitive to the level of detection threshold set, which is often potentially semi-arbitrary. An alternative approach relevant to instances where all steps in a series are expected to be of the same size is to convert the time-series into a *frequency domain* using a Fourier transform, and then detect the characteristic periodicity in the original trace by looking for the fundamental peak in the associated *power spectrum.* This has been used to good effect with certain molecular motors, such as kinesin, running on tracks, as well as for the estimation of molecular stoichiometry using step-wise photobleaching of fluorescent proteins (see section 8.3).

One recent improvement to objectifying single-molecule data is how the *distributions* of single-molecule properties (for example, step-size in terms of force or displacement, number of molecules measured in a complex, etc.) are rendered. Traditional approaches have used *histograms*, however these are notoriously sensitive to the size of the histogram bins and their positions and so is potentially prone to subjective bias, especially in the case of heterogeneity in these single-molecule properties. A more objective and general approach is to use *kernel density estimation* (KDE). In its simplest form the data, instead of being binned into semi-arbitrary histograms, are *convolved* with a Gaussian function whose width is an estimate of the *measurement error* for that property in that particular experiment, and whose height is then normalized such that the area under each unitary Gaussian curve is equal to precisely one (i.e. reflecting one detected event, but with the Gaussian indicating a realistic estimate for the actual *sample distribution* for that single event).

Where histograms with too few bins can hide heterogeneity, and those with too many bins potentially suggest there is more heterogeneity than there really is, a KDE will produce the most *objective* rendering of the distribution of any single-molecule property (Figure 6.2B). This is important not simply for a qualitative visual interpretation, but also the position of significant peaks in the KDE can then be robustly quantified to refer to positions of distinct single-molecule states.

KEY POINT

Any single-molecule experiment, if well characterized, should have an associated **measurement error** for each observed single-molecule property which can be estimated precisely. Therefore, there is **no excuse for using histograms at all**, apart from laziness – the distribution of every such experimentally measured single-molecule property can and should be **displayed as a KDE**!

6.3 Carving up the cell into a few sensible themes

Cellular processes span not only multiple length scales but also multiple time scales (see Chapter 1 and, for example, Spiller *et al.*, 2010), and there can be significant degrees of *cross-talk* not only between adjacent levels but also between broadly distant levels. This makes a description of cellular processes structured on the basis of either length or time scale fundamentally confusing to the reader since it is not possible to refer to real processes in the cell at any one particular length or time scale without reference to another scale at a different level. Similarly, discussing different processes on the basis of molecular type is also less than ideal since in general even the simplest biological process may involve molecules of several different types.

In the following three chapters we encounter multiple pioneering single-molecule case studies of native cellular systems or of test tube assays that have a very significant physiological relevance. These have been categorized more simply on the basis of where the principal biological process under study normally occurs *relative to the cell membrane*.

To a certain extent this categorization is somewhat arbitrary since there exist processes that in effect extend across multiple regions of the same cell. However, taking the cell, as encapsulated by a cell membrane, as the fundamental unit of life (see Chapter 2), what happens in the immediate vicinity outside the cell and involves some degree of transient binding to the cell surface (Chapter 7), what happens to molecules integrated in the cell membrane itself (Chapter 8), and what happens to molecules inside the cell away from the cell membrane (Chapter 9) is, on an *ontological level* at least, a not unreasonable approach for drawing together common themes of cell biology for the purposes of establishing a platform for learning and discussion.

Figure 6.3 is a schematic diagram which illustrates broadly where the chapter lines at least have been drawn. But the reader should not forget that the cell has no knowledge of such lines, and that in reality processes may be *integrated* significantly across different regions of the cell, with additional *cross-coupling* possible between different processes at

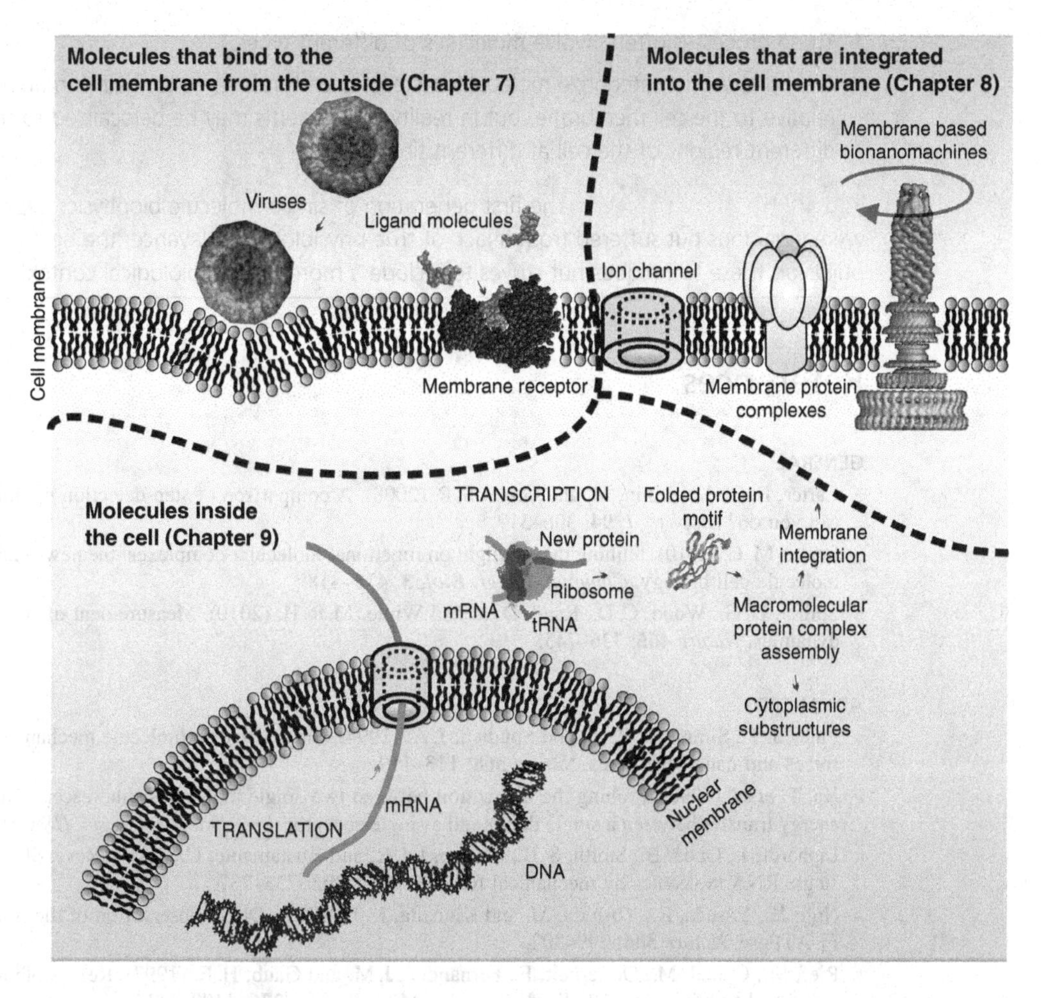

FIGURE 6.3 Schematic for carving up the molecular systems in a cell into broad themes for discussion (dotted lines): roughly, molecular systems that are predominantly in the immediate vicinity outside the cell membrane and bind to it transiently (Chapter 7), those that are integrated into the membrane itself (Chapter 8), and those that primarily occur within the cell innards (Chapter 9). (Virus images adapted with permission from Cornelia Büchen-Osmond, EM reconstruction of flagellar motor original data used with permission from David DeRosier.).

different points in time for different given processes. In short, for the experimentalist and theorist alike, trying to tie single-molecule biological data into the context of a functioning, living cell is an enormous challenge – one, however, very worthy of pursuit!

THE GIST

- First generation single-molecule biophysics experiments involved purified molecules in isolation, but succeeded in addressing many challenging biological questions which were intractable using conventional bulk ensemble averaging methods.
- The next generation single-biomolecule experiments involve increasing the complexity either by using multiple single-molecule methods in combination in a 'maximum information' approach, or by adding physiological relevance by increasing the number of relevant molecular components in vitro or by performing experiments in the living cell.
- Cellular processes span multiple length and time scales that can interact in non-trivial ways.

- These processes often involve molecules of different types.
- A simple way to categorize molecular systems is with reference to their primary location relative to the cell membrane, but in reality components may be delocalized to several different regions of the cell at different times.

TAKE-HOME MESSAGE The first generation of single-molecule biophysics experiments were ingenious but suffered from a lack of true physiological relevance; the next generation builds on these techniques but strives to include a more faithful biological context.

References

GENERAL

- Carter, B. C., Vershinin, M. and Gross, S. P. (2008). A comparison of step-detection methods: how well can you do? *Biophys. J.* **94**: 306–319.
- Leake, M. C. (2010). Shining the spotlight on functional molecular complexes: the new science of single-molecule cell biology. *Commun. Integr. Biol.* **3**: 415–418.
- Spiller, D. G., Wood, C. D., Rand, D. A. and White, M. R. H. (2010). Measurement of single-cell dynamics. *Nature* **465**: 736–745.

ADVANCED

- Finer, J. T., Simmons, R. M. and Spudich, J. A. (1994). Single myosin molecule mechanics: piconewton forces and nanometre steps. *Nature* **368**: 113–119.
- Ha, T. *et al.* (1996). Probing the interaction between two single molecules: fluorescence resonance energy transfer between a single donor and a single acceptor. *Proc. Natl. Acad. Sci. USA* **93**: 6264–6268.
- Liphardt, J., Onoa, B., Smith, S. B., Tinoco, I. J. R. and Bustamante, C. (2001). Reversible unfolding of single RNA molecules by mechanical force. *Science* **292**: 733–737.
- Noji, H., Yasuda, R., Yoshida, M. and Kinosita, K. J. (1997). Direct observation of the rotation of F_1-ATPase. *Nature* **386**: 299–302.
- Rief, M., Gautel, M., Oesterhelt, F., Fernandez, J. M. and Gaub, H. E. (1997). Reversible unfolding of individual titin immunoglobulin domains by AFM. *Science* **276**: 1109–1112.
- Sako, Y., Minoghchi, S. and Yanagida, T. (2000). Single-molecule imaging of EGFR signalling on the surface of living cells. *Nature Cell Biol.* **2**: 168–172.
- Smith, S. B., Cui, Y. and Bustamante, C. (1996). Overstretching B-DNA: the elastic response of individual double-stranded and single-stranded DNA molecules. *Science* **271**: 795–799.
- Svoboda, K., Schmidt, C. F., Schnapp, B. J. and Block, S. M. (1993). Direct observation of kinesin stepping by optical trapping interferometry. *Nature* **365**: 721–727.
- Yildiz, A. *et al.* (2003). Myosin V walks hand-over-hand: single fluorophore imaging with 1.5-nm localization. *Science* **300**: 2061–2065.
- Yu, J. *et al.* (2006). Probing gene expression in live cells, one protein molecule at a time. *Science* **311**: 1600–1603.

Questions

FOR THE LIFE SCIENTISTS

Q6.1. How, and why, might the elastic parameters that emerge from a single-molecule stretching experiment on DNA in isolation (for example, Smith *et al.*, 1996) be different to what might emerge if the experiment were performed in vivo? What challenges are there to doing single-molecule stretch experiments in a living cell?

Q6.2. Describe one of the many cellular processes that are localized to discrete regions of the cell. What biological benefits are there to having such a local sub-cellular architecture. How does this present a problem for test tube type single-molecule experiments?

Q6.3. With reference to specific examples, explain why the dotted lines marked in Figure 6.3 are artificial?

FOR THE PHYSICAL SCIENTISTS

Q6.4. Mechanical polymer properties such as persistence and contour length are intrinsic characteristics of each single stretched biological molecule (see for example Rief *et al.*, 1997; Smith *et al.*, 1996). Why then (see Q.6.1) should this be any different inside a living cell?

Q6.5. 'Maximum information' experiments strive to combine multiple single-molecule biophysics techniques on the same single molecule. (a) How does Shannon's information theory relate to the information content in one of these experiments? (b) What limits the fundamental amount of information that these experiments can extract?

Q6.6. (a) In a chain of first order biochemical reactions in which molecules 1, 2, 3,. . ., n react in effect irreversibly in the series $1 \rightarrow 2 \rightarrow 3 \rightarrow \ldots \rightarrow n$ with respective rate constants $k_1, k_2, k_3, \ldots, k_n$, what is the effective overall rate constant relevant to the process $1 \rightarrow n$. (b) What happens to this rate if the time scales for the intermediate processes are significantly different from each other? (c) What implications does this have for the local concentrations of the intermediates? (Hint: Such an intermediate is often referred to as the 'rate-limiting step'.)

Q6.7. In a stretch experiment utilizing the deflection of a thin glass needle attached to a single filament of F-actin, its Hookean stiffness was estimated to be $\sim 40\,\text{pN nm}^{-1}$. In a modified version of the dumbbell optical tweezers assay of Finer *et al.* (1994) using F-actin tethered between two optically trapped beads (see Chapter 5), the polymerization state of the F-actin was dynamically controlled to allow individual globular 'G-actin' monomer subunits to be added to one of the ends. (a) If the trapped beads were held in place using positional feedback optical tweezers clamps then the addition of more subunits resulted in filament buckling with an equal chance of buckling above or below the axis joining the two beads. Why? (b) If the G-actin diameter is ~3 nm, the filament is initially 10 μm long and held 1 μm above the top of a third surface-immobilized bead coated in myosin, and the G-actin spirals in a double-helix of pitch 36 nm composed of a 13 subunit repeat in the F-actin filament, how many G-actin subunits need to be added in order to detect a power stroke event? In a separate experiment the thermal fluctuations of single free fluorescently labelled F-actin filaments were measured using TIRF to estimate the bending stiffness, EI, also known as the flexural rigidity, at $\sim 7 \times 10^{-26}\,\text{N m}^2$, where E is the Young's modulus and I is the moment of inertia about the filament axis. (c) Will a typical optical trap be able to exert sufficient force on the ends of the buckled filament to allow the beads to be positionally clamped? You may assume that for a thin rod the torque acting about a given point (the bending moment) equals EI/r where r is the radius of curvature. (Hint: Approximate the F-actin buckle as an arc of a circle; this problem is essentially the same as welding in a small extra section into a straight train track which is fixed at both ends.)

FOR THOSE WHO HAVE NOT MADE UP THEIR MIND

Q6.8. Single-molecule experiments on the rotary F1-ATPase motor performed after the seminal study of Noji *et al.* (1997) suggest some shorter-lived angular states which were not picked up in the original investigation. Why is this?

Q6.9. The viscosity in the cell membrane is 100–1000 times greater than the viscosity in the cytoplasm. Does this mean that the time scale of molecular processes in the cytoplasm will be greater by this amount?

Q6.10. If an extra-cellular ligand binds to a receptor complex in the cell membrane, which then causes a conformational change in the receptor, which in turn triggers a biochemical cascade on the inside of the membrane in the cytoplasm, how should this receptor be classed in terms of being part of an external binding process to the membrane, or an example of an integrated membrane complex, or part of a process which involves molecular systems in the innards of the cell?

7 Molecules from beyond the cell

At present we are on the outside of the world, the wrong side of the door.

(C. S. LEWIS, THE WEIGHT OF GLORY, © C. S. LEWIS PTE LTD. 1941)

GENERAL IDEA

Here we explore some of the pioneering single-molecule experiments that have increased our understanding of the processes that essentially involve foreign molecules external to cells either binding to the cell surface or being internalized by cells. Such molecules interact directly with the semipermeable membrane of the cell, making it an exceptionally lively and dynamic environment.

7.1 Introduction

Highly efficient mechanisms exist that allow foreign molecules to bind to cell surfaces, ultimately evoking some form of *signal response*, and allow a variety of external molecules over a broad range of size, chemistry and charge to enter cells. At one level these mechanisms include processes involving *receptor* molecules embedded in the outer membrane of cells that bind to *ligand* molecules. Many of the stages of this detection and *signal transduction* process have canonical features, sometimes involving adapter molecules binding to the original ligand, as well as specific binding events and changes in molecular conformation which can often be transmitted over relatively long length scales of several nanometres spanning one or sometimes more lipid bilayer membranes and sometimes involving cooperative effects from other receptor molecules. Non-native particles which are internalized by cells range in size from single molecules to much larger heterogenous macromolecular complexes such as *viruses*. In this chapter we will encounter first-hand examples of how these processes have been investigated using single-molecule biophysics. These investigations reveal highly complex behaviours, indicating that the world outside the cell is just as important as the world on the inside.

7.2 Receptor molecules and ligands in the cell membrane

Receptor molecules and their associated ligand complexes have traditionally been prime targets for live-cell single-molecule research, principally because their location in the cell membrane makes them ideal for the contrast-enhancement single-molecule imaging method of total internal reflection fluorescence (TIRF) microscopy (see Chapter 3).

7.2.1 Receptors relevant to normal tissue assembly (and cancers when they go wrong)

The first truly single-molecule study involving definitively living cells used TIRF microscopy on a type of cultured cancer cell from the human vulva to monitor the so-called *epidermal growth factor receptor* (EGFR) complexes expressed in the cell membrane (Sako *et al.*, 2000). EGFR is part of a large family of cell surface receptors which possess intrinsic *tyrosine kinase* activity. Such receptor tyrosine kinases are membrane integrated enzymes found in many cellular processes involved in *signal transduction*. They typically possess an exposed extra-cellular domain which contains a binding site to a specific ligand, a connected domain which spans the cell membrane, and a further linked domain on the inside of the cell.

This intra-cellular domain functions ultimately to transfer a phosphate group from the amino acid tyrosine to a specific *target protein* in the cell cytoplasm, generally in response to the binding of some ligand molecule to the extra-cellular domain. In doing so, this phosphorylated protein then becomes *activated*, since the presence of the newly bound phosphate can then trigger a cascade of further chemical reactions with other molecules in the cell involved in a specific cellular *signal response*, for example to bring about a change in the level of expression from a multitude of genes, thus in effect *transducing* the original ligand chemical *signal* into a whole cell *response* of some form. In this sense, therefore, the tyrosine kinase functions as an on/off switch for this signal transduction.

EGF receptors are important in epidermal tissue *morphogenesis*. That is, they are involved in inter-cellular communication resulting ultimately in coordinated cell localization in space and time. This is very important in the development of several vital tissues and, ultimately, organs in the human body (for the life scientists, a good review can be found in Schlessinger, 2000). EGFR is a cell-surface receptor for members of the epidermal growth factor (EGF) family of extra-cellular protein ligand molecules. Since EGF receptors are involved directly in multi-cellular tissue formation and growth, it is unsurprising that genetic mutations affecting EGFR have been found to result in cancer in these tissues, since by definition a cancer involves uncontrolled cell division. These receptors have therefore been a focus of significant recent biomedical research.

In the seminal in vivo TIRF study by Sako *et al.* (2000), the researchers were able to observe individual molecules of the protein ligand EGF, tagged fluorescently using single molecules of either green (Cy3) or red (Cy5) organic dyes, as they bound to the cell membrane, putatively to membrane-integrated EGFR molecules to form a receptor–ligand complex (Figure 7.1A). Each emitting dye molecule was manifest as a point spread function intensity distribution, which equated to a blurry spot of light of width

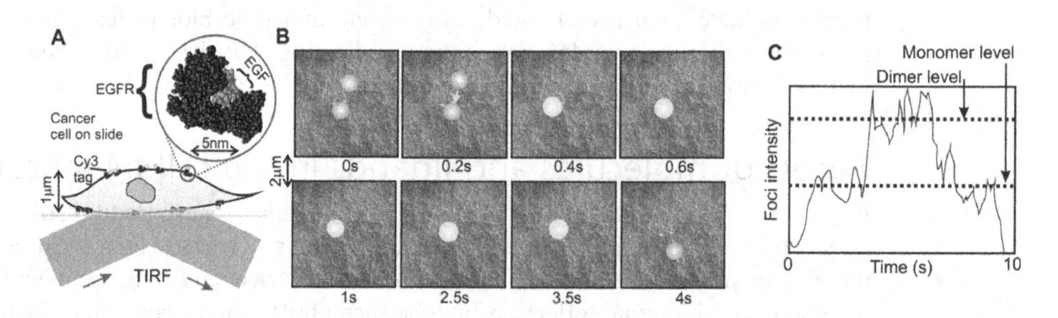

FIGURE 7.1 **(A)** Schematic of TIRF imaging of Cy3-labelled EGFR complexes in the membrane of cultured human cancer cells. **(B)** Schematic depiction of sequential TIRF images taken at different time points, showing fluorescent foci which mark the location of ligand-bound EGFR. Putative fusion of two dimmer foci (0–0.2 s), movement indicated by arrow, to a single brighter focus (0.4–1 s) then subsequently separation to dimmer foci again (2.5–4 s) is shown. **(C)** Schematic depiction of intensity versus time trace for typical EGFR foci, indicating possible dynamic monomer/dimer populations.

250–300 nm (see Chapter 3). By measuring the integrated fluorescence intensity from each spot in the cell, which putatively marked the positions of the receptor-bound EGF ligand conjugated to an EGFR complex in the cell membrane, the researchers were able to show that binding of the ligand was consistent with inducing neighbouring receptor complexes to *cluster* (Figure 7.1B, C).

The specific changes in fluorescence intensity observed indicated that receptors formed a stably bound *dimer* complex of two such EGFR molecules in response to a *single* molecule EGF binding event. Furthermore, the investigators were able to demonstrate that the ligand components of the monomer subunit complexes must be within a few nanometres distance of each other since they were observed to undergo appreciable FRET. This could indicate some level of collective behaviour in receptor complexes, in other words, the binding of one EGF molecule to one receptor complex may result in *spatial pooling* of receptors to regions of high local ligand concentration, thereby ultimately increasing the cellular response substantially by in effect conferring an amplification-adaptation to the initial signal.

Tyrosine kinases have also been studied using single particle tracking in bacteria. Using a YFP tag on the protein PleC, which is a membrane-integrated histidine kinase found in the bacterium *Caulobacter crescentus* and is implicated in the regulation of *cell division* and thought to be localized at cell poles, researchers were able to track single molecular complexes using widefield fluorescence imaging for several seconds, and established that two populations existed in terms of molecular mobility, one relatively immobile at the cell poles and another which exhibited relatively unhindered diffusion in the cell membrane (Deich *et al.*, 2004).

Single-molecule studies such as those described here have contributed greatly to our knowledge of tyrosine kinase receptors, in particular of this biomedically important EGF receptor. Not only do we now know that the binding of the EGF ligand causes receptors to cluster into dimers, but we also know that the initial phosphorylation activation of the tyrosine kinase on the EGF-bound receptor of EGFR in the receptor dimer actually results in *internal* phosphorylation of the other EGF receptor in the dimer which does not have EGF bound to it. This is a common device used in cellular signal transduction called *trans-autophosphorylation*. This phosphorylated receptor then acts as the *target protein* to initiate a specific *cascade* of chemical reactions inside the cell, so in other words, neighbouring receptors *communicate* to convey the signal to the cell.

7.2.2 Sticking cells together

The ability of certain cells to remain attached to each other is vital in higher organisms to allow many such cells to pool together to form integrated tissue structures. The principal mechanism for achieving this is through *focal adhesion complexes* assembled at the membrane junction between neighbouring cells. In a study that was the first to image single fluorescent protein molecules successfully in a living cell (Iino *et al.*, 2001), researchers investigated a membrane protein called *E-cadherin* in live mice cells called *fibroblasts*.

Fibroblasts are the most common cell type of *connective tissue* in animals, which secrete a material to generate the *extra-cellular matrix*. This is a viscoelastic, heterogeneous material composed of protein *fibrils* and modified polysaccharide biopolymers, most commonly including the protein *collagen*, which forms a jelly-like stabilizing structure between different organs of the body while still permitting the diffusion of essential molecules through the matrix such as water, diffused oxygen and carbon dioxide, as well as ions and small *metabolite* chemicals utilized in a variety of cellular processes.

E-cadherin is a special type of *adhesion molecule* which, along with other types of cadherins and other related molecules such as *selectins* and *integrins*, allows neighbouring cells to stick together, similar in action to each other. In this study, GFP was fused to one of the ends of the E-cadherin protruding into the inner cytoplasm of the cell. The investigators again used the summed pixel intensities from each observed distinct spot of green fluorescence, here reporting on the position of E-cadherin molecules, as a measure of the number of E-cadherin molecules bound together in a complex. They observed a heterogeneous distribution of brightnesses consistent with a mixed population of E-cadherin monomers, dimers, trimers and tetramers, and were able to perform single particle tracking on these spots to estimate the effective diffusion coefficients (see Chapter 3).

By modelling how they would expect the physical size of such oligomer complexes to vary with the number of monomer subunits, they could predict how they would expect the diffusion coefficient to vary with observed spot brightness on the basis of simple differences in subunit number resulting in a predictably different frictional drag coefficient. What they actually found was that the diffusion coefficient was substantially lower than expected for larger complexes. This could indicate that larger E-cadherin complexes were being *confined*, possibly by a cytoskeleton network of protein filaments.

BIO-EXTRA

The **cytoskeleton** is a protein based internal cellular scaffold, typically composed of multiple subunits of polypeptide monomers, forming stiff, fibrous-like structures that are implicated in the determination of **cell shape** and also act as **tracks** along which molecular cargoes can be transported. These were once considered to be unique to eukaryotes but have now been identified in several different prokaryotic cells. The eukaryotic cytoskeletal components include **microtubules**, made up of **protofilaments** of **tubulin** monomers (which has a homology of the protein **FtZ** in bacteria used in constricting cells during cell division), which are ~20 nm in diameter and can grow up to 25 μm long, each possessing a '+' and '−' end from which polymerization and depolymerization of tubulin monomers can occur respectively – motor proteins **kinesin** and **dynein** use microtubules as tracks (see Chapter 8), travelling principally towards + and − ends respectively, and convey a multitude of **molecular cargoes.** Other cytoskeletal components include **intermediate filaments** composed of a variety of subunit proteins including **vimentin**, which are 7–25 nm in diameter and can grow to lengths comparable to that of microtubules, and **microfilaments** composed of **actin** in eukaryokes (or **MreB** type proteins in bacteria) which have a narrower diameter of ~6 nm and are associated with the molecular motor protein myosin in various forms. At one extreme, the muscle system of animals can be considered a highly packed array of microfilaments which interact with a filamentous array of myosin molecules to bring about muscle contraction via molecular **power strokes.**

Such putative *corralling* effects could have implications beyond E-cadherin, and have now been observed in several different membrane protein complexes. The source of the confinement/corralling is unclear, and may indeed be the result of several different factors including the presence of a cytoskeleton matrix underneath the cell membrane, other crowding proteins in the membrane (typically 30–40% of the cell membrane is actually made up of proteins by surface area) or even *lipid raft* islands due potentially to differences in physical phases of lipids in the bilayer (see Chapter 8).

PHYSICS-EXTRA

Lipid bilayers are examples of systems that can exist at a local level in two different, stable physical phases, either a **gel phase** or a more compact, viscous **solid phase**. The probability of being in either phase depends on the specific compositions of lipids, but also on the temperature. Under certain circumstances it is possible to have **raft-like** islands composed of lipids, typically over a range of 10–200 nm effective diameter, of one phase surrounded by lipids of another phase. These islands are intrinsically unstable and may themselves diffuse *en masse*, and may also fluctuate in both size and shape with time, but might still result in transient zones of confined mobility for certain membrane protein molecules or complexes which are in effect trapped within the islands, since they need to overcome an appreciable free energy barrier of several k_BT to jump across a **phase-transition boundary** at the island's perimeter.

7.2.3 Signal transduction from activated membrane receptors

Ultimately the detection of an external ligand molecule by a receptor complex located in the cell membrane, which constitutes an initial cellular signal, needs to be transformed, or *transduced*, into another type of molecular signal which can then bring about some form of cellular response, typically involving a complex cascade of coupled chemical reactions. There are several examples of this signal transduction process which have been studied using single-molecule cellular biophysics techniques.

One common method of signal transduction in the cell uses a protein called *Ras*. Ras was first studied on a type of cancerous cell found in connective tissue called a *sarcoma*, and was first investigated in rats (hence *Rat sarcoma*, shortened to *Ras*). Ras is an intra-cellular signalling molecule that relays signals from activated receptor tyrosine kinase molecules, the family of receptors to which the EGF receptor discussed in the section above belongs. The cell membrane receptors act in several signal transduction pathways, with the hydrolysis of receptor-bound *guanosine triphosphate* (GTP) to *guanosine diphosphate* (GDP) generating a local energy source in much the same way as the hydrolysis of ATP to ADP (see Chapter 2). This most probably fuels a conformational change in the receptor–ligand complex that initiates a cascade of chemical reactions in the respective transduction pathway in the cell, of which Ras is a part. Genetic mutations of genes which express proteins that are involved at some level in these pathways often result in the formation of cancer cells, in a similar way to mutations of genes involved in the EGF pathway (such genes are referred to as *oncogenes*) and have therefore also been an intensive focus of much biomedical research.

During its normal biological function, a Ras molecule binds to an activated EGF receptor in the cell membrane. The EGF receptor then facilitates the activation of Ras through an adapter protein, known as Grb2 in mammalian cells, to a *Ras guanine nucleotide exchange factor*. Thus, Ras is a guanosine-nucleotide-binding protein, binding to GTP; the general class is thus more simply referred to as *G-proteins*.

In a seminal single-molecule investigation of the signal transduction of Ras proteins, researchers monitored Ras protein tagged with yellow fluorescent protein, YFP, in living cultured human cancer cells. Simultaneously, they monitored a lipid-soluble form of GTP tagged with a derivative of the orange organic dye rhodamine (see Chapter 3), which acts as a good FRET acceptor molecule for a donor molecule of YFP in a living cell (Murakoshi *et al*., 2004). By using highly sensitive single-molecule FRET measurements, the researchers discovered that, after first stimulating the cultured cells with the ligand EGF, GTP was found in the same positions as Ras proteins in the membrane roughly 30 s after the addition of EGF, with the distance between Ras and GTP being sufficiently small at a few nanometres or less to generate a measured FRET response.

Furthermore, the investigators observed that the brightness of the YFP spots increased following activation by EGF, and that the diffusion of these brighter spots showed signs of confined mobility in a similar way to that observed in the previous section for E-cadherin oligomers. Subsequent investigation by other researchers, again using single-molecule TIRF imaging on live cells of a similar type, showed that in fact what appeared to be oligomers were a mixture of monomers and dimers (Hern *et al.*, 2010). It is likely that such a monomer/dimer mix randomly distributed over the cell membrane surface will result in a certain proportion of these complexes being spaced at a distance which is less than the optical resolution limit, thus appearing as a single 'oligomeric' foci under TIRF illumination (Chapter 2), or it may be that the differences between the cell strains are actually real. This illustrates one of the common pitfalls in describing 'seminal' research papers in such a rapidly moving field, since within a decade someone else appears have obtained a 'more correct' answer! Even so, this suggests that not only does GTP bind preferentially to Ras proteins following EGF ligand binding to Ras, but also that this may subsequently initiate the formation of dimers of Ras, some of which may have confined mobility in the cell membrane, which result in local amplification of the transduced signal in a similar way observed for E-cadherin previously.

Chemotaxis is another prime example of signal transduction involving cell membrane receptors. It is the remarkable process by which whole cells can move ultimately either towards a source of food or away from noxious substances, by the detection of external chemo-attractants and chemo-repellents respectively. These attractants and repellants act as ligand molecules which bind either directly to receptor complexes on the cell membrane surface or first to adapter molecules which then in turn bind to the receptor complex (for a good early single-molecule study on the chemotaxis effects of changes in cell shape using different *spatial concentration gradients* of chemo-attractants see Watanabe and Mitchison, 2002). This binding event might, speculatively, result in a molecular conformational change that is transformed via some often remarkable coupled chemistry into a signal cascade inside the cell which, by a variety of other complex mechanisms, feeds into the cell motility control system, resulting in concerted, directional movement of the whole cell.

Prokaryotes such as bacteria live in a dynamic and often harsh environment in which other cells compete for finite food resources (for a good life scientist review see Wadhams and Armitage, 2004). Remarkable chemotaxis sensory transduction systems have evolved for highly efficient detection of, and cell movement in response to, external chemicals. The manner in which this is achieved is fundamentally different to the general operating principles of many biological systems in that no genetic regulation as such appears to be involved. That is to say, signal detection and transduction does not involve any direct change in the amount or type of proteins that are made from the genes, but rather utilizes in effect a network of proteins and protein complexes *in situ* to bring about this end.

In eukaryotes, the process of chemotaxis is critical in several systems which rely on concerted, coordinated multi-cellular responses from many such individual cells, including aspects of the immune response, patterning of cells in neuron localization and morphogenesis of complex tissue during the different stages of early development in an organism.

Fungal cells of *Dictyostelium discoideum* display a strong chemotaxis response to *cyclic adenosine monophosphate* (cAMP), which is mediated through a cell surface receptor complex and G protein-linked signalling pathway, as seen for the Ras protein above. Using fluorescently labelled Cy3-cAMP in combination with single-molecule

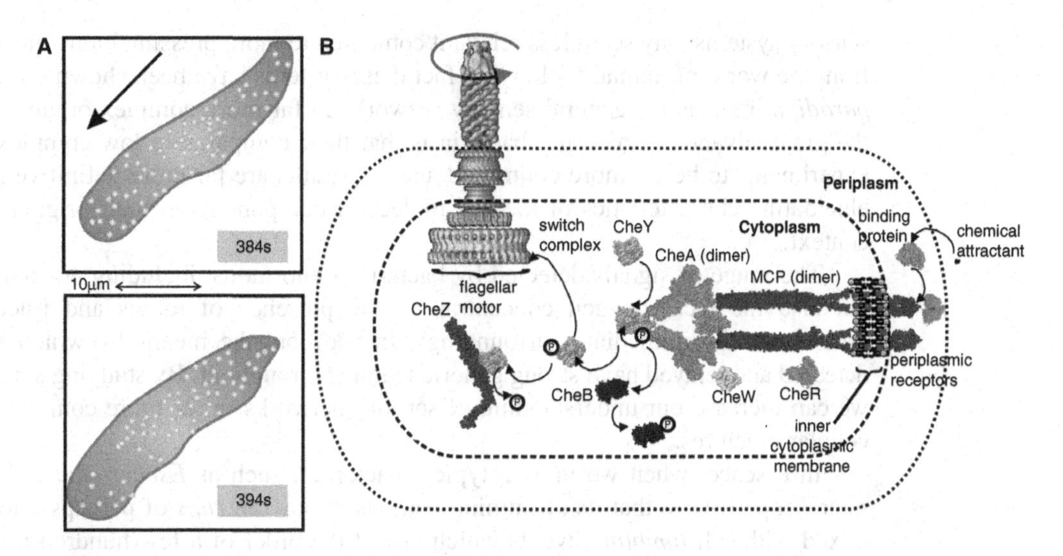

FIGURE 7.2 **(A)** Schematic depiction of receptor occupancy of a *Dictyostelium* cell under a gradient of Cy3-cAMP (arrow shows direction of source of Cy3-cAMP), with detected receptor positions marked (circles). The time after addition of the cAMP is given in seconds, indicating a cellular signal adaptation so that there is net diffusion of receptors away from the cAMP source. **(B)** Bacterial chemotaxis: cartoon (not to scale) of the most basic chemosensory pathway illustrated using known structures for chemotaxis proteins from *E. coli* and *S. enterica*, as well as the EM structure of the flagellar motor (original data with kind permission of David DeRosier, Brandies University, see Thomas *et al.*, 2006 for full description). Thousands of MCP receptors are clustered at the cell poles, and respond with a putative conformational change either upon binding of a specific protein attached to an attractant molecule, or to an attractant molecule directly. This initiates a biochemical cascade culminating in an activated form of the protein CheY (CheY-P) binding to the switch complex of the flagellar motor. This causes the motor rotation direction to switch, inducing tumbling of the whole of the cell.

TIRF imaging and single particle tracking on live *Dictyostelium* cells, researchers were able to monitor the dynamic localization of ligand-bound receptor clusters and to measure the kinetics of ligand binding in the presence of a chemo-attractant concentration gradient (Ueda *et al.*, 2001; Figure 7.2A).

In eukaryotic cells, such as *Dictyostelium*, the direction of a concentration gradient of an external chemical is detected by a complex mechanism that essentially compares the rate of binding of ligand by receptors on one side of a given cell compared to those on the other side. That is, it uses the physical length of the cell to generate probes in different regions of the concentration gradient such that on the side of the higher concentration there will be a small but significant and measurable increase in the rate of ligand binding compared to the opposite side of the cell. This is in effect *spatial sampling* of the concentration gradient. Prokayotic cells such as bacteria cannot use such a mechanism because their physical length of roughly a micrometre results in far too small a difference in ligand binding rates either side of the cell, at least for the relatively small concentration gradients in external ligand that the cell may need to detect. Instead, the cellular strategy evolved is one of *temporal sampling* of the concentration gradient.

The way in which bacteria sense and react to their environment has been well studied but only recently have tools been available to permit their investigation in living, functional cells at a single-molecule level. The ability of bacteria to swim up a concentration gradient of a chemical attractant is well known, so these cells can clearly detect their surroundings and act appropriately. Although at first glance the study of bacterial sensory systems may seem less relevant compared to more pressing biomedical questions from the world of human biology, in fact these systems have been shown to be excellent *paradigm* models for general sensory networks in far more complex organisms. One of the great advantages of using bacteria is that their comparative low complexity allows experiments to be far more controlled and the results are far more definitive in terms of elucidating characteristics of the key molecular components in their original biological context.

The range of signals detected by bacteria is enormous, including not only nutrients but also the local oxygen concentration, the presence of toxins and fluctuations in pressure in the immediate surroundings. In addition, the means by which signals are detected and relayed have strong generic features throughout. By studying single bacteria we can increase our understanding of sensory networks in far more complicated multicellular creatures.

In essence, when we view a typical bacterium such as *Escherichia coli* under the microscope we see that its swimming consists of smooth *runs* of perhaps a few seconds mixed with cell *tumbling* events which last of the order of a few hundred milliseconds. (According to Howard Berg, one of the true pioneers in the field of bacterial motility, the phrase *twiddle* was favoured initially for describing the random motion of the cells, but the consensus from the research team was that no one would really understand what they were talking about, and so *tumble* it became!) After each tumble the direction of swimming of the cell is randomized, so in effect each cell performs a three-dimensional *random walk*.

However, the key feature of bacterial chemotaxis is that if a chemical attractant is added to the solution then the rate of tumbling drops off – the overall effect is that the swimming phase, although still essentially randomized by tumbling, is then biased in the direction of an increasing concentration of the attractant; in other words this imparts an ability to move *closer* to a food source. The mechanisms behind this have been studied using optical microscopy on active, living cells, and single-molecule experiments are now starting to offer enormous insight into *systems-level behaviour* of these processes.

Much of our experimental knowledge comes from the chemosensory system exhibited by the bacteria *E. coli* and *Salmonella enterica*, and it is worth discussing this paradigm system in reasonable depth since it illustrates some remarkable general features of signal transduction regulation that are applicable to several different systems. Figure 7.2B illustrates a cartoon of our understanding to date based on these species in terms of the approximate spatial locations and key interactions of the various molecular components of the complete system. In a nutshell, these species are engineered to be intrinsically optimistic – if life is good, don't change it!

Travelling in the direction of the signal, i.e. from the outside of the cell in, the first sub-system we encounter concerns the primary detection of chemicals outside the cell. Here, we find many thousands of tightly packed copies of a protein complex which forms a chemoreceptor spanning the cell membrane. These complexes can undergo chemical modification by methyl groups and are thus described as *methyl-accepting chemotaxis proteins* or MCPs.

These MCPs are linked via a protein called CheW to the CheA protein. This component has a phosphate group bound to it, which can be shifted to another part of the same molecule, by the same process of trans-autophosphorylation encountered in the previous section on EGFR. It was found that the extent of this trans-autophosphorylation is increased in response to a *decrease* in local chemo-attractant binding to the MCPs.

Two different proteins known as CheB and CheY compete in binding specifically to this transferred phosphate group. Phosphorylated CheB (CheB-P) catalyses *demethylation* of the MCPs and controls receptor adaptation in coordination with CheR which catalyses MCP *methylation*, which thus serves as a *negative feedback* system to *adapt* the chemoreceptors to the size of the external chemical attractant signal. The phosphorylated CheY, CheY-P, diffuses through the cytoplasm and ultimately binds to a protein called FliM on the membrane-integrated *rotary flagellar motor* (the molecular machine which is ultimately responsible for causing the cell to swim). This binding to FliM causes the direction of motor rotation to reverse momentarily, which causes the cell to tumble. The CheZ protein is then required for *signal termination* by catalysing *dephosphorylation* of CheY-P back to CheY. The whole point of this rather elaborate biochemical scheme is that, in reducing the cellular pool of CheY-P, CheZ actually reduces the overall response time of the signal transduction system from ~20 s in the absence of CheZ to just a few milliseconds. In other words, the cell can respond far more rapidly to changes in concentration of local external chemo-attractant.

Much work has been done on the mathematical modelling of the bacterial chemotaxis system, both on whole populations and at single cell level (interested physical scientists should read Tindall *et al.*, 2008). Our experimental understanding of the systems involved in chemotaxis in living cells has come primarily from fluorescence microscopy. A recent study which has made significant inroads into our understanding of the switching behaviour investigated the *switch complex* of the flagellar motor at the single-molecule level (Delalez *et al.*, 2010). Here, single-molecule TIRF imaging was used on single live cells of *E. coli* bacteria which expressed genomically encoded YPet derivatives of the FliM protein at physiological levels (see Chapter 3).

Analysis of functional motors based on measuring the total fluorescence intensity from each motor revealed that the stoichiometry of each has a distribution from motor to motor in the range 10–40 FliM molecules, and in fact that two FliM populations coexist within each motor, one undergoing stochastic *turnover*, the other fixed. Surprisingly, exchange within the dynamic population appeared to rely on the presence of a *response regulator* protein linking the complex to the rest of the sensory pathway.

Cells could be observed using brightfield and laser excitation TIRF microscopy via a functional assay which conjugates the flagellar motor to a glass microscope slide (Figure 7.3A). Cell bodies could be seen to be either freely rotating around a point of tether attachment, or immobilized to the coverslip surface, which tells us immediately that the tethered motor complex is functional and the cell is alive. The measured width of these spots was 300–350 nm, consistent with a FliM ring of diameter ~50 nm convolved with the point spread function of a single YPet molecule. To measure the fluorescence intensity of each spot the investigators defined a circular region of interest of radius 400 nm centred on the motor and used automated software to separate out the intensity due to the bound membrane-integrated component of FliM-YPet from the local background intensity, which was a combination of diffusive cytoplasmic FliM-YPet, cellular autofluorescence and instrumental background (Figure 7.3B, C).

Continuous TIRF illumination indicated an *exponential decay* in spot fluorescence intensity which occurred in a *step-wise* manner of roughly integer multiples of a unitary spacing, consistent with the irreversible photobleaching of YPet (see Chapter 3) measured using a *Fourier spectral method* (Figure 7.3D). By calculating the initial intensity from an exponential fit for each trace, the stoichiometry could be estimated. This indicated ~34 FliM molecules per motor, in very good agreement to previously reported copy numbers of 33–35 FliM molecules from cryo-EM images.

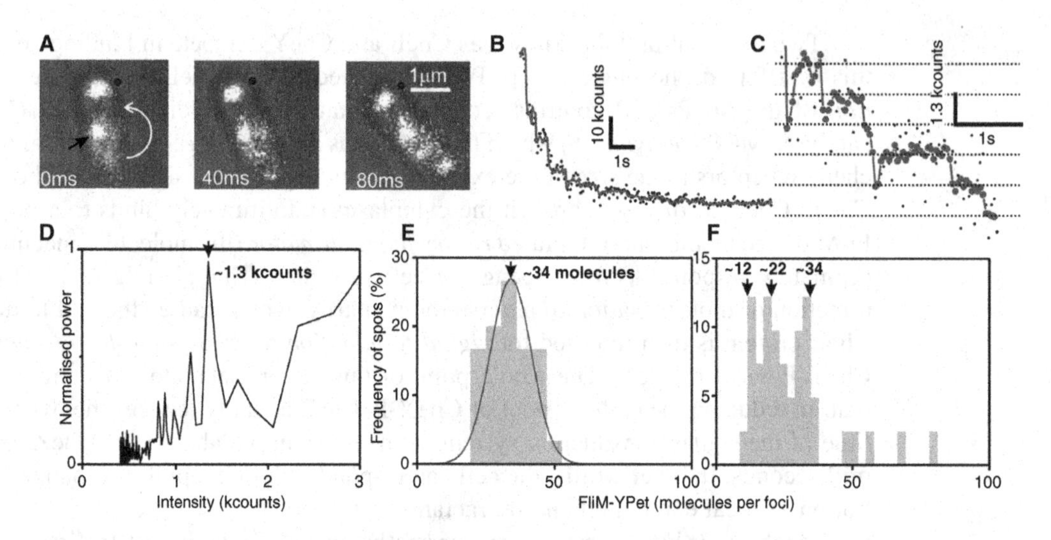

FIGURE 7.3 Using step-wise photobleaching to estimate FliM-YPet stoichiometry. **(A)** Sequential TIRF images of live cells plus overlaid brightfield images for a tethered, rotating FliM-YPet cell; the motor is at the centre of rotation, the direction of rotation is indicated (arrow). **(B)** Corresponding photobleach fluorescence intensity trace for a motor showing raw (dots) and filtered (line) motor intensity with **(C)** expansion of trace (grid lines at intervals of a single YPet molecule intensity, in this case each consisting of ~1300 counts on the camera detector used in the microscope), and **(D)** Fourier spectral analysis predicting the brightness of a single YPet molecule. Stoichiometry distributions for **(E)** tethered and **(F)** immobilized cells, peaks indicated (arrows). (For full details see Delalez *et al.*, 2010.)

These cells were also used to estimate the number of fluorescent FliM-YPet spots within the 1/e penetration depth of the TIRF evanescent field, indicating ~4 spots per cell. Since the TIRF evanescent field was calculated to encapsulate ~1/6 of the membrane area, this suggested a total number of fluorescent FliM-YPet complexes in the membrane of 20–30 per cell. An *E. coli* cell possesses 4–8 flagella on average, indicating that the number of fluorescent spots per cell is approximately four times greater than the number of functional flagellar motors, and thus may be indicative of *pre-assembly* states for this switch unit (Figure 7.3E, F).

The researchers were also able to monitor molecular turnover of FliM using both fluorescence recovery after photobleaching (FRAP) and fluorescence loss in photobleaching (FLIP) in normal cells, and also in strains which had deleted or elevated levels of the signalling protein CheY, which resulted in either *zero* turnover or *elevated* turnover rates respectively (Figure 7.4). This study suggests a mechanism by which signals originally detected from cell membrane receptors can be transformed into a *molecular accelerator* to modify the rate at which protein turnover in the whole cell motility system occurs.

7.3 Endocytosis and exocytosis

Endocytosis is the process by which eukaryotic cells internalize molecules from the outside by engulfing them. Exocytosis can be viewed in a naïve way as the reverse of this in that it involves the expulsion of molecules from the cell membrane via *secretory vesicles*. However, the molecular modes of action of each are distinct, and both have been studied in live, functional cells using single-molecule techniques. *Cargo sorting* of large molecules can be mediated by a process which is very similar to the mechanism of

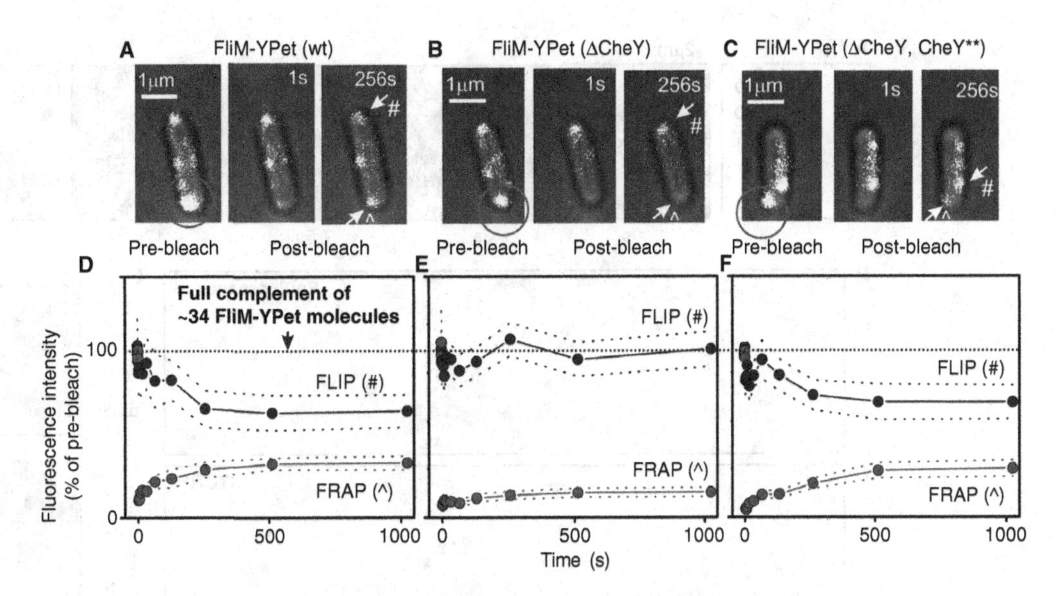

FIGURE 7.4 Turnover in the FliM-YPet complex. Pre- and post-focused laser bleach images for the three FliM-YPet strains: (**A**) wild-type (wt), a strain containing FliM-YPet at roughly native levels of concentration in the cell, (**B**) ΔCheY, in which the gene expressing the response regulator protein CheY has been deleted and (**C**) ΔCheY, CheY** where the gene expressing CheY has been deleted but then replaced with a mutated version known to have greater phosphorylase activity. Approximate extent of the bleach-zone (circle) is indicated and typical examples of complexes analysed for FRAP (^) and FLIP (#) are shown by arrows pointing to relevant fluorescent foci. (**D–F**) Corresponding mean FRAP and FLIP curves calculated using ~20 traces for each, with s.e.m. error bounds (dotted) indicated. (For full details see Delalez *et al.*, 2010.)

endocytosis, which we also encounter briefly here but return to subsequently in Chapter 9 when discussing how cargoes are trafficked specifically in the cell using molecular motors.

7.3.1 Live-cell imaging of clathrin-based endocytosis

Relatively large molecules such as proteins in general have some net electrostatic polarity and so will encounter a very large free energy barrier when crossing the cell membrane from the outside of the cell to the inside (see Chapter 2). Endocytosis allows large molecules to be internalized by a lower free energy path via an *invagination* of the cell membrane and budding-off of an internal vesicle of the molecules then encapsulated by a lipid bilayer. These molecules include signalling molecules such as hormones (see Chapter 2), lipid complexes such as *low-density lipoproteins* (LDLs), proteins including immunoglobulins (the basic structural motif from which antibodies are made) and *transferrin* (a glycoprotein, that is a complex formed from a sugar and a protein, which is used to carry atoms of iron), and even gigantic macromolecular complexes such as viruses (section 7.3). Earlier bulk ensemble-average biochemical evidence indicated that a protein *clathrin* was specifically involved in first coating these vesicles before they budded off.

Endocytosis was first formally studied in living cells at the single-molecule level using rhodamine-tagged transferrin, which was monitored using confocal fluorescence microscopy as it was engulfed into the cell in what was technically the first published single-molecule in vivo study for which the molecules studied were (at some point at

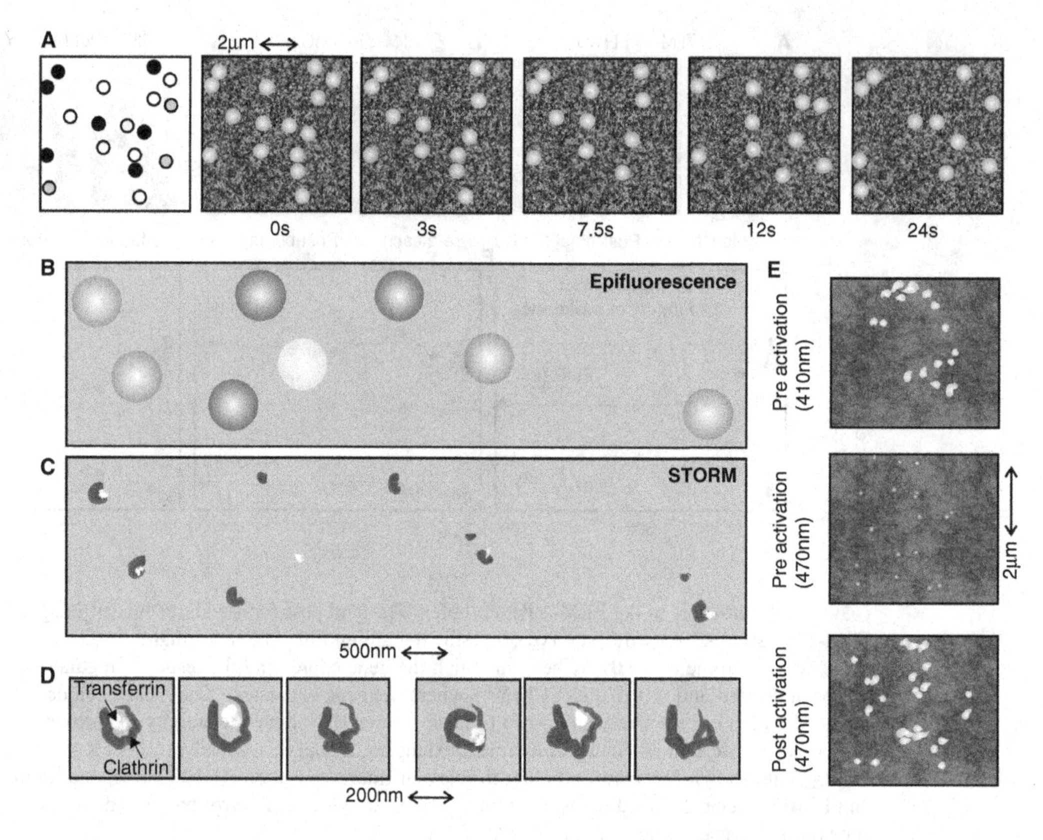

FIGURE 7.5 **(A)** Schematic depiction of sequential images for clathrin-GFP 'pits' forming at the start of endocytosis, indicating dynamic mobility behaviour of some spots which disappear with time (white circles, left panel), or appear with time (grey spot, left panel), or remain stationary and do not disappear with time (black spots, left panel). **(B)** Schematic of a conventional epifluorescence dual-colour image in which clathrin and transferrin have been separately labelled, with **(C)** schematic of associated STORM image and **(D)** schematic examples of expanded parts of the STORM image, transferrin often being engulfed by clathin-coated vesicles. **(E)** Schematic depiction of sequential images of a synapse junction between two live nerve endings (upper and lower regions of images), with pre-activation 410 nm image marking the position of all pHlourin, and the 470 nm images showing the regions of normal intra-cellular pH (positions of fusing vesicles) following nerve cell activation.

least) inside the cell cytoplasm as opposed to being just on the cell membrane surface (Byassee *et al.*, 2000). The first single particle live-cell experiment to investigate the dynamics of this process robustly involved tagging clathrin with GFP and observing foci clusters forming in the cell membrane using standard widefield epifluorescence (Gaidarov *et al.*, 1999).

These clusters were consistent with *clathrin-coated pits* of ~100 nm which initiate vesicle formation, and the researchers were able to perform single particle tracking (see Chapter 3) which indicated a potential confinement at the cell membrane related to the presence of actin components in the cytoskeleton just beneath the membrane (see section 7.2). Later experiments using live-cell confocal imaging (Figure 7.5A) utilizing similar fluorescently tagged clathrin allowed individual clathrin-coated vesicles to be tracked from their point of formation at the cell membrane to their ultimate loss of clathrin inside the cell prior to vesicle fusion with an organelle called the *endosome* (Kirchhausen *et al.*, 2005).

A recent study utilized SNAP-Tag technology in combination with dual-colour STORM to use bright organic dyes to separate labelled clathrin proteins and a transferrin endocytosis substrate (Jones *et al.*, 2011). Here, the investigators were able to label clathrin with the red dye Alexa647, and transferrin with the orange dye Alexa568. This allowed them to visualize both components simultaneously (Figure 7.5B–D) with a spatial precision of ~30 nm laterally and ~50 nm axially, requiring typically just a few seconds or less to acquire each reconstructed image. This clearly showed transferrin engulfed in a vesicular substructure coated with clathrin molecules.

Complex cells such as eukaryotes which are sufficiently large to possess multiple internal membrane-bound compartments, which in general are functionally distinct (see Chapter 2) and require a coordinated movement of molecules between compartments most especially of proteins and lipids, undergo cargo sorting in which potentially large molecules are ferried to specific compartments in the cell. One common way to achieve this is by forming vesicle buds from the lipid bilayer of a donor compartment in the same way as occurs with endocytosis at the cell membrane, with subsequent fusion of this vesicle with the lipid bilayer of an acceptor compartment.

The specific ways in which these vesicles are actually trafficked involves primarily the molecular motors kinesin and dynein as well as some forms of myosin, and are discussed later in Chapter 9. These and similar subsequent live-cell single-molecule fluorescence tracking studies have led to a vast increase in our understanding of the process of endocytosis and cargo sorting, indicating a complex requirement of several different accessory proteins for the clathrin-coat assembly-disassembly process.

7.3.2 Single-molecule fluorescence imaging of exocytosis in functional cells

Exocytosis involves the expulsion of secretory vesicles from the cell membranes of eukaryotic cells which, depending on the type of exocytosis involved, may or may not require an external stimulus. One of the commonest studied forms of exocytosis, partly because of its prime importance in biomedicine, is the chemical transmission of nerve impulses which occur in the spacing between adjacent nerve cell endings called the *synapse*. This is triggered by intra-cellular *calcium ions*, and involves the exocytosis of a variety of different neurotransmitter chemicals in phospholipid-bound vesicles, which then diffuse across the synapse gap from their point of departure in an electrically depolarized neuron cell at the so-called *pre-synaptic junction*, ultimately fusing with the cell membrane of the adjacent neuron of the *post-synaptic junction*, which subsequently results in a cascade of chemical reactions resulting in the depolarization of this second cell, hence transmitting the electrical nerve impulse.

Using live nerve endings, researchers have been able to perform single-molecule level fluorescence imaging on exocytosed vesicles secreted into the synapse junction between such adjacent nerve endings in cultured cells from mammalian brain tissue (Miesenböck *et al.*, 1998; later with significant improvements in detection sensitivity by Balaji and Ryan, 2007). Here, a derivative of GFP called *pHlourin* (see Chapter 3) was used as a single-molecule fluorescence tag, which had been modified to give it an increased brightness sensitivity to pH at long excitation wavelengths of ~470 nm while still being relatively insensitive to changes in pH when excited at shorter wavelengths of ~410 nm, making it very useful as a local *ratiometric* live-cell pH indicator, since the ratio of the fluorescence obtained when excited at 470 nm compared to 410 nm is a direct measure of the difference in local pH in the vicinity of the fluorescent protein (the intensity at 410 nm wavelength excitation in effect normalizes the signal for the total amount of pHlourin present).

The exocytosed synaptic lipid bilayer vesicles, which are budded off the cell membrane of one of the nerve endings, could be labelled by fusing this pHlourin at the level of the cell's DNA, using the same genetic encoding methodology as used for GFP, to one of the membrane proteins which are expressed in the phospholipid bilayer of these synaptic vesicles, inside of which is a neurotransmitter chemical which conveys the nerve impulse signal between the two nerve cells.

The pH within released vesicles was found to be *lower* than the cell itself (5–6 compared to ~7.4). The effect of this was to suppress dramatically the fluorescence of pHlourin molecules inside the vesicles when excited at a wavelength of ~470 nm. However, if vesicles fused with the membrane of either nerve ending then the local pH in the vicinity of the pHlourin molecules would *increase* dramatically, thereby making them fluorescent once more. This resulted in a dramatic decrease in local effective concentration of photoactive fluorophores in the vicinity of the synapse such that the average separation of labelled secretory vesicles at the point of fusing with either nerve end was then greater than the optical resolution limit, thus allowing single fusing vesicles to be observed distinctly (see Chapter 3). This then permitted real time single particle tracking on the fused vesicles (Figure 7.5E). In doing so it became evident that a significant subpopulation of exocytosed vesicles was being subsequently endocytosed by the same cell, in effect being *recycled*.

Similarly, standard GFP fusions, in combination with fluorescence imaging and single particle tracking, have been used to monitor other manifestations of cellular exocytosis at the single-molecule level. One very important case involves the expulsion of IgG protein complexes from the cell. IgG forms the basis of a variety of different classes of antibodies (see Chapter 2), and their exocytosis is an essential part of the eukaryotic cellular immune response. By labelling the Fc *non-variable* region of IgG using GFP, researchers were able to track the fate of individual IgG exocytosed vesicles (Ober *et al.*, 2004). Their observations were similar to those on the synapses of nerve cells – a significant proportion of vesicles often fuse back to their initial cellular point of release and are recycled within the cell, potentially resulting in a more controllable net rate of release if the cell can generate interventions to change the rate of re-uptake.

In another key investigation, relevant to the understanding of the disease *diabetes*, the exocytosis of single granules of the hormone *insulin* was studied using TIRF imaging of GFP-tagged insulin in live cells of the pancreas organ called β-cells, which normally function to produce the insulin protein. This added significant insight into our understanding of the process of insulin release from cells and their subsequent docking and fusion with the cell membrane of either a different or the same cell (Ohara-Imaizumi and Nagamatsu, 2006).

7.3.3 SNARE exocytosis proteins

Docking and fusion of exocytosed synaptic vesicles, as seen in the previous section, require the interaction of specialized complementary *SNARE* proteins both in the vesicle membrane and in the cell membrane of the destination nerve ending. Although technically still too challenging to probe in a live cell, these SNARE proteins can at least be reconstituted under physiologically relevant conditions in vitro, allowing the binding and unbinding interaction forces and kinetics of association and dissociation to be measured very precisely on a single-molecule basis using AFM spectroscopy (Yersin *et al.*, 2003).

This allows a powerful mechanistic understanding of the sequence of multiple SNARE interactions to be built up for the entire docking/fusion process.

In a separate study, GFP fusions to two of these complementary SNARE proteins were monitored using single-molecule fluorescence imaging in live cells (Knowles *et al.*, 2010). Interestingly, the researchers concluded that these proteins showed clear signs of being confined to nano-domains of ~100 nm length scale in the membrane, and by measuring the precise fluorescence intensity from the domains they estimated that 50–70 protein molecules of each pair were likely to be present. This potentially results in a very firm and rapid docking response, as there is likely to be significant cooperativity in binding since when a single SNARE pair binds the resulting confined motion of the vesicle will clearly increase the probability for other SNARE pairs in the local vicinity to bind.

7.4 Viral invasion

Viruses are remarkable *macromolecular machines*, sitting right on the fence which divides living and non-living matter (see Chapter 2), and are the causal agents of over 100 common diseases in humans. In 1892, by which time *germ theory* had already been established and the scientific community had begun to accept the notion that micro-organisms invisible to the naked eye were the cause of many common diseases, the first clear cut evidence was presented for the existence of an infecting agent smaller than individual cells. The Russian botanist Dmitri Iwanowski generated a sap extract from *tobacco* plants which were suffering from a common disease, causing a characteristic *mosaic* pattern on the leaves which ultimately resulted in their shrivelling and dying, in an attempt to isolate what he thought was the bacterial cause of the infection.

He passed this sap through a porcelain filter, which was small enough to trap all known bacterial cells, and inoculated fresh tobacco leaves with this filtered solution which he assumed should not result in any subsequent infection, thereby demonstrating that the pathogenic micro-organism had indeed been trapped by the filter. However, to his surprise he found that the plants all developed characteristic mosaic disease patterns in originally uninfected leaves. Iwanowski believed that a bacterium of exceptionally small size had managed to pass through the filter. It was another four decades before the source of this infection was correctly identified as being the now well characterized *tobacco mosaic virus* (for a comprehensive description of the history and developments of modern virology see Dimmock and Primrose, 2006).

Since then, enormous insights have been made into the properties of several different viruses infecting plants, animals and bacterial cells, especially into their structure, using EM, x-ray crystallography and NMR techniques. Viruses are remarkable nanoscale objects. They are essentially protein containers housing the genes necessary to hijack a cell's replication machinery to allow more viruses to be made, and they possess remarkable *thermodynamic* and *mechanical* properties (for a good review of the physics of viruses see Roos *et al.*, 2010).

They form highly stable protein *capsid* coats by spontaneous self-assembly inside cells, with AFM spectroscopy experiments on single viral capsids indicating enormously strong structures capable of withstanding internal osmotic pressures equivalent to several tens of atmospheres (physical scientists interested in the modelling of the way that these remarkable protein structures can withstand such stress should read Zandi and Reguera, 2005). Viral capsids need to withstand such high stresses because the high density of packing of nucleic acid inside the capsid results in exceptionally elevated electrostatic repulsion and bending energies. These can be estimated directly using AFM spectroscopy on single viral capsids under a variety of osmotic stress conditions to vary the stress exerted on the capsid by the nucleic acid (Ivanovska *et al.*, 2007).

Although the structure of viruses varies with type, with some possessing very specialized features allowing them to inject their genetic material into cells, typically they are classed as either *non-enveloped* or *enveloped*, with both characterized by having the protein capsid coat composed of multiple tessellating protein subunits in an array of different beautifully geodesic patterns including tightly packed near-spherical *icosahedral* structures a few tens of nanometres in diameter, and more extended *helical* type arrangements for example found in the tobacco mosaic virus. Inside these protein structures is a tightly packed nucleic acid, either of RNA or DNA depending on the specific type of virus. In enveloped viruses there is a second coat outside the first capsid consisting of a mix of sugars bound to proteins (i.e. glycoproteins) as well as phospholipids (see Chapter 2) which in some ways can be viewed as a primordial type of membrane encapsulating the viral particle.

The infection of a virus typically starts with the virus binding to specific molecular receptors on the outer surface of the cell membrane. Some non-enveloped viruses will then release their genetic contents by directly injecting nucleic acid across the cell membrane, most importantly viruses which infect prokaryotes such as bacteria. However, enveloped viruses will typically be internalized following surface attachment and entry essentially via endocytosis (section 7.3). The internalized genetic contents can be either RNA or DNA (see Chapter 2) depending on the virus type, and these can fuse with and in effect hijack the native cell's DNA, forcing the cell to manufacture proteins that are used to generate more virus particles in the cell. These viral particles can either exit the cell via exocytosis or build up in number within the cell; once a critical concentration is reached the cell literally splits open to spew out free viruses, and the cycle of infection can then start again with a different cell nearby.

Although viruses themselves are not single molecules as such, but rather are large macromolecular complexes composed primarily of multiple proteins and nucleic acids, a very powerful method for elucidating the various mechanisms involved in each stage of the viral infection process is the use of single-molecule techniques in the form of single particle tracking of fluorescently labelled virus particles in live cells. These began as relatively crude single colour two-dimensional tracking studies but now involve complex three-dimensional tracking using multi-colour fluorescence imaging (for a concise review see Brandenburg and Zhuang, 2007).

BIO-EXTRA

Arguably the most useful form of classification for viruses involves the **Baltimore classification system** in which each virus is formally classed into one of seven different groups ultimately centred on how the virus generates mRNA in the host cell. This depends on whether the virus contains RNA or DNA, whether this nucleic acid is single or double stranded, whether or not the virus utilizes the enzyme reverse transcriptase, and also for single-stranded RNA viruses whether they replicate in a sense or antisense fashion. There are currently around 3000 established viral species, with around the same number which remain as yet unclassified.

7.4.1 Tracking single viruses using single colour fluorophores

The first successful study that tracked single functional viruses investigated *reoviruses*, a family of viruses with a double-layer protein coat encapsulating viral RNA, which infect cells in the respiratory and gastrointestinal tracts. These were inoculated with host cells and were observed using EM on dead fixed samples and widefield epifluorescence on live cells. At that time, epifluorescence required at least ~500 molecules per virus of the organic dyes fluorescein or rhodamine (see Chapter 3) to produce enough fluorescence to observe single viruses (Georgi *et al.*, 1990). Although the researchers did not perform any robust tracking analysis, the technique permitted individual viral particles to be visualized accumulating on the cell membrane soon after inoculation of the host cells, and diffusing in the cytoplasm in a manner consistent with *trafficked* motion on microtubules similar to endocytosis.

A big development came with subsequent improvements in camera detection sensitivity which allowed single colour dye molecular level imaging for single tagged virus particles, and permitted a significant improvement in the time resolution of imaging (Seisenberger *et al.*, 2001), indicating that the process of viral infection of cells was much faster than previously assumed, but also offering insight into the hindered mobility of these viral particles inside cells.

The mobility of tracked viruses in living cells in one sense relates to the cellular mechanisms of organized trafficking of sub-cellular cargo (see Chapter 9) while in another sense it is indicative of the immediate viscoelastic and potentially crowded molecular environment in which the virus finds itself. As we will see in Chapter 9, cargo trafficking is mediated via the movements of molecular motors on tracks provided by the cytoskeleton, an *active* process, namely one which requires an *energy input*, generally provided by the hydrolysis of molecules of ATP (see Chapter 2). Such mobility results in *directed* motion of the virus particles when they are being actively trafficked. This motion along a track as a function of time t can be modelled by introducing an additional speed term v into the mean square displacement $\langle R^2 \rangle$ relation referred to previously in Equation 3.4 (typical speeds for viruses exhibiting such directed motion have been measured from single particle tracking to be in the range 200–1400 nm s^{-1}), under the assumption that particles exhibit one-dimensional Brownian type diffusion along a molecular track with a direction-dependent speed bias:

$$\langle R(t)^2 \rangle = 2Dt + v^2 t^2 \tag{7.1}$$

Here, D is the effective one-dimensional diffusion coefficient of the virus along the molecular track. Great care needs to be taken to minimize any sample drift when monitoring potential directed diffusion, since if uncorrected this could be manifest as an artifactual speed bias.

Some viruses when first entering the cell appear to exhibit *confined* diffusion, not along any specific track. Here, the viral particles appear to diffuse with Brownian motion over small length scales but are prevented from diffusing over longer length scales by some putative corralling effect. The source of this confinement is unresolved but may possibly be due to the mesh-like network of actin microfilaments that often occurs just beneath the cell membrane of eukaryotic cells. The mean square displacement in one dimension of such particles can be modelled using:

$$\langle R(t)^2 \rangle = \frac{L^2}{6} - \frac{16L^2}{\pi^4} \sum_{n=1(\text{odd})}^{\infty} \frac{1}{n^4} \exp\left\{ \frac{-t}{2} \left(\frac{n\pi\sqrt{2D}}{L} \right) \right\} \tag{7.2}$$

Here, L is a length parameter, such that at large time scales the mean square displacement converges to a value of $L^2/6$; in other words, the particle's mobility is confined to a specific domain in space. Similarly, viral particles may also exhibit mobility consistent with Brownian (i.e. *normal*) diffusion and *anomalous* (or *sub*-) diffusion, for which $\langle R^2 \rangle = 2Dt^{\alpha}$ for one-dimensional motion, where $\alpha = 1$ for Brownian diffusion and $0 < \alpha < 1$ for anomalous diffusion. Typically, virus particles in live cells exhibiting anomalous diffusion will have an α value of 0.7–0.8, which is found for many other unrelated diffusing particles in live cells where sub-diffusion is prevalent and is thus likely to be a feature of the underlying molecular environment surrounding the particle which hinders its motion, rather than being an intrinsic feature of the particle itself. For a comprehensive treatment of modelling different diffusive modes during single particle tracking see Qian *et al.* (1991); a more analytical treatment of anomalous diffusion for interested physical scientists can be found in Bouchard and Georges (1990).

Identifying the underlying diffusive *mode* of a particle using a combination of single particle tracking and advanced mobility analysis is enormously powerful, not just for viral research. This can reveal mechanistic details about not only motor proteins involved in cargo trafficking but also the local molecular architecture of the medium in which the particle is diffusing. The prime difficulty is that diffusion is ultimately a *stochastic process*, and so random movements of a given individual particle undergoing normal Brownian diffusion may suggest a different non-Brownian diffusive mode, for example giving the appearance of being confined, directed or anomalous. There is a distinct danger of subjectively selecting such particles as evidence of non-Brownian behaviour whereas in fact they are merely examples of outliers of normal Brownian mobility. However, with robust, automated *probabilistic criteria* for identifying different diffusive modes, such bias can at least be minimized.

7.4.2 Multi-colour fluorescence tracking of viruses

Use of dual-colour fluorescence labelling has substantially increased our understanding of the mechanisms of viral infection in cells. A seminal dual-label fluorescence study managed to perform proper fluorescence-based single particle analysis on tracked *adenoviruses*, DNA-containing viruses which account for several upper respiratory tract infections (Suomalainen *et al.*, 1999). The protein coat had been genomically labelled using GFP (Figure 7.6A, B), while the microtubule cytoskeletal network of the host cell

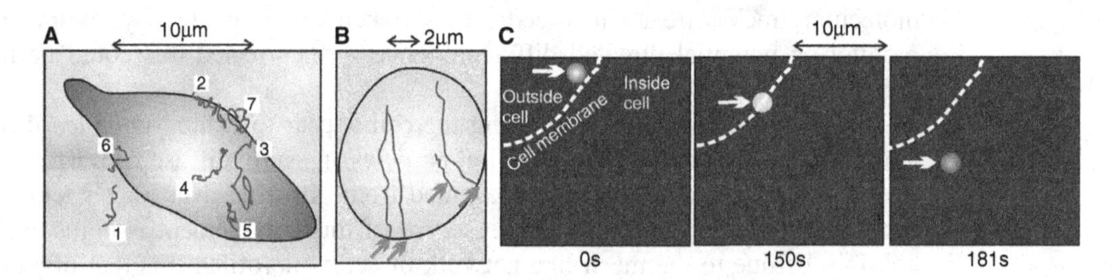

FIGURE 7.6 **(A)** Schematic depiction of single particle tracks of individual GFP-tagged adenovirus particles overlaid on a brightfield image of a single cell (7 numbered tracks shown in total), and **(B)** schematic of fluorescence image of tracks of four different adenovirus particles (arrows) inside the cell nucleus. **(C)** Schematic depiction of live-cell fluorescence dual-colour images from green and red channels merged showing cellular entry of a single influenza virus (tagged with a red dye, arrow), with green-labelled clathrin particles which co-localize with the virus as it is endocytosed (150 s time point) and are subsequently shed from the virus as it enters the inside of the cell (181 s time point), with seconds after the virus first makes contact with the cell indicated, and approximate position of the cell membrane shown by a dashed line.

had been labelled separately by tagging the tubulin protein monomers with the organic dye Texas Red, thereby establishing definitive evidence that viruses were trafficked inside the cell via microtubule tracks in much the same way as endocytosed particles, as well as the trafficking used in internal cellular cargo sorting. Similar techniques were later used to observe single tracked human immunodeficiency viruses (McDonald *et al.*, 2002), the infamous HIV.

Investigations have since become more elaborate, labelling multiple different components of more complex viruses with different colour fluorophores. For example, the *influenza* virus, a virus of a type possessing an outer complex protective coat composed of glycoproteins and phospholipids, was first specifically labelled using a lipid conjugating chemical called DiD, which in turn could be tagged with an organic fluorescent red dye CypHer5, while the protein capsid was labelled using the fluorescent green organic dye Cy3. The labelled viruses were then imaged using dual-colour widefield epifluorescence (see Chapter 3) to observe their progress infecting live cells (Lakadamyali *et al.*, 2003).

Further significant improvements were also made in labelling the clathrin protein component of cells, which is essential to primary entry mechanisms for endocytosis (section 7.2) while simultaneously observing virus particles tagged with a different colour fluorophore, again using the influenza virus as a model system (Rust *et al.*, 2004). This study clearly demonstrated that some viruses not only utilize the cell's own microtubule trafficking system but also definitively take over the initial clathrin-mediated entry system of normal endocytosis in order to gain entry to the cell in the first place (Figure 7.6C).

Similar dual-colour imaging was also later performed using the *polyoma* virus (viruses which cause a variety of diseases resulting from virus-mediated tumour formation) using both controlled artificial lipid bilayers and live-cell imaging (Ewers *et al.*, 2005) to characterize further the heterogeneous mobility behaviour of single viral particles. This was observed to consist of a complex mixture of Brownian and confined diffusive behaviour related again to the presence of cytoskeletal components. Similar dual-colour live-cell single-molecule imaging techniques have been applied to investigate the molecular mechanisms for the early stages of cellular infection of the pernicious *poliovirus*, another example of an RNA-containing virus which contains no secondary outer envelope aside from the relatively simple protein capsid (Brandenburg *et al.*, 2007).

These multi-colour fluorescence single-molecule imaging methods applied to investigate a variety of different virus types have dramatically increased our understanding of the number of distinct strategies which are exploited for the initial *entry* of the virus into the cell and the methods for *delivering* its genetic contents. For enveloped viruses, as we have seen, there can be some level of protein-mediated fusion of viral and cellular membranes which involves hijacking the clathrin-coated vesicle system of normal endocytosis. For non-enveloped viruses the mechanism appears to be slightly different and relies first upon proteins in the virus capsid to disrupt cell membranes mechanically, potentially to form nanopores through which viral nucleic acid can pass.

Viruses will often make contact with the surface of cells several times before successfully penetrating the cell membrane (somewhere in the range 1–10 times being not atypical), with some evidence indicating that successful penetration is related to the local *curvature* of the cell membrane. The typical number of such attempts can be approximated by relatively simple modelling using a *diffusion-to-capture* approach (an excellent treatment is given in Berg, 1993) by treating the cell membrane as a spherical shell of effective radius r_{cell}. It is assumed that viral particles are released a distance d

away from the centre of this effective shell (which can be crudely approximated by the distance from a neighbouring cell which has just undergone *lysis* to release fresh viruses), and move towards the shell at rate k_1 and away from the shell at rate k_2. It can be shown (see Q7.8) that the probability p that a virus will touch the cell membrane is:

$$p = k_1/(k_1 + k_2) \approx r_{cell}/d \tag{7.3}$$

It can also be shown that the average number of touches a given virus undergoes oscillating between the cell membrane and its point of release before diffusing away from the membrane and never returning is $\langle n \rangle \approx r_{cell}/(d - r_{cell})$.

KEY POINT

Although a virus in itself is not a single molecule, the techniques developed to study viral infection in live cells have moved forward the field of single-molecule cellular biophysics considerably.

THE GIST

- Receptor–ligand complexes in cell membranes are prime candidates for single-molecule fluorescence imaging using TIRF and TIRF-FRET because they are accessible to evanescent field illumination.
- Fluorescence intensity can be used to estimate the oligomeric state of membrane complexes
- TIRF imaging in combination with FRAP/FLIP can be used to probe the architecture and dynamics of bacterial chemotaxis at the single-molecule level.
- Endocytosis can be monitored in live cells by fluorescent tagging of the substrate being internalized or of the protein coat which makes up the endocytosed lipid vesicle.
- Exocytosis can be monitored using fluorescence imaging of tagged substrates, especially using pH sensitive fluorophores.
- AFM spectroscopy can be applied in the test tube to understand the docking and fusion processes of exocytosed vesicles in live cells.
- The mechanism of viral infection into living cells can be extensively studied using single and multi-colour single-molecule fluorescence imaging combined with single particle tracking.

TAKE-HOME MESSAGE Single-molecule methods, primarily utilizing either TIRF or confocal microscopy, in combination with single particle detection and tracking can extract substantial quantitative details from biological processes which involve external molecules interacting with the cell membrane.

References

GENERAL

- Berg, H.C. (1993). *Random Walks in Biology*, 10th edition. Princeton University Press, Princeton, NJ.
- Bouchaud, J.P. and Georges, A. (1990). Anomalous diffusion in disordered media: statistical mechanisms, models and physical applications. *Phys. Rep.* **195**: 127–293.
- Brandenburg, B. and Zhuang, X. (2007). Virus trafficking – learning from single-virus tracking. *Nature Rev.* **5**: 197–208.
- Dimmock, N.J. and Primrose, S.B. (2006). *Introduction to Modern Virology*, 6th edition. John Wiley & Sons.

- Qian, H., Sheetz, M. P. and Elson, E. L. (1991). Single particle tracking. Analysis of diffusion and flow in two-dimensional systems. *Biophys. J.* **60**: 910–921.
- Roos, W. H., Bruinsma, R. and Wuite, G. J. L. (2010). Physical virology. *Nature Phys.* **6**: 733–743.
- Schlessinger, J. (2000). Cell signaling by receptor tyrosine kinases. *Cell* **103**: 211–225.
- Wadhams, G. H. and Armitage, J. P. (2004). Making sense of it all: bacterial chemotaxis. *Nature Rev. Mol. Cell Biol.* **5**: 1024–1037.

ADVANCED

- Balaji, J. and Ryan, T. A. (2007). Single-vesicle imaging reveals that synaptic vesicle exocytosis and endocytosis are coupled by a single stochastic mode. *Proc. Natl. Acad. Sci. USA* **104**: 20576–20581.
- Brandenburg, B. *et al.* (2007). Imaging poliovirus entry in live cells. *PLoS Biol.* **5**: e183.
- Byassee, T. A., Chan, W. C. and Nie, S. (2000). Probing single molecules in single living cells. *Anal. Chem.* **72**: 5606–5611.
- Deich, J., Judd, E. M., McAdams, H. H. and Moerner, W. E. (2004). Visualization of the movement of single histidine kinase molecules in live Caulobacter cells. *Proc. Natl. Acad. Sci. USA* **101**: 15921–15926.
- Delalez, N. J. *et al.* (2010). Signal-dependent turnover of the bacterial flagellar switch protein FliM. *Proc. Natl. Acad. Sci. USA* **107**: 11347–11351.
- Ewers, H. *et al.* (2005). Single-particle tracking of murine polyoma virus-like particles on live cells and artificial membranes. *Proc. Natl. Acad. Sci. USA* **102**: 15110–15115.
- Gaidarov, I., Santini, F., Warren. R, A, and Keen. J, H. (1999). Spatial control of coated-pit dynamics in living cells. *Nature Cell Biol.* **1**: 1–7.
- Georgi, A., Mottola-Hartshorn, C., Warner, A., Fields, B. and Chen, L. B. (1990). Detection of individual fluorescently labeled reovirions in living cells *Proc. Natl. Acad. Sci. USA* **87**: 6579–6583.
- Hern, J. A. *et al.* (2010). Formation and dissociation of M-1 muscarinic receptor dimers seen by total internal reflection fluorescence imaging of single molecules. *Proc. Natl. Acad. Sci. USA* **107**: 2693–2698.
- Iino, R., Koyama, I. and Kusumi, A. (2001). Single molecule imaging of green fluorescent proteins in living cells: E-cadherin forms oligomers on the free cell surface. *Biophys. J.* **80**: 2667–2677.
- Ivanovska, I., Wuite, G., Jönsson, B. and Evilevitch, A. (2007). Internal DNA pressure modifies stability of WT phage. *Proc. Natl. Acad. Sci. USA* **104**: 9603–9608.
- Jones, S. A., Shim, S. H., He, J. and Zhuang, X. (2011). Fast, three-dimensional super-resolution imaging of live cells. *Nature Methods* **8**: 499–508.
- Kirchhausen, T., Boll, W., van Oijen, A. and Ehrlich, M. (2005). Single-molecule live-cell imaging of clathrin-based endocytosis. *Biochem. Soc. Symp.* **72**: 71–76.
- Knowles, M. K. *et al.* (2010). Single secretory granules of live cells recruit syntaxin-1 and synaptosomal associated protein 25 (SNAP-25) in large copy numbers. *Proc. Natl. Acad. Sci. USA* **107**: 20810–20815.
- Lakadamyali, M., Rust, M. J., Babcock, H. P. and Zhuang, X. (2003). Visualizing infection of individual influenza viruses. *Proc. Natl. Acad. Sci. USA* **100**: 9280–9285.
- McDonald, D. *et al.* (2002). Visualization of the intracellular behavior of HIV in living cells. *J. Cell Biol.* **159**: 441–452.
- Miesenböck, G., De Angelis, D. A. and Rothman, J. E. (1998). Visualizing secretion and synaptic transmission with pH-sensitive green fluorescent proteins. *Nature* **394**: 192–195.
- Murakoshi, H. *et al.* (2004). Single-molecule imaging analysis of Ras activation in living cells. *Proc. Natl. Acad. Sci. USA* **101**: 7317–7322.
- Ober, R. J. *et al.* (2004). Exocytosis of IgG as mediated by the receptor, FcRn: an analysis at the single-molecule level. *Proc. Natl. Acad. Sci. USA* **101**: 11076–11081.
- Ohara-Imaizumi, M. and Nagamatsu, S. (2006). Insulin exocytotic mechanism by imaging technique. *J. Biochem.* **140**: 1–5.
- Rust, M. J., Lakadamyali, M., Zhang, F. and Zhuang, X. (2004). Assembly of endocytic machinery around individual influenza viruses during viral entry. *Nature Struct. Mol. Biol.* **11**: 567–573.
- Sako, Y., Minoghchi, S. and Yanagida, T. (2000). Single-molecule imaging of EGFR signalling on the surface of living cells. *Nature Cell Biol.* **2**: 168–172.
- Seisenberger, G. *et al.* (2001). Real-time single-molecule imaging of the infection pathway of an adeno-associated virus. *Science* **294**: 1929–1932.

- Suomalainen, M. *et al.* (1999). Microtubule-dependent plus- and minus end-directed motilities are competing processes for nuclear targeting of adenovirus. *J. Cell Biol.* **144**: 657–672.
- Thomas, D. R., Francis, N. R., Xu, C. and DeRosier, D. J. (2006). The three-dimensional structure of the flagellar rotor from a clockwise-locked mutant of *Salmonella enterica* serovar Typhimurium. *J. Bacteriol.* **188**: 7039–7048.
- Tindall, M. J. *et al.* (2008). Overview of mathematical approaches used to model bacterial chemotaxis I: the single cell. *Bull. Math. Biol.* **70**: 1525–1569.
- Ueda, M. *et al.* (2001). Single-molecule analysis of chemotactic signaling in *Dictyostelium* cells. *Science* **294**: 864–867.
- Watanabe, N. and Mitchison, T. J. (2002). Single-molecule speckle analysis of actin filament turnover in lamellipodia. *Science* **295**: 1083–1086.
- Yersin, A. *et al.* (2003). Interactions between synaptic vesicle fusion proteins explored by atomic force microscopy. *Proc. Natl. Acad. Sci. USA* **100**: 8736–8741.
- Zandi, R. and Reguera, D. (2005). Mechanical properties of viral capsids. *Phys. Rev. E* **72**: 021917.

Questions

FOR THE LIFE SCIENTISTS

Q7.1. The cytoskeleton of eukaryotic cells is composed primarily of microtubules, intermediate filaments and microfilaments. Would it be more efficient to have a system based on just a single protein type to perform all the roles of the cytoskeleton?

Q7.2. 10 ml of a bacteriophage virus culture was prepared from the cellular extract of bacteria at the peak of viral infection and the solution was divided into 10 volumes of 1 ml. Nine of these were incubated with nine separate fresh uninfected bacterial cultures; all of these subsequently developed viral infections. The 10th volume was then added to fresh culture medium to make up to the same 10 ml volume as the original virus culture, and mixed well. This diluted viral solution was then divided up into 10 equal volumes as before, and the previous procedure repeated up to a total of 12 such dilutions. In the first nine dilutions all nine fresh bacterial cultures subsequently developed viral infections. In the 10th dilution only five of the nine fresh bacterial cultures developed viral infections, in the 11th dilution only one of the nine bacterial cultures developed viral infections, while in the 12th dilution none of the nine bacterial cultures developed viral infection. (a) Estimate the molarity of the original virus culture. (b) If this culture consisted of virus particles tightly packed such that the outer coat of each virus was in physical contact with that of its nearest neighbours, estimate the diameter of the virus. (Hint: a virus culture is likely to cause a subsequent infection of a bacterial culture if there is at least one virus in the solution.)

Q7.3. A fluorescently labelled virus was tracked inside a living cell for several seconds and then an uncoupling reagent was added to the cell and the virus was tracked subsequently for a few more seconds. The mean square displacement appeared to be roughly parabolic in shape with respect to time both before addition of the uncoupler and afterwards. Upon investigation it was found that the sample holder in the stage was a little loose. After tightening and performing the experiment again on a different cell the mean square displacement of the tracked virus appeared to be parabolic in shape with respect to time before addition of the uncoupler as before but afterwards it now appeared to be roughly linear. Explain these observations.

Q7.4. If bacterial chemotaxis control can be achieved in the absence of gene regulation, why do other cell types often employ gene regulation strategies as part of signal transduction control?

FOR THE PHYSICAL SCIENTISTS

Q7.5. Crystallographic data suggest that a certain receptor molecule expressed in the cell membrane may occur as both a monomer and a dimer. A single-molecule GFP fluorescent tag was fused to the receptor to perform live-cell fluorescence imaging. TIRF microscopy was used on live cells indicating typically 10 fluorescent foci per cell in each illuminated patch of cell membrane. The TIRF microscopy was found to illuminate ca. 1/6 of the total extent of the cell membrane per cell, with the total membrane area being ca. 20 μm^2 per cell. On the basis of intensity measurements of the foci (see for example Sako *et al.*, 2000 and Iino *et al.*, 2001) this indicated that ca. 80% of the foci were monomers, ca. 10% were dimers and the remaining 10% were consistent with being oligomers. Are the oligomers likely to be real or simply monomers or dimers which are very close to each other? (Hint: If two fluorescent molecules are separated by less than the optical resolution limit we may interpret them as a single spot of light.)

Q7.6. Bacteria contain a tough cell wall composed of proteins and sugars (called 'peptidoglycan') in addition to a phospholipid bilayer, which allows the cell to withstand build up of osmotic pressures up to ca. 15 bar but is semi-permeable so allows a variety of small molecules, including water, to pass through. (a) If the virus particle is ca. 50 nm in diameter, is it likely that the reason why each bacterial cell ultimately splits open is due to the pressure build up from the large number of virus particles? If each virus consists of a maximum of ca. 1500 protein monomer subunits in its capsid coat which split apart from each other spontaneously, would this make any difference to your answer? (Hint: Treat the viruses as an ideal gas whose concentration is limited by tight-packing.)

Q7.7. Using the same methods of Q7.3, a different virus was fluorescently labelled and monitored in a live cell at consecutive sampling times of interval 40 ms up to 1 s. The root mean square displacement was calculated for two such viral particles, giving values of [61, 75, 81, 95, 107, 112, 128, 131, 158, 167, 181, 176, 177, 182, 183, 178, 177, 180, 182, 184, 181, 179, 180, 178, 180, 182] nm and [59, 65, 66, 60, 64, 63, 58, 62, 63, 61, 64, 60, 59, 64, 62, 65, 61, 60, 63, 66, 62, 58, 60, 57, 61, 62] nm. What might this imply about the diffusion of each particle?

Q7.8. (a) Show that the probability that a virus will touch a cell membrane can be approximated by r_{cell}/d and that the mean number of touches a virus undergoes oscillating between the membrane and its release point before diffusing away completely is $\sim r_{cell}/(d - r_{cell})$. (b) If a cell extracted from a tissue sample infected with viruses resembles an oblate ellipsoid of major axis ~10 μm and minor axis ~5 μm with a measured mean number of membrane touches is 4.1±3.0 (±standard deviation) of a given virus measured using single particle tracking from single-molecule fluorescence imaging, estimate how many cells there are in a tissue sample of volume ~1 ml. (c) If the mean effective Brownian diffusion coefficient of a virus is 7.5 $\mu m^2 s^{-1}$, and the incubation period (the time between initial viral infection and subsequent release of fresh viruses from an infected cell) is 1–3 days, estimate the time taken to infect all of the cells in the tissue sample, stating any assumptions that you make. (Hint: Treat the distribution of number of virus touches on the cell membrane as binomial.)

FOR THOSE WHO HAVE NOT MADE UP THEIR MIND

Q7.9. In the original single-molecule live-cell receptor–ligand studies the ligand molecule was labelled with a fluorescent tag but not the receptor. Why?

Q7.10. Is it acceptable to include virus tracking as an example of a single-molecule investigation?

Q7.11. Endocytosis, exocytosis and cargo sorting appear to be found primarily in eukaryotic cells and not in prokaryotes. What are the advantages of such processes, and why do most prokaryotes not utilize them?

8 Into the membrane

Be like a duck. Calm on the surface, but always paddling like the dickens underneath.

(ATTRIBUTED TO MICHAEL CAINE, BRITISH ACTOR, BORN 1933)

GENERAL IDEA

In this chapter we encounter some key examples of single molecules and molecular complexes which are primarily integrated into the cell membrane performing a range of essential biological functions, and of the surrounding molecular architecture of the phospholipids, that have been studied extensively using exemplary single-molecule biophysics techniques

8.1 Introduction

Roughly 30% of all proteins are integrated into the membranes of cells, a significant proportion which is indicative of the importance of molecular processes which occur at the surfaces of cells. The cell membrane is an *enormously* important structure. It provides a *physical support* for seeding a vast array of complex surface chemistry reactions, as well as acting as the site for *molecular detectors*, *pumps*, *channels* and *motors*, not to mention its obvious function as a physical *boundary* to the cell. In the previous chapter, we discussed some of the important biological systems that deal with molecules and molecular complexes which spend significant periods outside the cell membrane boundary, and how single-molecule methods have dramatically improved our understanding of these processes. In this chapter, we will discuss several of the biological systems which are primarily associated directly with the cell membrane itself, and how single-molecule techniques have probed many of these in physiologically relevant settings, including both *protein complexes* integrated into the membrane and the makeup of the *phosopholipid bilayer*. These include molecular complexes which *transport* molecules across cell membranes, such as *ion channels* and protein transport *nanopores*, as well as some remarkable molecular machines which are involved in *cell motility* and *cellular fuel manufacture*.

8.2 Molecular transport via pores, pumps and carriers in membranes

Cells have the ability to *translocate* not only relatively large molecules and particles across the cell membrane, or between sub-cellular membrane-bound compartments, as mediated through endocytosis (see Chapter 7) or related intra-cellular cargo sorting processes (see Chapter 9), but also small molecules such as ions, small sugars, amino acids, peptides, nucleosides and even single protein molecules. In doing so, the cell encounters three primary problems.

(i) The *relative concentrations* inside and outside the cell. Net diffusion of molecules is from a region of high to low concentration, whereas if the molecules in question are to be transported in the opposite direction, as is often the case since a cell embodies a *confined volume* but often needs to capture molecules from a much larger effective external volume from the outside world, then this counter-diffusion represents a local *entropic decrease* which therefore needs to be energized in some way if it is to occur, i.e. it must involve *active transport* in some form to *pump* the molecules.

(ii) The phospholipid bilayer of the membrane constitutes an *enormous* free energy barrier for the crossing of charged molecules and ions (see Chapter 2). Water and other small molecules with relatively low polarity such as ethanol can permeate through the bilayer, as can essential metabolite gases such as oxygen and carbon dioxide, whereas small ions and larger polar molecules require passage either via some form of protein based membrane pore which has a suitable *hydrophilic inner wall* thereby allowing the solvated molecules to pass through without encountering a hydrophobic free energy impediment, or via some form of *protein carrier* located in the membrane that has a hydrophilic binding site for attachment of the translocating molecule. In the case of net translocation occurring from a high local concentration immediately outside the cell to a lower concentration inside the cell on the other side of the membrane, these membrane-integrated carriers or pores permit *facilitated diffusion*; this is in principle a *passive* process (i.e. requiring no external energy input).

(iii) Such a pore or carrier in the membrane may result in problems relating to either the entry of other types of molecules in the cell, or similarly the exit of other molecules out of the cell. This is particularly problematic in the case of bacteria, but also in the case of the membrane-bound mitochondria and chloroplast organelles of eukaryotic cells, since the fundamental source of cellular energy is derived from the *electrochemical potential* established from a *gradient of protons* across the membrane (see section 8.3), thus allowing protons to pass indiscriminately through a membrane pore or to be ferried by a carrier molecule would very quickly eradicate this electrochemical gradient thereby killing the cells rapidly. Thus, an essential requirement for pores and carriers is *molecular specificity*, often combined with some form of *controlled gating*.

In this section, we discuss some key single-molecule studies, either on living cells or in test tube environments which have been designed to have significant physiological relevance, which have improved our understanding of these transmembrane molecular transport processes dramatically. Many of these processes are relevant primarily to the cell membrane, but some are also pertinent to membrane-bound organelles within cells.

BIO-EXTRA

Biochemists often classify membrane-integrated protein pores/channels and carriers into three broad classes: **uniporters**, **antiporters** and **symporters**. Uniporter translocation involves **facilitated diffusion** down a transmembrane concentration gradient and is implicitly **unidirectional**. Several types of small molecules are translocated in this way, ranging from potassium ions in bacteria to glucose molecules in mammalian cells, but each individual uniporter is highly specific to a single type of molecule. Antiporters (also known as **exchangers**) translocate two or more different molecules in opposite directions across the membrane by **coupling the free energy** embodied in the transmembrane electrochemical potential of one molecule down a concentration gradient to the movement of the other molecule in the opposite direction against its respective concentration gradient. For example, the sodium/calcium antiporter in cardiac muscle cells uses the Na^+ gradient to pump Na^+ ions into the cells coupled to pumping Ca^{2+} ions out of the cell against its concentration gradient and has the crucial role of maintaining low Ca^{2+} concentrations in cardiac muscle cells. Symporters (also known as **co-transporters**) translocate two different molecules, one of which is usually either H^+ or Na^+, across a membrane simultaneously in the same direction; the transmembrane electrochemical potential generally energizes the translocation of the second molecule against its respective concentration gradient. For example, the **sodium/glucose co-transporter** translocates one molecule of glucose against a glucose concentration gradient coupled to the translocation of two Na^+ ions in the same direction but down a Na^+ concentration gradient.

8.2.1 Small sugar molecule carriers

The translocation of small sugar molecules, mainly in the form of glucose or more rarely fructose, utilizes multiple classes of *sugar-carrier proteins*, and in mammalian cells this is primarily carried out by 14 different sugar-carrier transporters, GLUT1 through to GLUT14 (simpler eukaryotes such as yeast and prokaryotic cells use a variety of different but related sugar-carriers). These transporters are variously expressed in different cell types with different glucose/fructose binding strengths, though their fundamental mode of action is believed to be driven by a *flip-flop mechanism* in which the sugar binds to the carrier on the cytoplasmic side of the cell membrane and then the carrier–sugar complex flips orientation to drag the sugar across the membrane and ultimately into the cell cytoplasm.

The most well studied of these mechanisms using single-molecule approaches has involved the *glucose-transporter protein* GLUT4, primarily because it is sensitive to external concentration levels of the hormone insulin and so has direct biomedical relevance for investigating causes of diabetes in insulin-sensitive muscle and fat cells in which the normal method of glucose storage inside these cells has been impaired. GLUT4 has been investigated using several different single and dual-colour TIRF imaging investigations at the multi-molecule level over the past decade (Huang and Czech, 2007). In a recent study (Xu *et al.*, 2011) single *GLUT4 storage vesicles* (GSVs) were studied by tagging a membrane protein VAMP2 expressed in the phospholipid bilayer of the GSVs with the pH-sensitive fluorescent protein pHlourin, as used previously in exocytosis systems (see section 7.3). This allowed much clearer visualization of single vesicles than earlier studies, permitting robust quantitative analysis of the size of these nano-vesicles under different states of insulin intervention in the cells, as well as pinpointing their cellular localization very precisely.

Single-molecule AFM imaging has also been applied to glucose transport in studying the sodium/glucose co-transporters in live cells (Puntheeranurak *et al.*, 2006). Here, researchers were able to monitor the topology and conformational changes of these co-transporter complexes in the presence of antibodies that variously disrupt the translocation action. Furthermore, combining this approach with AFM spectroscopy using a glucose-coated AFM tip indicated directly that, in the presence of Na^+ ions, a glucose binding site would appear in the transporter, but was absent at lowered Na^+ levels.

8.2.2 Ion and water channels

There are hundreds of different ion channels, varying in specificity to the type of ion and in the strategies used to open and close (i.e. *gate*) the channel controllably (see Chapter 4). A seminal investigation on single ion channels in live cells involved utilizing a known cell toxin called *hongtoxin* which produced toxicity effects from its ability to bind specifically to *voltage-gated* potassium ion channels (Schütz *et al.*, 2000a). This toxin was labelled with the red organic dye Cy5 (see Chapter 3) and incubated with live so-called *Jurkat* cells, which are an *immortalized* cell strain derived from *T-lymphocytes* (cells used in the immune response in mammals), often used in studying cell signalling and immunodeficiency diseases such as leukaemia and HIV.

BIO-EXTRA

An **immortal cell** is derived from a multi-cellular organism which, under normal circumstances, would not proliferate indefinitely but, due to genetic mutation, is no longer limited by the so-called **Hayflick limit** – a limit to future cell division which is set either by DNA damage or by shortening of sub-cellular structures called **telomeres** (repetitive DNA sequences capping chromosomes which normally become shorter with each subsequent division such that at a critical length cell death is triggered by a complex process called **apoptosis**). Immortal cells can continue to undergo cell division and can be grown under cultured 'test tube' conditions for prolonged periods, making them very useful for studying a variety of cellular processes, since otherwise only short duration studies could be performed. **Cancer cells** are natural examples of immortalized cells, but they can also be prepared using a variety of biochemical methods. Common immortalized cell lines include variants of the **CHO** (Chinese hamster ovary), **HEK** (human embryonic kidney), **Jurkat** (T-lymphocyte) and **3T3** (mouse fibroblasts from connective tissue) cells, with the oldest and most commonly utilized human cell strain being **HeLa** cells (epithelial cervical cells which were originally cultured from a cancerous cervical tumour of the patient **Henrietta Lacks** in 1951, who ultimately died as a result of the cancer, but left a substantial scientific legacy for future biomedical research in the form of her tumour cells). Although there are possible limitations to their use after having undergone potentially several mutations from the original 'normal' cell source, they are still of vital use to biomedical research, increasingly so at the single-molecule level.

Using a *narrow-field* form of laser excitation (see Chapter 3) the researchers were able to perform fluorescence imaging on single dye-tagged ion channels with a very fast lateral time resolution of 5 ms per image, allowing the position of the ion channels to be measured using Gaussian fitting of the point spread function of the emitting dye molecule to ~40 nm precision (Figure 8.1A). With a little modification this also permitted proper three-dimensional imaging of the whole relatively large ~10 μm diameter cell in a rapid 300 ms. This resulted in very fast single-molecule level imaging using a probe that was relatively small (~1 nm) compared to other fast imaging techniques such as laser dark-field that uses gold-coated probes of the order of tens of nanometres in diameter (see Chapter 3), and thus was physiologically much more relevant.

It indicated that multiple, immobile ion channels per cell localized to the cell membrane. They could be investigated with a very high imaging signal-to-noise ratio of ~10, which paved the way for subsequent functional studies to use a variety of chemical interventions to the ion channels as well as genetic mutation to investigate their mode of action.

Another common form of ion channel gating is *ligand-activation*. These channels will open or close upon binding of a specific ligand, which may be either intra-cellular or extra-cellular depending on the specific ion channel. They are similarly highly specific for the type of ion transported through the channel, most commonly including Na^+, K^+, Ca^{2+} and Cl^-, and these channels are often associated with a *neurotransmitter ligand*. The binding of a chemical messenger in the form of a neurotransmitter molecule may cause the channel to open/close, with the mode of action thought to be *allosteric* (meaning that the binding of the ligand at a site away from the channel itself affects the conformation of the whole receptor complex so as to change the open/closed state of the channel).

One such example are the *AMPA glutamate receptors* (AMPARs) which are ligand-activated cation (i.e. positive ion) channels concentrated in the post-synaptic area, the region of the *acceptor* neuron (i.e. nerve cell) just after the synapse junction (see section 7.3). Using

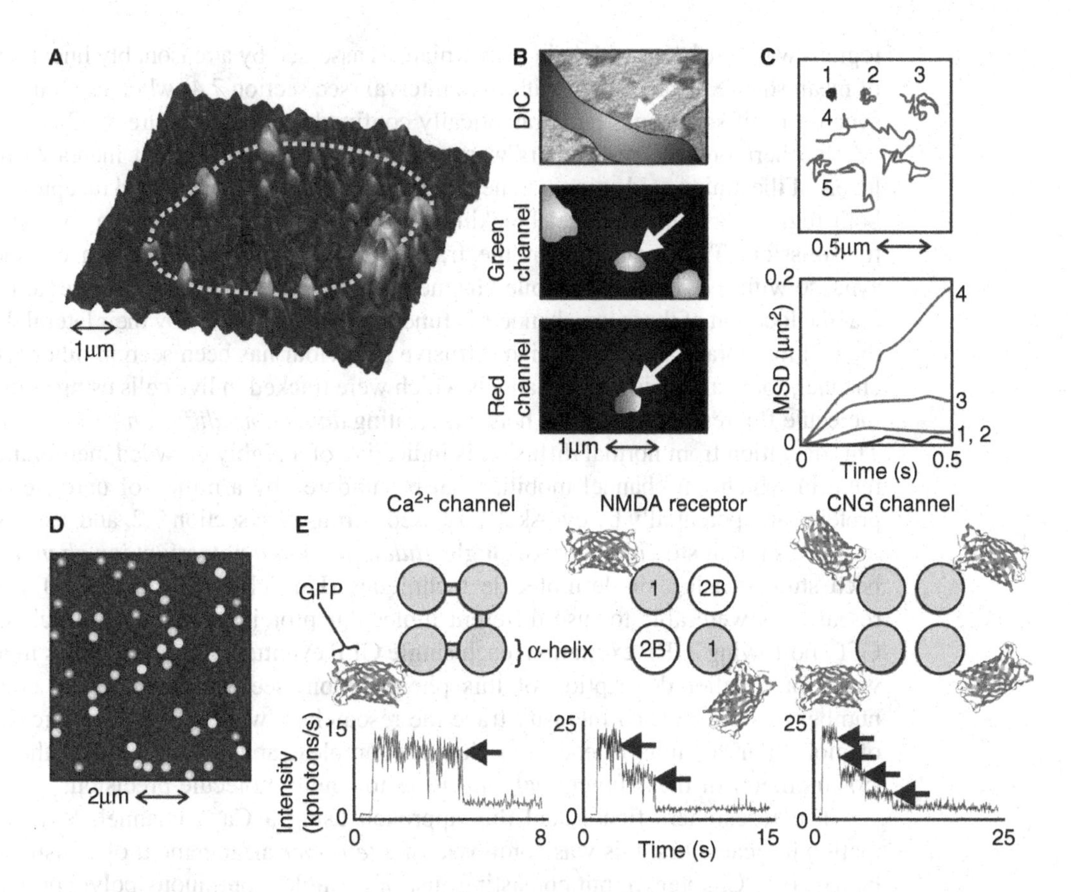

FIGURE 8.1 (A) Intensity contour plot for fluorescently labelled K^+ channels expressed in the cell membrane (Schütz *et al.*, 2000a), perimeter of the cell indicated by a dashed line (data courtesy of Gerhard Schütz, Vienna University of Technology). (B) Schematic depiction of brightfield DIC image (upper panel) of live nerve cell, intensity in the green fluorescence channel indicating the position of the synapse (middle panel), intensity in the red channel indicating the AMPA-Cy5 K^+ ion channel (lower panel), shown with arrows, and (C) schematic trajectories in the focal plane (upper panel) of single tracked AMPA-Cy5 fluorescent foci either inside (tracks 4 and 5) or just outside (tracks 2 and 3) the synapse, compared with a surface-bound AMPR-Cy5 complex conjugated to a glass coverslip via a specific antibody (track 1), with corresponding mean square displacements (MSD) of these tracks (lower panel). (D) Schematic of a TIRF image of a cell membrane expressing GFP-tagged K^+ channels, and (E) schematic (upper panel) of a Ca^{2+} and two K^+ channels (NMDA, consisting of two α-helix subunits each of NR1 and NR2B, and CNG consisting of four subunits of the same type) with corresponding intensity traces indicating step-wise photobleaching of single GFP molecules (arrows). (Original photobleach data courtesy of Maximilian Ulbrich, Freiburg University.)

a red Cy5 or Alexa647 dye tagged antibody which bound to a specific region of the glutamate receptor of the accessible outer surface of the complex, the researchers were able to perform single-molecule fluorescence imaging of live neurons using again a narrow-field approach to permit high signal-to-noise ratios for single fluorophore imaging (Tardin *et al.*, 2003).

Using an additional green stain the researchers could also identify the location of the synapse itself simultaneously (Figure 8.1B). They were able to perform single-molecule tracking on individual tagged ion channels using iterative Gaussian fitting of the point spread function as performed in the earlier study, but here it allowed the mobility of the receptor complexes to be interrogated in either the synapse zone or the *extra-synapse* region as indicated by the green stain. This showed that diffusion in the extra-synapse regions was consistent with being Brownian, as assessed by a reasonably linear dependence of mean square displacement with time interval (see section 7.4) whereas complexes in the synapse itself sometimes had dramatically confined mobility (Figure 8.1C).

Furthermore, the researchers were able to correlate subsequent increases in external levels of the amino acid anion (i.e. negative ion) *glutamate* to increased receptor mobility in both regions, as did chemical blocking of a known inhibitory system to nerve impulse transmission. This regulation of the fraction of immobile/confined ion channels in the synapse with the relatively mobile channels in the nearby extra-synapse regions indicates that the location of these ion channels is fundamentally regulated by their lateral diffusion in the cell membrane. Non-Brownian diffusive behaviour has been seen in other types of ion channel, for example in Ca^{2+} channels which were tracked in live cells using similar single-molecule fluorescence imaging, hence indicating *anomalous diffusion* (Harms *et al.*, 2001). This deviation from normal diffusion is indicative of a highly crowded membrane environment in which ion channel mobility can be hindered by a milieu of unrelated crowding proteins and potentially by cytoskeletal-based corrals (see section 7.2, and see Q8.7).

The subunit stoichiometry of single *glutamate-gated potassium ion channels* has also been studied using single-molecule techniques (Ulbrich and Isacoff, 2007). Here, the researchers were able to fuse different molecular protein subunits of the channels with GFP. Following TIRF excitation, each single GFP eventually photobleaches in a step-like way (for a fuller description of this phenomenon, see section 8.3). By counting the number of steps in each intensity trace the researchers were able to measure the number of molecular subunits in each ion channel complex, and hence estimate the molecular stoichiometry of these functional *multimers* to single-molecule precision.

The researchers first tested this approach using a Ca^{2+} channel. Structural information indicated that this was composed of a *tetramer* arrangement of transmembrane α-helices (see Chapter 2), but consisting just of a single, continuous polypeptide. The vast majority of GFP fluorescent spots observed displayed one bleaching step, consistent with the channel being composed of just one polypeptide. This method was then applied on K^+ channels (for example either so-called CNG or NMDA receptor complexes) which again were known to be arranged as a tetramer of transmembrane α-helices, but now expressed as separate subunits either of two different types (the NMDA receptor) or of just one type (the CNG channel). This revealed mostly either two or four steps in the photobleach trace, thus clearly demonstrating that the method could be used to estimate single-molecule precise stoichiometry in functional multimer type complexes, such as ion channels, expressed in the cell membrane (Figure 8.1D, and see Q8.9).

The actual cellular *mechanism* of ion transport has been investigated using single-molecule cellular biophysics approaches. Here, both live cell as well as elegant test tube approaches have been very successful in using *combinatorial* single-molecule techniques of both single-molecule fluorescence imaging plus single ion channel conductance measurements on the same single channel (see Chapter 4 and Ide *et al.*, 2002 for a good description of a typical setup). These are highly sensitive experimental approaches which permit *voltage clamping* across the membrane in combination typically with TIRF measurements on fluorescently tagged components of the functional ion channel. They have been used to investigate dynamic conformational changes in both K^+ (Sonnleitner *et al.*, 2002) and Ca^{2+} ion channels (Harms *et al.*, 2004), and have indicated both possible *cooperative* features of subunits in these channels as well as important *molecular rearrangements* of the channel components which can be made without the need to affect the open or closed state.

Furthermore, much work has been done in studying cell membrane channels using the model protein *gramicidin* (for a good review see Kelkar and Chattopadhyay, 2007).

Gramicidin is an antibiotic which is technically a mix of six related linear peptides which can each spontaneously self-assemble into prototypical ion channels in a cell membrane, specific to *monovalent* cations. Its bactericidal nature results from allowing inorganic ions such as Na^+ and K^+ to translocate unrestrictedly across the cell membrane in affected bacterial cells, thereby destroying the respective transmembrane ion gradient on which membrane transport processes rely. For example, using physiologically relevant artificial lipid bilayers, gramicidin channel formation and action have been studied using simultaneous electrical and fluorescence measurements, including smFRET to monitor specific molecular conformational changes during the channel's functioning (Harms *et al.*, 2003).

Such studies have indicated a heterogeneous distribution of FRET efficiencies consistent with a broad range of different molecular conformational states, i.e. the mode of action in transporting a single cation through a typical ion channel may be the result of several *transient* molecular conformational states corresponding to different transit positions of the cation in the channel. This is also consistent with atomic level structural data in which the cation has been captured in the channel at different effective transit points.

Ultimately, gating of an ion channel requires some form of *mechanical coupling* between different components of the channel, and in fact there is one other class of ion channel where the gating is explicitly mechanical in nature, commonly referred to as *stretch-activated* or *mechanosensitive* channels, found in the cell membranes of a number of different tissues but especially prevalent in certain nerve cells, and typically requiring a local *deformation* of the cell membrane in the vicinity of the channel leading to opening or closing. Recent developments in mechanical-sensitive single-molecule FRET-based probes are likely to have a large impact in studying such mechanical gating at the single channel level in the near future (see Chapter 3 and Stabley *et al.*, 2011). A recent finding has indicated that one of the largest ion channels yet discovered, which has an enormous molecular weight in excess of 1 MDa and is composed of subunits of a protein called *Piezo*, is mechanically activated (Coste *et al.*, 2012). Here, the investigators were able to monitor single ion channels in live cells using patch clamping (see Chapter 4), and could activate the channels by applying tiny *pressure pulses* via the end of a nano-pipette. Furthermore, they could also reconstitute the ion channel complex in liposomes and count how many subunits each complex contained, again using step-wise photobleaching, as described previously.

Water-specific channels, or so-called *aquaporins*, have some features similar to ion channels since transported ions are in a water-solvated environment, but they function exclusively to regulate water flow across the cell membrane. Although some water permeates across the hydrophobic free energy barrier of the phospholipid cell membrane via *osmosis*, this is a relatively minor effect compared to the regulated flow that occurs through aquaporin water channels. They conduct water both into and out of the cell while being *impermeable* to ions, including protons. They also exclude a variety of non-charged solutes, though there is a class of *aquaglyceroporins* that will, in addition to water, transport some non-charged small solute molecules such as glycerol and a variety of, primarily, waste products, including ammonia, urea and carbon dioxide.

There are several identified classes of aquaporins but they are all characterized by being integrated membrane protein complexes with the pore itself formed from a wall of six transmembrane α-helices which give a pore diameter only just large enough to permit the passage of a single water molecule at any one time. Single-molecule studies have recently been employed to monitor such channels, most prevalently in the aquaporin-4 variant Structural evidence from EM had indicated that either this can exist in a relatively sparsely expressed state in the membrane, or it can coalesce in what appear to be orthogonal arrays. Using QD and GFP tagging in combination with TIRF and FRAP

(Crane *et al.*, 2010; Tajima *et al.*, 2010) researchers were able to apply single particle tracking to individual aquaporin channels to observe the array assembly process and demonstrate consequent deviations from Brownian diffusive behaviour, in addition to estimating the molecular stoichiometry of the subunits of functional channels using step-wise photobleaching of single GFP molecules.

BIO-EXTRA

Voltage-gated ion channels are a broad class of transmembrane protein pore complexes that allow specific passage of usually one of three types of ion, either calcium, sodium or potassium, but have open or closed states of the central channel which is dependent on the local transmembrane voltage potential difference. They are common in many cell types but especially important for **nerve impulse conduction** in neurons and the activation of **muscle contraction**. The mode of action has been studied using crystallographic structural studies (see Chapter 1) on voltage-gated potassium ion channels, indicating a molecular conformational change on charged transmembrane helices that make up a **paddle motif**, which is highly conserved across many species (Alabi *et al.*, 2007). Gating by **ligand-activation** of ion channels can involve binding of an extra-cellular ligand molecule, generally with neurotransmitter action, for example GABA (**gamma-aminobutyric acid**), **glutamate** and **acetylocholine**, or may involve binding of an endogenous intra-cellular ligand which is often a variant of nucleoside bases, such as **cyclicAMP** and **cyclicGMP** which are used in the cellular **second messenger signal application response**. ATP may also be utilized as an intra-cellular ligand for some channels. One particular ATP-dependent channel which transports anions is called the **cystic fibrosis transmembrane conductance regulator** (CFTR), since mutations of the associated genes which express the channel's protein components disrupt this anion transport and may lead to the disease cystic fibrosis.

8.2.3 Protein translocation

In eukaryotic cells, protein export across the cell membrane is the endpoint of highly complex transport and packaging processes in the cell that translocate proteins from their point of synthesis largely from the endoplasmic reticulum through to the Golgi apparatus (see Chapter 2). Here they are chemically modified by adding sugar groups prior to insertion into intra-cellular phospholipid vesicles which are then trafficked to the cell membrane via the cytoskeleton (discussed more fully in Chapter 9), with ultimate transport across the cell membrane being a secretory exocytosis process (see Chapter 7).

In bacteria, the transport of protein molecules across the cytoplasmic membrane is very similar to the process evolved for the membranous thylakoid system of chloroplasts in plant cells, and involves principally one of two mechanisms, either the general secretory (*Sec*) pathway or the twin-arginine translocase (*Tat*) system, both energized by the *transmembrane protonmotive force* (see Chapter 2, and section 8.3). Sec requires proteins first to be *denatured*, tagged with a specific Sec-pore recognition sequence, and then *threaded* in this non-functional *unfolded* state through the specific and very narrow Sec nanopore in the cytoplasmic membrane. Conversely, the Tat nanopore is substantially larger, and can translocate whole proteins in their native *folded* conformation.

Such Tat-mediated translocation presents certain advantages to the cell. Principally, the protein is in a fully functional state following translocation, since in general a protein's function is linked intrinsically to its final native folded conformation, with a requisite normal secondary and tertiary structure (see Chapter 2). Thus it does not require subsequent refolding which is either a very slow spontaneous process or requires the cell to use energy in manufacturing so-called *chaperone proteins* which can assist in the protein folding process, to reach its native functional state once more. However, the presence of such a

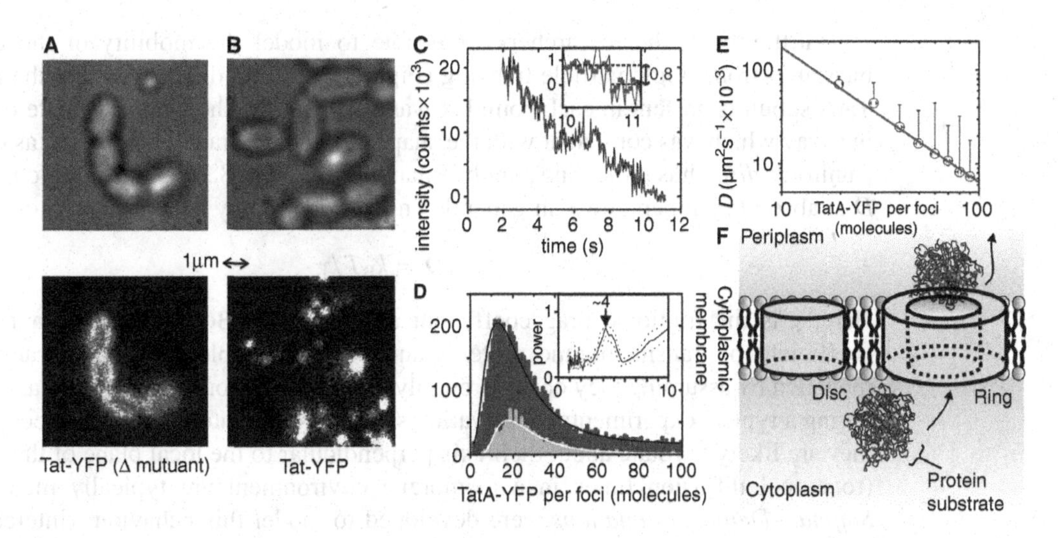

FIGURE 8.2 Protein transport across bacterial cytoplasmic membranes via Tat. Brightfield (upper panels) and epifluorescence (lower panels) images of live bacteria for **(A)** a mutant in which the TatB component has been deleted, and **(B)** a non-deletion strain in which all three essential Tat protein components are present. **(C)** Intensity versus time trace for a typical tracked Tat focus from **(B)** with an expanded section showing a single YFP photobleach step (inset). **(D)** Distribution of the number of TatA subunits per focus (dark grey), and the same but with addition of an uncoupler chemical that destroys the protonmotive force (light grey) with power spectrum for the periodicity in the stoichiometry distribution (inset). **(E)** Variation of estimated effective diffusion coefficient D with number of TatA subunits per focus (standard deviation error bars) with log-log fit. **(F)** Schematic for disc and ring models for the TatA complex. (Original data, author's own work, for full details see Leake *et al.*, 2008.)

large Tat nanopore, which can translocate a variety of protein substrates over a range of molecular weights, whose effective diameters vary in the range 1–8 nm, presents a significant challenge to the cell in that this represents a very large hole at the length scale of single ions, most especially protons, and any passive *ion leakage* by diffusion down a transmembrane concentration gradient is likely to be highly detrimental to the cell, especially of protons which would destroy the protonmotive force very rapidly, thereby killing the cell.

To investigate the behaviour of functional Tat complexes researchers applied single-molecule real time fluorescence imaging to live *E. coli* bacterial cells in which the component which makes up the nanopore wall, a protein called TatA, was fused to YFP (Leake *et al.*, 2008). The Tat system also requires two other protein components in order to function, TatB and TatC which were believed to be involved in recruitment of the protein substrates to be translocated. The researchers were able to generate mutant varieties of the YFP cell fusions which had either or both of these additional components deleted, which resulted in diffuse halo images in epifluorescence consistent with a membrane localization but no multimeric assembled TatA complexes (Figure 8.2A).

In variants with no such deletions the images consisted of multiple, mobile fluorescent foci of a range of brightness (Figure 8.2B). By using single particle tracking and iterative Gaussian fitting (see Chapter 3), the researchers were able to construct a *dynamic* photobleach trace for each fluorescent focus (Figure 8.2C), and by utilizing step-wise photobleaching of the YFP molecules the investigators were able to estimate the number of TatA molecules associated with each focus. This indicated that the stoichiometry was peaked at intervals of roughly every ~4 TatA molecules with a mean of 20–30 but covering a very wide range of values up to 100 molecules and beyond, with no obvious dependence of this stoichiometry on protonmotive force (Figure 8.2D).

Furthermore, the researchers were able to model the mobility of the complexes measured from single particle tracking (Figure 8.2E) and discovered that the number of TatA subunits present in each complex was correlated to the complex's rate of diffusion in a way which was consistent with the shape in the membrane being a *ring* as opposed to a uniform *disc*, thus indicating a mobile nanopore (Figure 8.2F). The diffusion coefficient D at absolute temperature T in general can be modelled by the Einstein–Stokes relation:

$$D = k_B T / \gamma \tag{8.1}$$

Here, γ is the frictional drag coefficient and k_B is the Boltzmann constant. The drag coefficient of transmembrane proteins and protein complexes in membranes is often modelled by assuming they occupy roughly a *cylindrical* conformation and assuming that during a typical experimental observation sampling time interval of milliseconds or more they are likely to rotate about their axis perpendicular to the local plane of the membrane (rotational diffusion times in a membrane environment are typically microseconds). *Saffman–Delbrück equations* were developed to model this behaviour (interested physical scientists should read Saffman and Delbruck, 1975 and Hughes *et al*., 1981). In this model, γ is expected to vary roughly with radius of the effective cylinder (see Q8.6).

The single-molecule observations of Tat were consistent with a system requiring some degree of spontaneous self-assembly of Tat complexes which did not require an external energy source. The nanopores assembled in tetramer subunits of TatA, resulting in a range of different sized diameters which could cater to the precise size of the substrate, resulting in a very tight fit and hence minimizing ion leakage through the pore.

There is also evidence that other membrane protein complexes in bacteria act to minimize further any such general proton leakage. One example is *phage-shock protein A* (PspA) which can assist in maintaining the transmembrane voltage by suppressing proton leakage, involving the association of multiple copies of individual PspA molecules. These molecules bind both to the cytoplasmic membrane and to the cell wall and it may be that their mode of proton leakage suppression involves a change in physical properties of the attached membrane. In a single-molecule epifluorescence study on live *E. coli* cells (Lenn *et al*., 2011), investigators labelled PspA with GFP and were able to count the number of individual PspA proteins in each complex by a similar method involving step-wise photobleaching of the GFP, which indicated a broad range of stoichiometry over the range 4–40 molecules per complex.

8.3 Rotary motors – the rise of the (bionano) machines

Rotary molecular motors in the natural world, most especially the *flagellar motors* of bacteria, have been viewed by some as evidence of higher design principles that conflict with natural selection models of evolution. It seems remarkable, to some, that natural selection mediated through random genetic mutations over multiple generations could have resulted in such efficient miniature machines, baffling in their complexity. This view is largely based on subtleties of scientific misunderstanding and is better discussed by evolutionary biologists and philosophers elsewhere (Dawkins, 2011), but one matter on which there is consensual agreement is that these machines truly are remarkable, and they have excited much recent interest in potential bionanotechnological/synthetic biology applications, though as yet with no resounding success in terms of realistic commercial exploitation.

The type of membrane-integrated molecular machine which causes whole bacterial cells to swim is this flagellar motor (for a good modern review see Sowa and Berry, 2008). It is a true rotary machine which in many ways can be seen as one of the remarkable *primordial molecular wheels* of the natural world, the others being two belonging to the FoF1-ATP synthase complex (see Chapter 2) which generates the universal cellular fuel of ATP. Neither rotary motor as a whole can be classified as a 'single molecule' as such, but rather consists of multiple subunits of single molecules which interact in a highly coordinated way.

For example, the flagellar motor is an assembly of ~13 different proteins into a structure ~50 nm in diameter, and although the FoF1 motors complex is smaller at ~10 nm diameter it too still contains eight different proteins expressed at different relative stoichiometries, making both machines relatively large compared to the length scale of single proteins of just a few nanometres. They both contain the two essential components you would expect from a motor: a ring of *stator units* which generate torque to rotate an inner *rotor* connected to a central shaft running through the cell membrane.

There have been several exemplary in vitro studies investigating a variety of rotary motors, including those that rotate on DNA, for example using magnetic tweezers (see Chapter 5) and using fluorescence imaging of fluorescently labelled F-actin filaments to monitor rotation of the F1-ATP synthase component of the FoF1 complex (see my personal top-ten list of single-molecule papers of Chapter 6), as well as methods using optical tweezers on fused microsphere pairs (see Chapter 5).

However, studying rotary motors in living cells to date has utilized the bacterial flagellar motor as the experimental system of choice, though exciting new experiments are now monitoring FoF1 in vivo at the single-molecule level. In this section we will discuss some of these pioneering single-molecule investigations, but also include discussion of some of the more physiologically relevant studies of the FoF1 motors which, albeit not in a true living cell, are moving in the direction of providing single-molecule information which is close to the native biological context.

8.3.1 Counting molecular subunits in a bacterial flagellar motor

In the model bacterium *E. coli*, the flagellar motor consists of a combination of four *concentric rings* stacked between the outer and inner cell membranes, centred on a rotary shaft component. The rotor shaft is connected to the outside world through a number of protein molecule *adapters* to a semi-rigid helical *filament*, and it is the rotation of this which ultimately brings about propulsion of the cell through its watery environment using basically a *corkscrew* action (Figure 8.3A). The fundamental generation of force by these stator units is fuelled by a constant flux of ions across the cell membranes into the cell from the outside. These ions, either sodium ions, Na^+, in some genera such as the marine dwelling *Vibrio*, or simply H^+ (i.e. protons) in the case of bacteria such as *E. coli*, flow down both a charge and a concentration gradient depending on the particular species of bacterium, hence known as the sodium-motive or protonmotive force respectively which equates to the free energy of that ion per unit charge. But the precise details of the molecular mechanisms by which the machine generates its rotary function are still far from clear.

A recent study combined true single-molecule imaging on living bacterial cells to investigate the biological system of whole cell motility (Leake *et al.*, 2006). The researchers used the model bacterium *E. coli* and used genetic engineering to place a GFP tag onto a protein called MotB, which was known to be one of the required components of each stator unit in the motor (Figure 8.3B). TIRF imaging was used to monitor single GFP–MotB molecules in the living cell.

To see the GFP–MotB protein construct at the flagellar motor, each cell was tethered to the microscope slide using antibodies that bind specifically to the helical flagellar filament and stick the bacterial cell down to the glass surface. Since the motions of the filament are

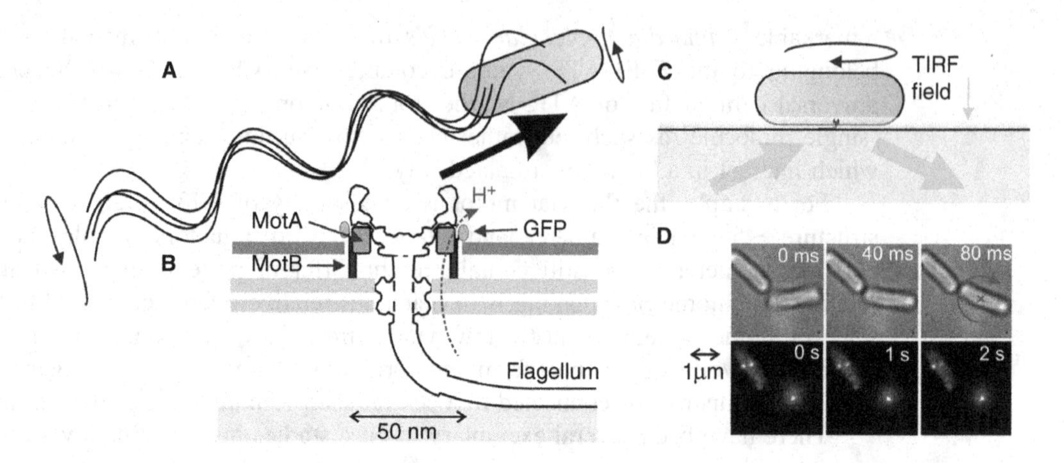

FIGURE 8.3 Visualizing a rotary motor molecular machine. **(A)** A bacterium swimming (in the direction of the straight arrow) by rotation of 4–8 filaments each powered by a single flagellar rotary motor (note the cell body will counter-rotate in the opposite direction by conservation of angular momentum). **(B)** Cartoon cross-section based on EM structural information showing the likely position of green fluorescent protein (GFP) fused to the motor protein MotB. **(C)** Tethered cell showing extent of the evanescent field in TIRF. **(D)** Consecutive brightfield (top) and TIRF (bottom) images. (Original data, author's own work, for full details see Leake *et al.*, 2006.)

then constrained, as the motor rotates instead of swimming the cell body will rotate about its point of tether attachment on the slide. When looking at such cells using TIRF, bright spots of fluorescence were seen which corresponded to the centre of cell rotation (Figure 8.3C), as we saw previously for a single-molecule study of bacterial chemotaxis (see section 7.2), indicating a high density of GFP–MotB protein molecules at the flagellar motor (Figure 8.3D). In observing the rotation of the cell body at the same time as GFP fluorescence in this way it was clear that the molecular machine under observation was *functional*; for live-cell single-molecule investigations there are currently only a handful of such examples where this claim can definitively be made (see Q8.2).

BIO-EXTRA

Bacteria such as *E. coli* are classified as **Gram-negative** because they do not retain a commonly used cellular dye marker of **crystal violet**, whereas **Gram-positive** bacteria will retain this dye. This is primarily because the Gram-negative bacteria possess an additional second outer membrane made of lipids and sugars (i.e. **lipopolysaccharides**) which prevent the dye from staining the **peptidoglycan cell wall** beneath. The flagellar motors of Gram-positive bacteria possess two ring-like systems centred on a rotor component which ultimately is connected to a filament in the outside world, one ring just beneath the cell membrane and one above it located in the cell wall. Gram-negative species have an additional two rings associated with the outer lipopolysaccharide membrane and the peptidoglycan wall. However, the essential molecular details which ultimately result in rotation of the rotary shaft are very similar across different classifications of bacteria.

When the actual intensity values of these bright spots were measured frame-by-frame under continuous TIRF illumination they were observed to decay with time in a roughly exponential fashion. However, on closer inspection, especially towards the end of each trace, the decay curves could be seen to be composed of many small steps either of roughly similar size, or of approximately *integer multiples* of this (Figure 8.4A). This is due to the phenomenon of photobleaching of individual GFP molecules (see Chapter 3).

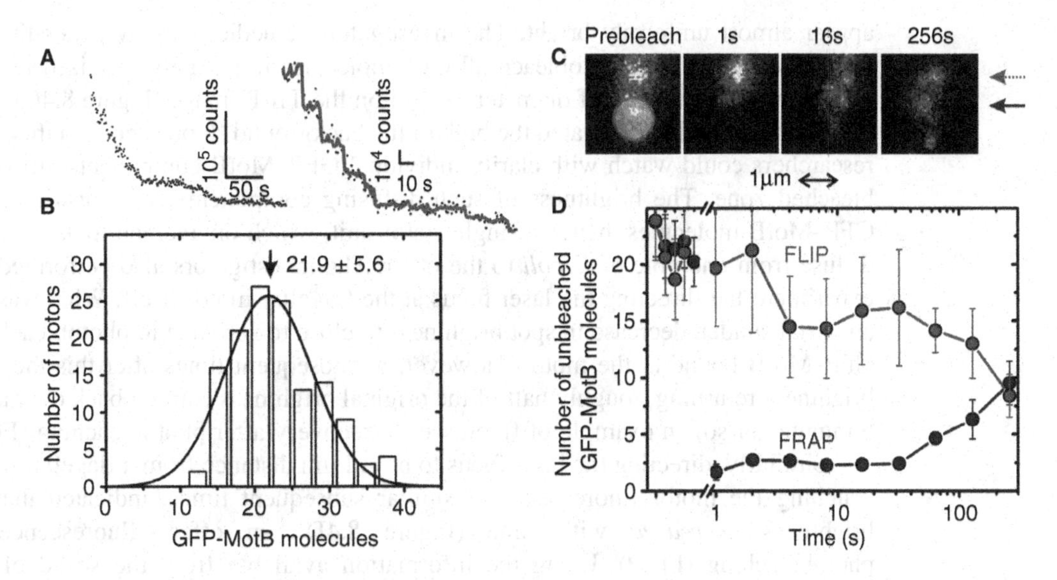

FIGURE 8.4 Counting and turnover. (A) TIRF photobleach of a flagellar motor in a living bacterial cell. (B) Raw photobleach trace (left panel) with expansion of a section including filtered data showing distinct steps due to step-wise photobleaching of single GFP molecules. (C) Distribution for stoichiometry of GFP-MotB per motor for many flagellar motors, mean ± standard deviation indicated. (D) FRAP/FLIP following focused-laser bleaching (circles). One of the complexes shows recovery (solid arrow), another distant from the bleach shows loss (dotted arrow). (Original data, author's own work, for full details see Leake *et al.*, 2006.)

Many fluorescent molecules such as GFP emit on average a roughly constant stream of light photons but then, after a mean period of time characteristic of that molecule, will stop emitting light generally owing to *free-radical* chemical damage induced by the action of the excitation light on the surrounding water. This means that the size of the small steps we observe is a measure of the brightness of a single GFP molecule or, less frequently, two or perhaps three GFP molecules which photobleach during the time window of each frame. By measuring this very accurately the investigators could extrapolate back to the brightness at the beginning of each trace to estimate precisely how many photoactive GFP molecules were present in each spot, and therefore how many MotB molecules are present in the flagellar motor.

This led to an estimate of ~22 molecules (Figure 8.4B); additional data suggested that there are two of these MotB molecules per stator unit, therefore indicating a ring in the functional motor of only ~11 stator units. This number was consistent with later experiments in which stator units were chemically *resurrected* in a modified bacterial cell strain, resulting in typically 11–13 *quantized* increments in motor speed (Reid *et al.*, 2006). But this value is just a snapshot of the mean number of MotB molecules present in the motor at any one time – there is a danger in thinking that because this number does not appear to change from snapshot to snapshot then everything must be static. It turns out that these molecules are far from stationary.

In *E. coli* there are typically 4–8 flagellar motors in each cell, randomly located in the cell membrane, each containing an average of around 22 MotB molecules. However, there is also a pool of roughly 200 MotB molecules not present in the motors but free to diffuse between them in the cell membrane. This could be estimated using the fluorescence intensity in the non-motor regions of the cell. The presence of these additional fluorescent tags makes tracking any individual molecule difficult since they cause the membrane to appear almost uniformly bright. The investigators remedied this by generating a highly intense laser focus to photobleach all GFP molecules in a highly localized region of the cell, manifest as a circle of diameter ~1 μm on the TIRF image (Figure 8.4C).

Having thereby eradicated the bulk of the background fluorescence in this region, the researchers could watch with clarity individual GFP–MotB components diffuse into the bleached zone. The brightness of such diffusing components was consistent with two GFP–MotB molecules, hence a single stator unit, which on average took ~2 minutes to diffuse from one pole of *E. coli* to the other. The investigators also performed the same experiment but directing the laser focus at the flagellar motor itself. What was seen was an initial sudden decrease in spot brightness to close to zero due to photobleaching of the GFP–MotB bound at the motor. However, at subsequent times after this they observed brightness returning, roughly half of the original brightness coming back over a period of 5 minutes or so, an example of fluorescence recovery after photobleaching (FRAP).

Similarly, directing the laser focus to be ~1 μm distance from a flagellar motor, then watching the motor fluorescence at similar subsequent times, indicated that the spot brightness *decreased* with time (Figure 8.4D), in effect fluorescence loss in photobleaching (FLIP). Using the information available from the speed of diffusing stator units in the cell membrane these data suggested that each individual stator unit will spend on average only about half a minute in each flagellar motor. In other words, the force-generating components of a functional motor are constantly on the move – truly remarkable if one considers attempting to change one of the wheels of a car while driving down a motorway at the same time.

8.3.2 Measuring nanometre-scale rotary movements of the flagellar motor

Our understanding of the way in which this nanoscale biological machine of the flagellar motor is fundamentally driven at the molecular level has been advanced significantly by the relatively recent ability to monitor its rotation in a living cell with an exceptional time resolution of less than a millisecond (Sowa *et al.*, 2005). The main problem with monitoring the rotation of a filament-tethered cell, as we saw previously in Figure 8.2, is that the cell body has a relatively large frictional drag associated with it. This means that relatively fast changes occurring in the molecular makeup of the motor remain hidden to the observer since the cell body cannot respond quickly enough.

To overcome this problem, the investigators turned the standard tethered cell assay on its head, so that now the cell body was immobilized on a microscope slide but one of the flagellar filaments was projecting above the cell free to rotate (Figure 8.5A). They further reduced the drag due to the filament itself by first truncating it down by *mechanical shearing* to a stub of just a few hundred nanometres length, and marked the angular position of the motor using a very small latex bead a few hundred nanometres in diameter. In one set of experiments the bead was made fluorescent and detected by a high efficiency camera using fluorescence imaging. In this scenario the relatively high fluorescence excitation induced photodamage in the motor, causing failure of some of the stator units that generate motor torque. By optimizing the conditions the researchers were able to tailor the extent of photodamage such that only a few, typically one to three, functional stator units were present in the motor – under this environment discrete steps in motor speed could be observed with the transition itself lasting less than a few milliseconds (Figure 8.5A, upper panel) indicating single molecular *stator-stepping* events.

In another set of experiments, the bead position was monitored by the method of *back focal plane detection* (see Chapter 4) in which a weak laser was focused onto the

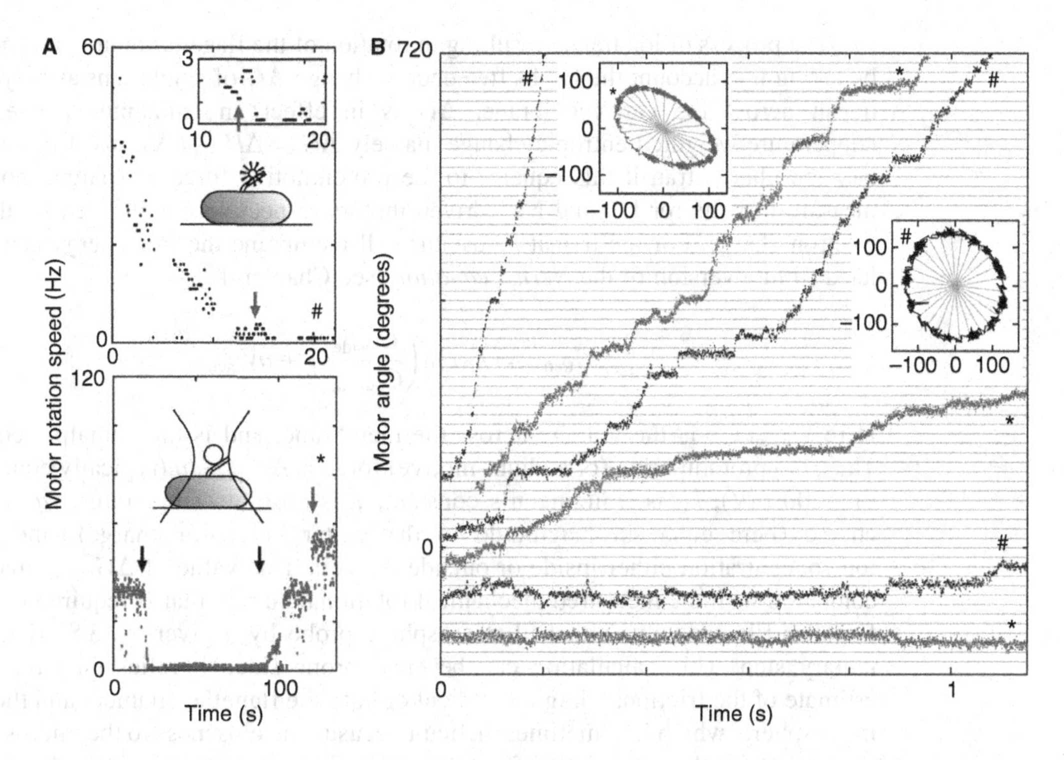

FIGURE 8.5 Monitoring molecular stepping in motor rotation. (A) (Upper panel) Reducing motor rotation using photodamage, combined with fluorescence detection. (Lower panel) Reducing motor rotation by lowering the concentration of sodium ions (5 mM to 0.1 mM and back, black arrows), speed measured using back focal plane detection. Doublings in motor rotation speed in both traces (grey arrows) indicate a change from one to two force-generating stator units. (B) Typical motor rotation speed traces using fluorescence (#) and back focal plane detection (*) using low stator number motors indicating discrete stepping of the flagellar motor. Grey lines indicate (360/26)° in both main and inset figures, suggesting a total of 26 steps per motor rotation. (Original data, author's own work, for full details see Sowa, 2005.)

bead and its movement though the laser focus detected using photodiodes split into quadrants so as to yield independent x and y positions of the bead centre which could be converted into the speed of rotation of the motor. The cell strain itself had been cleverly modified to contain stator parts of both a proton driven motor from one species and a sodium driven component from another.

By reducing the external Na^+ concentration, the researchers found that these hybrid stator units became destabilized, and conditions were again optimized so that only a few functional stator units remained (Figure 8.5A, lower panel). Back focal plane detection of the bead then indicated similar discrete motor speed steps as seen previously using the fluorescence method. Both approaches indicated that the size of these unitary steps in terms of angle about the motor centre was roughly constant at 1/26th that of a single rotation (Figure 8.5B). This 26-fold periodicity was the same as that of a protein called FliG known to be located in the rotor part of the motor. These steps were therefore *direct observations* of single molecular jumps of around ~10 nm in length of individual stator units around the rotor axis of the motor.

Thus a direct connection could be made between the ultimate organism-level system property of whole cell motility at a length scale of several thousand nanometres over a time scale of seconds to that of single, molecular-level, discrete events of stator unit motor stepping which occur over milliseconds and nanometres.

The process of ion transit resulting in rotation of the flagellar motor can be modelled by taking into account the Gibbs free energy change ΔG of single ions as they make the transit across the cell membrane. ΔG is in effect an enthalpic change minus a temperature-weighted entropic change, namely $\Delta G = \Delta H - T\Delta S$, and for a single unitary ion charge transit this equates to the protonmotive force or sodium-motive force measure in volts for H^+ and Na^+ driven motors respectively multiplied by the unitary electron charge. For ion transit across the cell membrane the free energy can be calculated from a version of the *Nernst equation* (see Chapter 4):

$$\Delta G_{\text{transit}} = k_B T \ln\left(\frac{C_{\text{inside}}}{C_{\text{outside}}}\right) + qV_{\text{membrane}} \tag{8.2}$$

Here V_{membrane} is the voltage across the membrane, and is the enthalpic component. Thus, the protonmotive (or sodium-motive) force is $\Delta G_{\text{transit}}/q$ (typically equal to -150 to -200 mV), k_B is Boltzmann's constant, T is absolute temperature, q is the ionic charge (here equal in magnitude to the unitary electron charge), and C is the ion concentration either inside or outside the cell. This value of $\Delta G_{\text{transit}}$ may then be compared with the measured mechanical rotational energy that is required to rotate the flagellar filament plus attached microsphere probe by a given ~13.8° (i.e. 360/26°) unitary step. This calculation can be error prone since it relies on having a good estimate of the frictional drag coefficient of both the flagellar filament and the attached microsphere, which is sometimes difficult becasue the closeness to the microscope slide surface affects the surrounding fluid dynamics. The most accurate estimates for viscous drag in rotary molecular machines have been made using the *fluctuation theorem* (interested physicists may wish to read Hayashi *et al.*, 2010). The best estimates currently suggest that only 1–2 ions may be required to bring about a unitary step of the flagellar motor.

PHYSICS-EXTRA

Although the Second law of thermodynamics predicts that, as an ensemble, a system will converge over large times to net values of entropy change which are positive or (in the case of equilibrium processes) zero, the **fluctuation theorem** of statistical mechanics predicts for a finite non-equilibrium system in a finite time (such as during a typical single-molecule experiment on a rotary motor), there is a small non-zero probability that entropy flows in a direction opposite to that indicated by the Second law; in the case when such systems are measured over very long times then the fluctuation theorem is essentially identical to detailed balance (see Chapter 5). Put mathematically, for the case of rotary motors, we can state that if $P(\Delta\theta)$ is the probability distribution for a small unitary angular fluctuation $\Delta\theta$ which results in net entropy increase and $P(-\Delta\theta)$ is the corresponding probability distribution for the reverse direction fluctuation which results in net entropy decrease, then $P(-\Delta\theta)/P(\Delta\theta) = \exp(-\Delta G_{\text{rot}}/k_B T)$ where ΔG_{rot} is the rotational Gibbs free energy change of the angular displacement, k_B is the Boltzmann constant, all at absolute temperature T. For a process in which torque τ is constant, as suggested from rotary motor single-molecule experiments, we can then say that the associated single-step power is $\tau\omega$ at angular speed ω, hence $\Delta G_{\text{rot}} = \tau\omega\,\Delta t = \tau\,(\Delta\theta/\Delta t)\,\Delta t = \tau\Delta\theta$, where the dwell time between unitary step-like angular changes is Δt. Thus, $P(-\Delta\theta)/P(\Delta\theta) = \exp(-\tau\Delta\theta/k_B T)$. Therefore, by making accurate measurements of $\Delta\theta$, and the associated forward and reverse probability distributions from single-molecule experiments, the free energy change and the torque can be measured without recourse to using potentially inaccurate estimates for frictional drag coefficients.

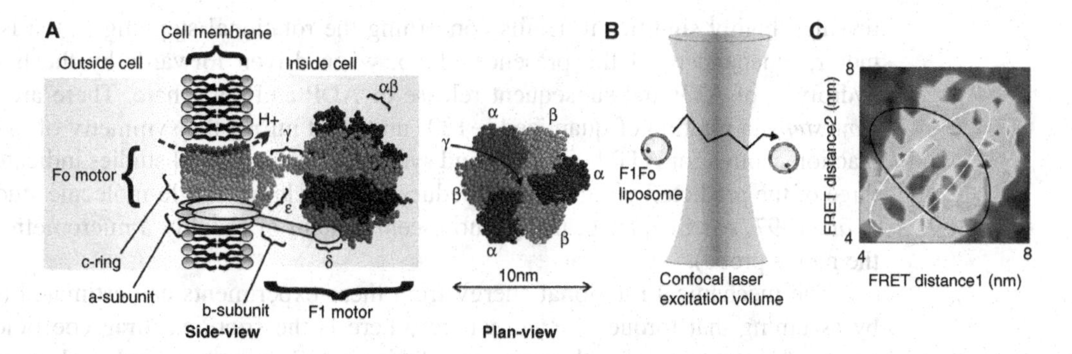

FIGURE 8.6 (**A**) Side (left panel) and plan (right panel) views of FoF1, based on crystal structure of F1 ATP synthase PDB ID:1QO1, with other components lacking in the experimentally determined structure added schematically. (**B**) Fluorescently labelled FoF1 complex reconstituted in a liposome which diffuses through a confocal laser excitation volume, in which a-subunits are GFP tagged (FRET donor) and c-subunits are Alexa568 tagged (FRET acceptor). (**C**) Plot of FRET distances for experiment of (**B**) between GFP and Alexa568 before (distance 1) and after (distance 2) each detected rotational step, with overlaid simulated models for a 36° (white), to be compared against a 120° step model (black) which would be expected from the slow rotational movement of the γ shaft relative to the ε subunit of F1, as opposed to the rotation of the a-subunit relative to the c-subunit for Fo, which clearly indicates that the rotation steps observed are associated with Fo and not F1 (original data courtesy of Michael Börsch, University of Jena).

8.3.3 Unravelling the mechanism of the FoF1-ATP synthase

The FoF1-ATP synthase is a ubiquitous protein complex in living cells, being expressed in bacteria, chloroplasts and mitochondria, and performs the key function of manufacturing ATP, a substance which is in effect a universally accepted *credit card* of chemical energy whose locked free energy can be released as heat upon hydrolysis to ADP and inorganic phosphate and *spent* by being *coupled* in generally highly coordinated ways to myriad different biological processes at the molecular level, in effect fuelling them (see Chapter 2).

The complex itself consists of two coupled rotary motors (for a good modern review see Okuno *et al.*, 2011). The inner *water-soluble* F1 motor is normally exposed to the cytoplasm of the cell and consists of a stiff rotor shaft γ protein subunit terminated with the ε protein subunit, with the γ shaft surrounded by three α and three β alternating stator units in a hexagonal arrangement. The outer *hydrophobic* Fo motor is linked to the rotor shaft through the protein δ via a second flexible shaft composed of b-protein subunits, which in turn links a- and c-protein subunits which embed the Fo complex into the cell membrane (Figure 8.6A).

Under certain conditions, for example in oxygen-starved bacteria, the free energy released during ATP hydrolysis to ADP and inorganic phosphate causes the F1 motor to rotate in a forward direction which is then coupled to the pumping of protons via Fo across the cell membrane out of the cell; the FoF1 therefore acts as a *transmembrane proton pump*. But, under more normal conditions, the Fo motor couples the chemiosmotic energy stored in the proton gradient across the cell membrane lipid bilayer to the rotation of the F1 motor in the *reverse* direction, which results in ATP being *synthesized* from ADP and phosphate. However, as yet very little is known about the exact mechanistic details of these molecular coupling processes.

Furthermore, the vast majority of single-molecule experiments on FoF1 have been performed in the test tube, many solely on the water-soluble F1 component. These have revealed highly significant details concerning the rotational stepping of the isolated F1 motor, independent of the presence of Fo, when driven forwards by the binding and hydrolysis of ATP and subsequent release of ADP and phosphate. There are relatively *slow steps* in rotation of quantized ~120° units that mirror the symmetry of the catalytic reaction centres for ATP hydrolysis and synthesis that structural studies indicate reside at three of the predicted αβ interfaces (deduced in the classic single-molecule study of Noji *et al.*, 1997, see Chapter 6, using fluorescent F-actin of roughly a micrometre length as the motor probe).

The mechanical rotational energy from these experiments can estimated (see Q8.8) by assuming that torque $\tau = \gamma\omega$, where γ here is the frictional drag coefficient of the F-actin filament and ω is the angular velocity in radians per second, and energy is then $\tau\omega$, assuming a highly damped low Reynolds number environment (see Chapter 5) in which all the rotational energy is *dissipated* in moving the surrounding water. The γ value can be estimated analytically by modelling the F-actin filament as a perfect cylinder of length L and radius r in water environment of viscosity η, which gives:

$$\gamma = \frac{4\pi L^3}{3\eta}\left(\ln\left[\frac{L}{2r}\right] - 0.447\right) \tag{8.3}$$

On closer observation, using subsequent faster and more intricate single-molecule methods of laser darkfield with a much smaller 40 nm nanogold probe (Yasuda *et al.*, 2001, see Chapter 3), in a very elegant combinatorial single-molecule assay using fused microsphere pairs attached to the rotor shaft monitored by back focal plane detection (see Chapter 4) while simultaneously imaging the binding of a fluorescent Cy3-tagged derivative of ATP using TIRF excitation combined with polarization imaging (Nishizaka *et al.*, 2004), these ~120° were found to be composed of two smaller sub-steps of 30–40° and 80–90°.

Although not performed in the living cell, such studies have dramatically improved the physiological relevance of the single-molecule data due to the ability to perform measurements either over a fast millisecond time scale much closer to the true biological context, or to use powerful combinatorial single-molecule *maximum information* techniques on the same single complex simultaneously.

This has led to a cyclical model of ~100% efficiency for the molecular action of the F1 motor, requiring cooperativity between the three ATP binding sites.

(i) There is an empty ATP-waiting state for one of the sites brought about by the orientation of the rotor shaft while the other two sites are each bound to either an ATP or ADP molecule.
(ii) ATP binding to this empty site drives an 80–90° rotation step of the rotor shaft, such that the wait time before the step is dependent on ATP concentration with a single exponential distribution indicating that a single ATP binding event triggers the event.
(iii) This binding then triggers ATP hydrolysis and possibly also phosphate release, but leaving an ADP molecule still bound in this site.
(iv) Unbinding of ADP from this site is then coupled to a further rotor shaft rotation of 30–40°, such that the wait time does not depend on ATP concentration but has a peaked distribution indicating that it must be due to at least two sequential steps (see Q8.9) of ~1 ms duration each (assumed to be correlated with the hydrolysis and product release in some way).
(v) The cycle (i)–(iv) above may then repeat.

This process of F1-mediated ATP hydrolysis can be represented by a relatively simple chemical reaction in which ATP binds to F1 with rate constants k_1 and k_2 for forward and reverse processes respectively to form a complex ATP.F1, which in turn then reforms F1 with ADP and inorganic phosphate as the products in an irreversible manner with rate constant k_3.

This reaction scheme is well known to biochemists in that it follows so-called *Michaelis–Menten kinetics*. Here, a general enzyme E binds reversibly to its substrate to form a metastable complex ES, which then breaks effectively irreversibly to form product(s) P while in principle regenerating the original enzyme E. In this case, F1 is the enzyme and ATP is the substrate. If the substrate ATP is in significant excess over F1 (which in general is true since only a small amount of F1 is needed as it is regenerated at the end of each reaction and so can be used in subsequent reactions for other molecules of ATP) this results in effectively steady-state values of the metastable complex F1.ATP, and it is relatively easy to show that this results in a net rate of reaction J given by the *Michaelis–Menten equation*:

$$J = J_{\max}[\mathrm{ATP}]/(K_{\mathrm{m}} + [\mathrm{ATP}]) \tag{8.4}$$

Here, $K_{\mathrm{m}} = (k_1 + k_2)/k_1$, and $J_{\max}$ is the maximum rate of reaction given by k_3[F1.ATP], and square brackets indicate a respective concentration. In the case of a non-zero load on the F1 complex (for example, it is dragging a large F-actin filament) then the rate constants are weighted by the relevant Boltzmann factor $\exp(-\Delta W/k_{\mathrm{B}}T)$ where ΔW is the rotational energy that is required to work against the load to bring about rotation (see Q8.9). The associated free energy released by hydrolysis of an ATP molecule to ADP and the inorganic phosphate Pi can be estimated as:

$$\Delta G = \Delta G_0 + k_{\mathrm{B}}T\ln\{[\mathrm{ADP}][\mathrm{Pi}]/[\mathrm{ATP}]\} \tag{8.5}$$

Here, ΔG_0 is the standard free energy for hydrolysis of a single mole of ATP molecule at pH 7, equivalent to roughly −50 pN nm per molecule (note the minus sign to indicate a spontaneously favourable reaction), and square brackets '[]' indicate molar concentrations.

The difficulty with observing the reverse process of ATP manufacture is that this requires the coupled presence of Fo which is not easy to generate since it needs to be stabilized by the presence of surrounding lipids. When performed in the test tube this requires using subtle quantities of detergents in challenging experimental assays. The most physiologically relevant studies to date have involved reconstitution of the full FoF1 complex in liposomes (Düser *et al.*, 2009). Here, researchers were able to fuse GFP to the a-protein subunits and the bright organic orange dye Alexa568 to the c-protein subunits to act as a FRET acceptor for the GFP fluorescence when monitored using a confocal laser excitation volume (see Chapter 3) for the GFP on freely diffusing liposomes in solution (Figure 8.6B).

The smFRET traces indicated distinct steps in the measured FRET efficiency with time, and hence the interpolated FRET distance. By plotting the pre-step FRET distance versus the post-step FRET distance the researchers were able to show that such events were consistent with a rotational stepping of the Fo motor in ~36° units, which mirrored the ~10-fold symmetry of binding sites of the c-ring seen from earlier structural studies (Figure 8.6C). A drawback of using liposomes in this way was that there was no independent control over membrane potential voltage, making it impossible to correlate changes in measured Fo rotational speed. Furthermore, the maximum likely pH difference across the liposome membrane was estimated to be ~4, which compares with more like 0.7–0.8 in living cells.

Understanding the true dynamics of the FoF1 complex in actual living cells at the single-molecule experimental level is still in its infancy, but developments both in reconstituted liposomes in systems permitting stable control over membrane potential voltage and in living cells in using brighter and more stable genomically encoded fluorophores than the standard fluorescence proteins, such as by application of SNAP and CLIP-tags (see Chapter 3) is likely to address many mechanistic questions in the near future (Seyfert *et al.*, 2011).

8.4 Energizing the cell

The principal processes by which cells generate their own 'fuel', namely converting some external source of energy into a chemical form which can be locked into a molecule until it is needed elsewhere in the cell, can be divided broadly into those that convert electromagnetic energy directly from the sun into chemical energy stored in the covalent bonds of sugar molecules (*photosynthesis*) and those that break down nutrient molecules ultimately to generate the universal fuel molecule ATP (*oxidative phosphorylation*).

8.4.1 Manufacturing ATP via oxidative phosphorylation

Although there are several peripheral methods that cells can use to generate ATP which are well known to biochemistry students (glycolysis, the Krebs citric acid or TCA cycle, as well as fermentation in plant cells), the primary method of ATP production, generating 80–90% of all cellular ATP, is oxidative phosphorylation, or OXPHOS. This is specifically the process by which nutrients in the cell are broken down to energize the pumping of protons across a membrane, and this electrochemical *chemiosmotic* proton gradient, essentially a form of electrostatic potential energy stored across the membrane in much the same way as an electrical capacitor stores charge across a dielectric compound, is used ultimately to bring about the rotation of the FoF1-ATP synthase which generates molecules of ATP, which we discussed in the previous section. These molecules of ATP lock in the electrochemical energy as chemical energy, thereby acting as a universal cellular fuel (see Chapter 2).

There are differences in how this is achieved between prokaryotic and eukaryotic cells, but they all utilize multiple proteins integrated into a phospholipid membrane, either of the cell membrane itself in the case of prokaryotes, or in the organelle membranes of mitochondria in eukaryotes (see Chapter 2). The basic process involves multiple enzymes which break down nutrient molecules and in the process generate electron-reduced states, most commonly of the reduced nucleoside NADH. Such molecules in effect contain one or more electrons with a high electrostatic potential. These molecules can then enter the *electron transport chain* which involves transferring the high energy electrons between several different *electron-carrier proteins* via a process of *quantum tunnelling*, resulting in sequential oxidation and reduction states for the carrier molecules as electrons are lost to another carrier down the chain or subsequently gained from a carrier up the chain respectively.

At each stage in this electron transfer process, the electrostatic potential energy of the electron goes down, with the deficit energy coupled to transmembrane proton pumps which pump protons across the membrane to establish a protonmotive force. This protonmotive force is then coupled to the rotation of the FoF1-ATP synthase in the membrane to generate ATP.

Mitochondria are difficult targets for single-molecule methods because of their small size and compartmentalized membranous inner architecture. They are roughly rod-like with a diameter of 200–500 nm and length typically ~1000 nm, and contain at least four physically distinct membrane-bound compartments separated by distances of only a few tens of nanometres. This is an order of magnitude smaller than the standard optical diffraction limit and so an obvious choice of functional single-molecule imaging is super-resolution microscopy (Brown *et al.*, 2010). A promising recent study applied structured illumination to fluorescently labelled whole mitochondria in live cells, and was able to image dynamic movement of these organelles at frame rate of ~1 Hz to a super-resolution precision of ~100 nm (Hirvonen *et al.*, 2009).

This is still an area of research in its early stages, but a seminal recent investigation on an individual mitochondrial channel protein was the first single-molecule study to label a functional mitchondrial target protein definitively with a fluorescent protein tag (Kuzmenko *et al.*, 2011). Here, the researchers were able to fuse the integral membrane of the *Tom-group* of proteins. The Tom-group of outer membrane proteins are not directly involved in OXPHOS – Tom stands for 'translocase of the outer membrane', and these are a class of channel proteins that imports a variety of molecules across the outer mitochondrial membrane. OXPHOS proteins are located in the inner mitochondrial membrane, but the mobility of molecules between the two membranes means that Tom proteins do give indirect information as to the activity of OXPHOS.

Tom40 was labelled with the photoactivatable protein *Dendra2* by genetic encoding at the level of the mitochondria's own DNA, and the investigators used PALM in combination with Gaussian fitting of the point spread function of each activated and then excited fluorophore intensity profile (see Chapter 3) to track each tagged molecule of Tom40 to a precision of ~40 nm over a sampling time of just 5 ms per image. In this way they could measure the mobility of the Tom40 molecules, indicating very dynamic behaviour.

In a later study researchers were able to monitor another Tom-group protein Tom20 (Appelhans *et al.*, 2012), in addition to OXPHOS components directly in the form of the OXPHOS enzymes ATP synthase and succinate dehydrogenase using dual-colour oblique epifluorescence/HILO illumination (see Chapter 3). The strategy employed in this study was to use a much brighter rhodamine-based *HaloTag* (see Chapter 3) in combination with photactivatable GFP, which generated a superior precision of ~15 nm following single particle tracking and iterative Gaussian fitting. This could reveal the molecular mobility characteristic of the different protein complexes, in addition to the presence of restricted mobility zones for both Tom and OXPHOS components.

Preliminary studies such as these are likely to pave the way towards many future single-molecule investigations on functional mitochondria, including those of OXPHOS proteins.

Bacteria are in several ways better generic experimental models for studying OXPHOS. They are slightly larger than mitochondria which makes optical imaging less challenging, they are far easier to cultivate, they posses none of the highly compartmentalized phospholipid sub-cellular features which might impede functionality in the presence of bulky fluorescent protein tags, and they are easier to manipulate genetically using fluorescent protein fusions. Even so, this is still an area of developing research, but a recent key study involved monitoring one of the electron transport carrier proteins expressed in the cell membranes of live *E. coli* cells called CytdB tagged with GFP (Lenn *et al.*, 2008a), in combination with oblique epifluorescence, FRAP and single-molecule TIRF imaging (see Chapter 3).

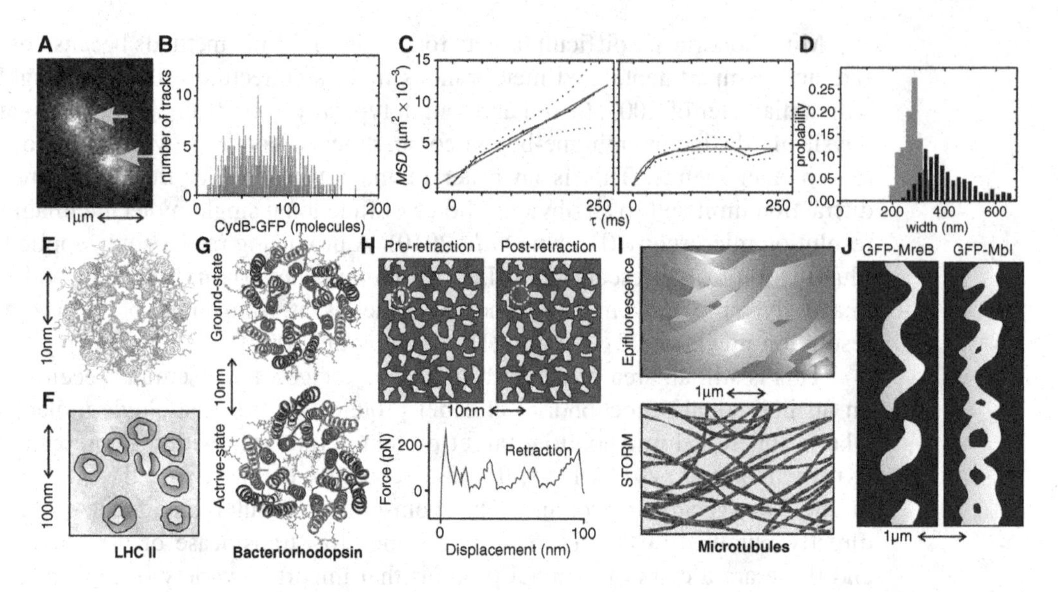

FIGURE 8.7 **(A)** TIRF imaging of single CytdB-GFP foci (arrows) in live *E. coli* cell, with **(B)** distribution of subunit number, **(C)** variation of mean square displacement (MSD) with time interval τ for Brownian (left panel) and confined (right panel) trajectories, and **(D)** variation of foci width (black) compared against single molecules of surface-immobilized GFP (grey) indicating diffusion in large membrane patches as opposed to compact complexes (original data, author's own work, for full details see Lenn, 2008a). **(E)** Plan-view head-on into the membrane of the structure of light-harvesting complex II (LHCII) from bacteria (assembled multimers of PDB ID: 1LGH), and **(F)** schematic depiction of AFM images of single LHCII complexes in reconstituted membranes on a mica surface. **(G)** Plan-view looking head-on into the membrane of the structure of bacteriorhodospin before (upper panel, assembled from PDB ID:1C3W) and after (lower panel, assembled from PDB ID:1DZE) photon absorption. **(H)** Schematic depiction of AFM image of BR membrane patch before (left panel) and after (right panel) AFM spectroscopy tip retraction, 'excised' BR subunit circled (dashed line), with corresponding schematic of a retracting-tip force-extension trace (lower panel). **(I)** Schematic depiction of conventional widefield epifluorescence image of fluorescently labelled microtubules in a live mammalian cell (upper panel), compared with a schematic for the same region of interest imaged using STORM (lower panel). **(J)** Schematic of kymograph images (compiled multiple separate images showing average path of tagged proteins) in live *Caulobacter crescentus* cells for GFP-tagged MbI (a cell wall protein, upper panel) compared with GFP tagged MreB (lower panel) indicating circumferential movement in the shape of helices around the cell.

Here, the researchers observed multiple mobile fluorescent foci in the cell membrane (Figure 8.7A), whose subunit stoichiometry was measured using step-wise photobleaching of single GFP molecules (Figure 8.7B, and see section 8.3) indicating a broad range peaking at around ~70 CytdB molecules per focus. Single particle tracking of these foci in combination with super-resolution Gaussian fitting of their point spread function intensity profiles suggested that foci were either freely diffusing or confined in domains of effective diameter ~60 nm (Figure 8.7C), with the width of the foci themselves larger than that expected from the point spread function width of single GFP molecules by typically 100–300 nm (Figure 8.7D).

This has led to speculation that these fluorescent foci are actually small, separate *patches* of cell membrane a few hundred nanometres in diameter which contain multiple copies not only of CytdB but potentially of other OXPHOS components in localized areas of fuel production called *respirazones* (Lenn *et al.*, 2008b), and can potentially diffuse *en masse* in the rest of the membrane, but sometimes with restricted mobility which may again have biological significance. Much of these speculations still remain to be tested, but it may be the case that key OXPHOS components are in effect pooled together in membrane-based *nanoreactor* vessels, which confine the key electron-carrier proteins, thereby generating substantially higher rates of electron-transport reaction than would be expected from free, unconfined diffusion in the cell membrane.

8.4.2 Using light to generate cellular energy

Photosynthesis is the primary process by which plant and certain bacterial cells convert energy from visible-light photons into the manufacture of small molecules of sugar (the photochemistry is well characterized at a bulk ensemble-average level, and interested life scientists should read Berg *et al.*, 2002). These cells represent essentially the lowest rungs in the food-chain – barring some rare bacteria that live in thermal springs and derive their energy from the heat of lava, photosynthetic cells form the key link between the primary ultimate source of free energy for all forms of life, sunlight, and all other cells, organisms and biological systems in general.

The first key stage in photosynthesis is mediated through a molecular enzyme complex called *Ribulose-1,5-bisphosphate carboxylase oxygenase*, *RuBisCO*, which, because of the significant mass of biological material (i.e. *biomass*) which can perform photosynthesis, has been estimated to be the most abundant protein on Earth. RuBisCo catalyses the fixation of carbon dioxide into a chemical precursor of sugars such as glucose called 2-phosphoglycerate in a process known as the *Calvin cycle*, requiring energization through ATP hydrolysis (whose generation is ultimately fuelled from solar energy). The enzymatic activity of RuBisCo is relatively low and so, to increase the chemical efficiency, such reactions are often carried out in specialized sub-cellular organelles called *carboxysomes*. Carboxysomes have their due place in evolutionary history since they are in fact one of the few clear examples of a distinct sub-cellular organelle which can be found in a prokaryotic cell.

Single-molecule methods have led to significant progress in the study of photosynthesis down to the level of the single pigment–protein molecular complexes involved (for a review, see Vacha *et al.*, 2005). The proteins which gather the light are expressed in phosholipid membranes, either directly in the cell membranes of photosynthetic bacterial cells called *cyanobacteria*, or in a highly invaginated membranous *thylakoid* system of chloroplast organelles in the eukaryotic cells of plants (see Chapter 2). The precise mechanisms of action differ between bacteria and chloroplasts, but essentially these *light-harvesting complexes*, are remarkable multi-molecular complexes that act as highly efficient antennae to absorb photons of visible light in combination with pigmented molecules such as *carotenoids* and *chlorophylls* (Figure 8.7E). This results in a *funnelling* effect for photon absorption, with subsequent transfer of energy to surrounding molecules, most probably by a FRET type effect, ultimately resulting in excitation of high energy electrons from a ground state focused on a *photosynthetic reaction centre*.

Quantum tunnelling of these high energy electrons then occurs in a complex series of electron-transfer reactions to electron carriers, similar to that found in OXPHOS, with the drop in electron energy coupled at each stage to a series of complex biochemical reactions, which in essence results in the overall effect of pumping protons across a membrane. The consequent electrochemical energy is used to fuel the reaction of carbon dioxide with water, ultimately producing small sugar molecules, which lock up the energy of the originally excited electrons into high energy covalent chemical bonds, and oxygen.

Substantial work has been done in the test tube at the single-molecule level at low, unphysiological temperatures, for example to unravel the complex electronic structure of the pigment–protein complexes (van Oijen *et al.*, 1999). Significant research has been done, albeit at a single cell but not yet at a single-molecule level, on live bacteria using FRAP, as well as on functional chloroplasts, to characterize the mobility of proteins and membranes associated with photosynthesis. Single-molecule studies performed at physiological temperatures include single-molecule TIRF-based spectroscopy characterization of light-harvesting complexes known as *chlorosomes* isolated from a branch of prokaryotes called *green sulphur bacteria* (Saga et *al.*, 2002).

Phycobilisomes (*PBsomes*) are structurally similar to chlorosomes and are the principal light-harvesting complex assemblies in the *blue-green* cyanobacteria, as well as in *red algae*. Similar single-molecule spectroscopy experiments of PBsomes indicate subtle energy *decoupling* mechanisms at high sunlight intensities, by which photodamage/saturation is minimized in reaction centres (Liu *et al.*, 2008). In addition, FRAP studies on single PBsomes have indicated that the recovery of fluorescence is due to intrinsic photon processes in the PBsomes as opposed to resulting from molecular diffusion processes (Liu *et al.*, 2009).

This process of energy decoupling appears to involve *quenching* of energy from the light-harvesting complexes to other nearby molecules which is dissipated as heat instead of resulting in the excitation of ground state electrons. This avoids a condition in which there are no remaining ground state electrons to excite at very high light intensities (i.e. saturation) as well as minimizing the risk of free-radical formation which would result in non-specific irreversible chemical damage (see Chapter 2). Note, however, the ultimate by-product of such energy decoupling is to transform a significant proportion of incident sunlight into heat elsewhere, and so in this sense photosynthesis is a relatively inefficient process in converting sunlight into locked chemical energy in the covalent bonds of sugar molecules (the theoretical maximum is estimated at ~25%, but in practice there are additional losses due to reflection and scatter of sunlight from leaves of plants etc., such that the real overall efficiency is more like ~5%).

AFM single-molecule imaging has revealed fine topographical details both of individual light-harvesting complexes in reconstituted membranes (Scheuring *et al.*, 2001) and of native photosynthetic membranes (Sturgis *et al.*, 2009). This indicates a very highly organized, tightly packed arrangement of reaction centres optimized for highly efficient light absorption (Figure 8.7F). Also, the forces of attachment in the cell membrane of light-harvesting complexes found in bacteria have recently been studied in fine detail using single-molecule AFM spectroscopy (Liu *et al.*, 2011).

Furthermore, by using an ABEL trap (see Chapter 5) researchers were able to monitor single molecules of the important photosynthetic *antenna* protein, *allophycocyanin*, in solution (Goldsmith and Moerner, 2010). Although again a test tube study, the researchers were able to monitor the molecular conformational changes of single allophycocyanin molecules in response to exposure to visible light over relatively long periods of several seconds. Such observations may prove to be highly relevant to future single-molecule imaging studies in live cells which are likely to develop in the near future.

Certain *purple bacteria* use light in a slightly different way. Whereas photosynthetic cells use light-harvesting complexes to funnel light onto mainly chlorophyll-type molecules, which in turn pump protons across either the cell membrane in the case of photosynthetic bacteria, or the thylakoid membrane in the case of chloroplasts, purple bacteria use the membrane protein *bacteriorhodopsin* (BR) both to absorb sunlight and primarily to pump protons across the cell membrane, thereby maintaining the protonmotive force (which is utilized for subsequent ATP manufacture via bacterial oxidative phosphorylation in a similar manner to that discussed previously in section 8.3).

BR is a remarkable light-driven ion pump (related molecules called *halorhodopsins* are also able to use light to pump halide ions across the membrane in bacteria that live in very salty conditions, thereby maintaining cellular levels of these ions). Rhodopsin contains a protein component *opsin* plus a smaller light-absorbing component in the form of the molecule *retinal* (which is the same light-absorbing molecule found in the retina of the human eye, see Chapter 3). BR undergoes a molecular conformational change upon the absorption of a single photon of light, which can be seen clearly by comparing two atomic-level crystallographic structures obtained pre- and post-absorption of a single photon of ~green light (Figure 8.7G).

The BR molecule has generated significant interest in terms of bionanotechnological application, for example as a potential miniaturized photoreceptor in which it has been used in conjunction with synthetic nanodroplets (see section 4.4, and Holden *et al.*, 2007). In a seminal single-molecule AFM study (Oesterhelt *et al.*, 2000) a membrane patch containing a tightly packed array of individual molecules of BR was first imaged using AFM (see Chapter 4), then the AFM tip was pressed hard in an approach to the sample surface and subsequently retracted to measure the resultant force–extension relation (see Chapter 5), and the same membrane patch was imaged using AFM once more (Figure 8.7H). This approach indicated that where the AFM tip had been retracted there was clearly no longer a BR complex present on the patch. Thus, it had been pulled free from the surface, indicating that the force–extension trace really did correspond to a single molecule, and so these before-and-after images constitute another example of a *molecular signature* (see section 5.4). In this way the researchers were able to correlate the observed forces and extension with the unravelling of key transmembrane α-helices in the BR complex – very important information, since although the atomic level crystal structures of Figure 8.7G are very powerful they do not necessarily indicate the true physiological structure in the actual native membrane.

KEY POINT

The primary forms of fuel production in the cell, **photosynthesis** and **oxidative phosphorylation**, both rely intrinsically on establishing, and subsequently feeding off, the **electrochemical potential energy** stored in a charge and concentration gradient of protons across a **phospholipid dielectric bilayer**.

8.5 Architecture and shape of the cell surface

The shape of a cell is maintained primarily by a variety of filamentous proteins that constitute the cytoskeleton (see section 7.2), as well as a cell wall in bacterial and plant cells, in addition to the local architecture of the phospholipid bilayer itself. Single-molecule methods have been applied in depth in several recent studies to investigate these structural and architectural factors in the cell.

8.5.1 The cytoskeleton

The cytoskeleton is an extensive network of protein filaments, including *microtubules*, *intermediate filaments* and *microfilaments* in eukaryotes with a variety of equivalent filaments in prokaryotes, which help to maintain cell shape but also, in the case of eukaryotic cells, allow cells to produce dynamic, directed movement of its cellular boundary, for example by protruding out arm-like processes to extend further into the external environment. The cytoskeletal protein filaments in cells appear to form essentially a *tensegrity* structure, an architectural term applied to, for example, modern tents, where the principal struts which hold the tent together do so under tension. This is a clever method for achieving strength of an integral structure using minimum quantities of structural material.

The source of the tension is twofold. Firstly, the osmotic pressure of the cells pushes the cytoskeleton outwards. This high osmotic pressure is a consequence of there being a large concentration inside the cell compared to outside for a multitude of biological solutes, and thus with the cell membrane acting as a semipermeable water barrier there will be a net diffusive flux of water towards the more concentrated inner region of the cell manifest as an increase in outward pressure on the membrane and attached cytoskeletel components, given by the *Van't Hoff equation*:

$$\Delta P = k_B T \,.\, \Delta n \tag{8.6}$$

Here ΔP is the pressure difference across the cell membrane, where the molar concentration difference of the solute in question is Δn. Secondly, there is a tension effect on the cytoskeleton from the contractile activities of motor proteins that use the cytoskeleton as a molecular track. The cytoskeleton is intrinsically involved in the trafficking of internal cargo in eukaryotic cells using such motor proteins (see Chapter 9).

The fine topographical details of the cytoskeleton have been a target for single-molecule AFM imaging on flat surfaces in vitro, since their location beneath the cell membrane surface makes them inaccessible to AFM tip probes thus not allowing live-cell imaging. This technique has revealed physiologically relevant intermediate structures in the assembly of microtubules (Hamon *et al.*, 2010). Also, the mechanical properties of the cytoskeleton have been investigated using single-molecule biophysical methods. These have involved both AFM imaging and force spectroscopy to measure the elastic response of intact cells, including *epithelial cells* and *lymphoctyes*, by pushing and retracting the AFM tip deep into the cell membrane to create a driving force for cytoskeletal filament deflection (Kuznetsova *et al.* , 2007). Optical tweezers have been used to stretch intact cells, most famously either whole *red blood cells* or their *ghosts*, which are cells that have been manipulated so as just to contain the cytoskeleton, here composed mainly of a mesh of *spectrin* protein molecules (Sleep *et al.*, 1999).

Although such ingenious cellular mechanical investigations do not involve direct experiments on single molecules of the cytoskeleton, they do employ single-molecule methods and can be related by suitable modelling implicitly back to the mechanical parameters of single molecules.

The cytoskeleton has been an exceptionally popular target for single-molecule fluorescence imaging methods in live cells. This is partly because the relatively high subunit stoichiometry in cytoskeletal filaments and substructures result in dense fluorophore labelling and hence high signal-to-noise ratios, making it a highly visible target substructure. *Fluorescent speckle microscopy*, or FSM (see Chapter 3), has been used to monitor tubulin dynamics of microtubules in live cells (Waterman-Storer *et al.*, 1998) as well as investigating the turnover of actin molecules in cytoskeletal substructures called *lamellipodia* that generate arm-like extrusions of the perimeter of eukaryotic cells, as a cellular chemotaxis response (Watanabe and Mitchison, 2002, and see Chapter 7).

Microtubule imaging in particular has emerged as a benchmark for the quality of fluorescence super-resolution imaging techniques, since the diameter of a single filament is ~20 nm which is comparable to the resolution obtainable from the emerging super-resolution fluorescence imaging methods (see Chapter 3). A good example of this is the application of multi-colour STORM (Bates *et al.*, 2007), which has been used to generate high resolution images to within a few tens of nanometres precision of microtubules in live mammalian cells even in very typical scenarios of microtubules being densely bundled (Figure 8.7I), with similar experiments performed using PALM and STED. One of the leading developments in the field utilizes a combination of STED with BiFC (see section 3.9) thereby permitting both super-resolution images of microtubule filaments in live cells in addition to molecular co-localization information for the binding of a *microtubule-associated protein* MAP2 (Lalkens *et al.*, 2011).

Furthermore, microtubules have been imaged using *blinking assisted localization microscopy* or BaLM (see section 3.9) which uses stochastic photoblinking of dye-tagged tubulin molecules to generate super-resolution reconstructed images (Burnette *et al.*, 2011). This approach has also been utilized to generate super-resolution live-cell images of *podosomes* (Cox *et al.*, 2011). Podosomes are cytoskeletal structures composed of an actin core surrounded by a ring of proteins such as *vinculin* and *talin* that are known to associated with integrins and thus are associated with cell adhesion complexes (see section 7.2), and here researchers were able to monitor the dynamics of podosomes to within ~50 nm precision over a sampling time of ~4 s per image.

Also, the fact that microtubules act as tracks for the transport of cargo in the cell (Chapter 9 explores the mechanisms of actual trafficking) has been utilized to map out the track itself by labelling the cargo (in this case *endosomes*) with individual QDs, then performing iterative Gaussian fitting on the observed point spread function for the fluorescence intensity profile of each QD as the cargo is trafficked, thereby mapping out with super-resolution precision an image of the track itself (Mudrakola *et al.*, 2009). This method has the advantages of not requiring an expensive specific super-resolution microscope. Recent advances have also been made in the use of Bessel beams for fluorescence imaging (see section 5.2) to obtain greater three-dimensional information from cellular samples, which have been applied to imaging components of the cytoskeleton, including microtubules involved in the mitosis process by which non-germinal eukaryotic cells divide, as well as imaging the actin mesh which results in spiky *filopodia* extrusions to the cell (Planchon *et al.*, 2011).

Super-resolution fluorescence imaging methods have also been applied to the cytoskeleton of prokaryotes. In a study related to the BaLM investigation above, researchers fused the intermediate filament protein *crescentin* with YFP. In the first instance they combined photobleaching of the YFP with stochastic reactivation following ~400 nm wavelength laser excitation in a PALM-like manner to generate super-resolution reconstructed images (Biteen *et al.*, 2008). In a second study the researchers then used photoblinking reconstruction, but utilizing the same longer wavelength of excitation as used for the actual imaging of YFP to map out its extent just beneath the cell membrane in single cells of the bacterium *Caulobacter crescentus* (Lew *et al.*, 2001), in a method which they dubbed SPRAIPAINT (see section 3.9).

The actin-like protein MreB found in the cytoskeletons of bacteria has been a target of several single particle tracking studies in live cells. By tagging MreB with YFP, researchers were able to monitor the dynamics of MreB in live *Caulobacter* cells using widefield epifluorescence illumination with single-molecule sensitivity to track at a sample rate of ~15 ms per image. This indicated a directed diffusive motion consistent with *helical* tracks just beneath the cell membrane, which suggested a possible *treadmilling* behaviour of MreB as a result of polymerization and depolymerization at opposite respective ends of each MreB filament (Kim *et al.*, 2006).

However, by subsequent fluorescent protein labelling of a marker for the cell wall in bacteria as well as MreB, it was later discovered that the whole of the cell wall actually performs a *circumferential* motion as it grows (Garner *et al.*, 2011), which results in shuttling the MreB most probably in grooves in the cell wall as opposed to MreB dynamics being due to intrinsic molecular treadmilling. This may have important dynamic significance for all bacterial cytoskeletal components (Figure 8.7J).

8.5.2 The micro- and nano-architecture of the membrane

The standard structural description of biological membranes for 40+ years has been the *fluid mosaic model* (Singer and Nicolson, 1972) which envisages proteins integrated in a surrounding phospholidid bilayer, and so unless proteins are immobilized by other interactions (for example with cytoskeletal proteins just beneath the cell membrane) they are free to diffuse among the surrounding lipid molecules. However, emerging evidence, primarily from single-molecule studies, now indicates that there is significantly more heterogeneity in the local micro-architecture of biological membranes than a simple fluid mosaic picture would suggest. When most biological membranes are cooled to ~10 °C below their growth temperature there is a step-like *decrease* in the fluidity of the lipid bilayer, which has been attributed to a phase transition from the fluid liquid crystalline state to the more viscous gel state. This has been observed by measuring the mobility of a variety of reporter molecules embedded in the lipid bilayer, having pronounced effects on both the translational and rotational diffusion of lipids.

Evidence for the additional presence of small quasi-crystalline *island* regions of lipids existing in a *sea* of more mobile lipids, essentially compartmentalized *lipid microdomains* with enriched content of hydrophobic molecules such as *cholesterol* and *sphingolipids* (see section 2.7) which are commonly referred to as *lipid rafts* (Simons and Ikonen, 1997), has emerged from bulk level biophysical studies, for example using calorimetry on different lipid mixtures, as well as x-ray diffraction (see Chapter 1). These putative rafts have been speculated as functioning as mobile platforms for lipids and proteins, especially for those involved in intra-cellular signalling, with potentially highly dynamic stability (for an accessible review, see Lingwood and Simons, 2010).

However, the existence of lipid rafts is still a subject of debate, primarily because historically the methods used in support of their existence were not performed on live cells and required recourse to chemical fixation, detergent solubilization of lipids as well as chemical extraction of cholesterol components, all of which potentially could result in raft-like artifacts. However, recent non-invasive single-molecule methods have provided invaluable support for the lipid raft hypothesis in exploring the properties of these heterogeneous membrane features, most especially using small fluorescent tags attached either to membrane proteins or to lipid molecules themselves, and then performing fluorescence imaging often in combination with single particle tracking.

The first single-molecule evidence came from single particle tracking in reconstituted synthetic membranes, for which the variable lipid composition could be finely tuned (Schütz *et al.*, 1997). Using narrow-field illumination (see Chapter 3) the researchers were able to track single lipid molecules labelled with the organic dye rhodamine, and perform iterative Gaussian fitting to a super-resolution precision of ~100 nm sampling at ~5 ms per image. To determine whether or not a given tracked lipid molecule was performing Brownian motion they estimated the probability p for finding each given tracked molecule at a distance r from its point of origin after a time t, and compared this with what is expected from simple Brownian diffusion. On average, Brownian-diffusing particles obey *Fick's second law of diffusion* (also known more commonly as the *diffusion equation*), which is for the radial dimension coordinate r:

$$\frac{dp}{dt} = D\frac{d^2p}{dr^2} \tag{8.7}$$

It can then be shown that the probability P for finding a single particle which starts off at the origin inside a circle of radius r at time t is given by:

$$P = 1 - \exp(-r^2/\langle r^2 \rangle) \tag{8.8}$$

Here, $\langle r^2 \rangle$ is the mean square displacement after time t (see Q8.12). This could be used as a criterion for deciding whether a molecule's mobility is due to underlying Brownian diffusion or not, which indicated evidence both for Brownian-like diffusion in some molecules, and for confined diffusion in corrals of size 100–200 nm in diameter in others, consistent with lipid rafts.

In an extensive live-cell investigation on Chinese hamster ovary cells (Nishimura *et al.*, 2006), researchers were able to label lipids using fluorescent lipid analogue molecules, as well as key membrane protein components, and then apply single-molecule fluorescence imaging combined with single particle tracking to investigate the effects of sample temperature, and the disruption of the cytoskeleton, as well as the effects of removing cholesterol from the cell membrane, on the prevalence of lipid raft regions. Interestingly, the researchers observed that removing cytoskeletal components had relatively little impact, whereas either reducing the temperature or reducing the local membrane concentration of cholesterol significantly increased the surface density for the number of putative mobility-confinement zones in the membrane.

Similar methods could resolve fluorescently labelled lipids in native membranes of muscle cells (Schütz *et al.*, 2000b), indicating again the presence of a sub-population of lipid molecules confined to microdomains of around a few tens of nanometres to ~200 nm diameter. Similar later studies on cell membranes of live T-lymphocyte cells (a white blood cell found in mammals) suggested that raft components may assemble in the vicinity of membrane receptor proteins associated with cell signalling in the immune response only after binding of the correct ligand has brought about activation of the receptor (Drbal *et al.*, 2007). In the non-active resting state of the receptor, membrane compartmentalization by lipid rafts may facilitate the separation of effector signalling molecules from their transmembrane receptor substrates, whereas upon activation this segregation is removed which results in efficient signal transduction. In other words, there may be distinct functional significance to the presence of lipid rafts, as opposed to their being an intriguing but not physiologically significant consequence of transitions between different lipid phase states.

Several other studies using single-molecule imaging combined with single particle tracking on fluorescently labelled membrane proteins already mentioned in this chapter and previously (see sections 7.2, 7.4, 8.2 and 8.4) show evidence for putative zones of confined diffusion in the cell membrane. There is analytical evidence from modelling experimental data from reconstituted fluorescently labelled lipid vesicles that the formation, size and distribution of lipid rafts can be sensibly predicted from critical fluctuation behaviour involving two or more different lipid phase states near to the phase transition temperature, that is the characteristic mean temperature at which a lipid bilayer of a given composition will change a gel-like to a crystalline-like state (interested physical scientists should see Veatch *et al.*, 2008). This may indicate that, although the formation of lipid raft structures in the membrane is likely to be a physical phase-transition effect, the living cell may have evolved ingenious strategies to utilize this phenomenon, either to separate one set of molecules transiently from another (as with regulation of signal transduction), or to keep interacting protein components compartmentalized together (as with the speculative respirazones of OXPHOS).

KEY POINT

Biological membranes not only mark the boundary of even ostensibly simple cells or sub-cellular organelles but have highly **complex** micro- and nano-**architecture** of significant physiological relevance.

THE GIST

- Single-molecule fluorescence imaging can give key details to the localization of protein-based channels, pores and carriers in the cell membrane.
- Many membrane protein-complexes are mobile, and single particle tracking can generate super-resolution precise information for molecular mobility in live-cell membranes over a time scale of a few milliseconds.
- By fluorescence imaging of both membrane protein and lipid components a dynamic nano-scale picture of the cell membrane can be built up, indicating significant local architecture with functional zones of compartmentalization.
- Using step-wise photobleaching of fluorescent proteins the number of subunits expressed in functional molecular machines in the cell membrane can be estimated to a single-molecule precision.
- AFM spectroscopy and imaging can be applied to single protein complexes expressed in native cell membrane patches.
- Super-resolution fluorescence imaging methods can yield very fine detail for the localization of single cytoskeletal protein filaments in the living cell.

TAKE-HOME MESSAGE The cell membrane is an enormously important structure, and aspects of both its structural details and the multitudinous cellular process that rely on it can be studied extensively using a broad range of different single-molecule biophysics techniques.

References

GENERAL

- Berg, J. M., Tymoczko, J. L. and Stryer, L. (2002). *Biochemistry*. Chapter 19: The light reactions of photosynthesis, 5th edition. W. H. Freeman, New York.
- Dawkins, R. (2011). *The Blind Watchmaker*. Penguin new edition.
- Huang, S. and Czech, M. P. (2007). The GLUT4 glucose transporter. *Cell Metabolism* **5**: 237–252.
- Kelkar, D. A. and Chattopadhyay, A. (2007). The gramicidin ion channel: a model membrane protein. *Biochim. Biophys. Acta* **1768**: 2011–2025.
- Lingwood, D. and Simons, K. (2010). Lipid rafts as a membrane-organizing principle. *Science* **327**: 46–50.

- Okuno, D., Iino, R. and Noji, H. (2011). Rotation and structure of FoF1-ATP synthase. *J. Biochem.* **149**: 655–664.
- Singer, S. J. and Nicolson, G. L. (1972). The fluid mosaic model of the structure of cell membranes. *Science* **175**: 720–731.
- Sowa, Y. and Berry, R. M. (2008). Bacterial flagellar motor. *Q. Rev. Biophys.* **41**: 103–132.
- Vacha, F., Bumba, L., Kaftan, D. and Vacha, M. (2005). Microscopy and single molecule detection in photosynthesis. *Micron* **36**: 483–502.

ADVANCED

- Alabi, A. A. *et al.* (2007). Portability of paddle motif function and pharmacology in voltage sensors. *Nature* **450**: 370–375.
- Appelhans, T. *et al.* (2012). Nanoscale organization of mitochondrial microcompartments revealed by combining tracking and localization microscopy. *Nano Lett.* Jan 13. [Epub ahead of print]
- Bates, M., Huang, B., Dempsey, G. T. and Zhuang, X. (2007). Multicolor super-resolution imaging with photo-switchable fluorescent probes. *Science* **317**: 1749–1753.
- Biteen, J. S. *et al.* (2008). Super-resolution imaging in live *Caulobacter crescentus* cells using photoswitchable EYFP. *Nature Methods* **5**: 947–949.
- Brown, T. A., Fetter, R. D., Tkachuk, A. N. and Clayton, D. A. (2010). Approaches toward super-resolution fluorescence imaging of mitochondrial proteins using PALM. *Methods* **51**: 458–463.
- Burnette, D. T. *et al.* (2011). Bleaching/blinking assisted localization microscopy for superresolution imaging using standard fluorescent molecules. *Proc. Natl. Acad. Sci. USA* **108**: 21081–21086.
- Coste, B. *et al.* (2012). Piezo proteins are pore-forming subunits of mechanically activated channels. *Nature* **483**: 176–182.
- Cox, S. *et al.* (2011). Bayesian localization microscopy reveals nanoscale podosome dynamics. *Nature Methods* **9**: 195–200.
- Crane, J. M., Tajima, M. and Verkman, A. S. (2010). Live-cell imaging of aquaporin-4 diffusion and interactions in orthogonal arrays of particles. *Neuroscience* **168**: 892–902.
- Delalez, N. J. *et al.* (2010). Signal-dependent turnover of the bacterial flagellar switch protein FliM. *Proc. Natl. Acad. Sci. USA* **107**: 11347–11351.
- Drbal, K. *et al.* (2007). Single-molecule microscopy reveals heterogeneous dynamics of lipid raft components upon TCR engagement. *Int. Immunol.* **19**: 675–684.
- Düser, M. G. *et al.* (2009). 36° step size of proton-driven *c*-ring rotation in F_oF_1-ATP synthase. *EMBO J.* **28**: 2689–2696.
- Garner, E. C. *et al.* (2011). Coupled, circumferential motions of the cell wall synthesis machinery and MreB filaments in *B. subtilis*. *Science* **333**: 222–225.
- Goldsmith, R. H. and Moerner, W. E. (2010). Watching conformational- and photo-dynamics of single fluorescent proteins in solution. *Nature Chem.* **2**: 179–186.
- Hamon, L., Curmi, P. A. and Pastré, D. (2010). High-resolution imaging of microtubules and cytoskeleton structures by atomic force microscopy. *Methods Cell Biol.* **95**: 157–174.
- Harms, G. S. *et al.* (2001). Single-molecule imaging of l-type Ca(2+) channels in live cells. *Biophys. J.* **81**: 2639–2646.
- Harms, G. S. *et al.* (2003). Probing conformational changes of gramicidin ion channels by single-molecule patch-clamp fluorescence microscopy. *Biophys. J.* **85**: 1826–1838.
- Harms, G., Orr, G. and Lu, H. P. (2004). Probing ion channel conformational dynamics using simultaneous single-molecule ultrafast spectroscopy and patch-clamp electric recording. *Appl. Phys. Lett.* **84**: 1792.
- Hayashi, K., Ueno, H., Iino, R. and Noji, H. (2010). Fluctuation theorem applied to F1-ATPase. *Phys. Rev. Lett.* **104**: 218103.
- Hirvonen, L. M., Wicker, K., Mandula, O. and Heintzmann, R. (2009). Structured illumination microscopy of a living cell. *Eur. Biophys. J.* **38**: 807–812.
- Holden, M. A., Needham, D. and Bayley, H. (2007). Functional bionetworks from nanoliter water droplets. *J. Am. Chem. Soc.* **129**: 8650–8655.
- Hughes, B. D., Pailthorpe, B. A. and White, L. R. (1981). The translational and rotational drag on a cylinder moving in a membrane. *J. Fluid Mech.* **110**: 349–372.

- Ide, T., Takeuchi, Y. and Yanagida, T. (2002). Development of an experimental apparatus for simultaneous observation of optical and electrical signals from single ion channels. *Single Mol.* **3**: 33–42.
- Kim, S. Y. *et al.* (2006). Single molecules of the bacterial actin MreB undergo directed treadmilling motion in *Caulobacter crescentus. Proc. Natl. Acad. Sci. USA* **103**:10929–10934.
- Kuzmenko, A. *et al.* (2011). Single molecule tracking fluorescence microscopy in mitochondria reveals highly dynamic but confined movement of Tom40. *Sci. Rep.* **1**: 195.
- Kuznetsova, T. G. *et al.* (2007). Atomic force microscopy probing of cell elasticity. *Micron* **38**: 824–833.
- Lalkens, B., Testa, I., Willig, K. I. and Hell, S. W. (2011). MRT letter: nanoscopy of protein colocalization in living cells by STED and GSDIM. *Microsc. Res. Tech.* **75**: 1–6.
- Leake, M. C. *et al.* (2006). Stoichiometry and turnover in single, functioning membrane protein complexes. *Nature* **443**: 355–358.
- Leake, M. C. *et al.* (2008). Variable stoichiometry of the TatA component of the twin-arginine protein transport system observed by in vivo single-molecule imaging. *Proc. Natl. Acad. Sci. USA* **105**: 15376–15381.
- Lenn, T., Leake, M. C. and Mullineaux, C. W. (2008a). Clustering and dynamics of cytochrome bd-I complexes in the *Escherichia coli* plasma membrane in vivo. *Mol. Microbiol.* **70**: 1397–1407.
- Lenn, T., Leake, M. C. and Mullineaux, C. W. (2008b). Are Escherichia coli OXPHOS complexes concentrated in specialized zones within the plasma membrane. *Biochem. Soc. Trans.* **36**: 1032–1036.
- Lenn, T. *et al.* (2011). Measuring the stoichiometry of functional PspA complexes in living bacterial cells by single molecule photobleaching. *Chem. Commun. (Camb).* **47**: 400–402.
- Lew, M. D. *et al.* (2011). Three-dimensional superresolution colocalization of intracellular protein superstructures and the cell surface in live *Caulobacter crescentus. Proc. Natl. Acad. Sci. USA* **108**: E1102–E1110.
- Liu, L. N. *et al.* (2008). Light-induced energetic decoupling as a mechanism for phycobilisome-related energy dissipation in red algae: a single molecule study. *PLoS One.* **3**: e3134.
- Liu, L. N. *et al.* (2009). FRAP analysis on red alga reveals the fluorescence recovery is ascribed to intrinsic photoprocesses of phycobilisomes than large-scale diffusion. *PLoS One.* **4**: e5295.
- Liu, L. N. *et al.* (2011). Forces guiding assembly of light-harvesting complex 2 in native membranes. *Proc. Natl. Acad. Sci. USA* **108**: 9455–9459.
- Mudrakola, H. V., Zhang, K. and Cui, B. (2009). Optically resolving individual microtubules in live axons. *Structure* **17**: 1433–1441.
- Nishimura, S. Y. *et al.* (2006). Cholesterol depletion induces solid-like regions in the plasma membrane. *Biophys. J.* **90**: 927–938.
- Nishizaka, T. *et al.* (2004). Chemo-mechanical coupling in F1-ATPase revealed by simultaneous observation of nucleotide kinetics and rotation. *Nature Struct. Mol. Biol.* **11**: 142–148.
- Noji, H., Yasuda, R., Yoshida, M. and Kinosita, K. J. (1997). Direct observation of the rotation of F_1-ATPase. *Nature* **386**: 299–302.
- Oesterhelt, F. *et al.* (2000). Unfolding pathways of individual bacteriorhodopsins. *Science* **288**: 143–146.
- Planchon, T. A. *et al.* (2011). Rapid three-dimensional isotropic imaging of living cells using Bessel beam plane illumination. *Nature Methods* **8**: 417–423.
- Puntheeranurak, T. *et al.* (2006). Ligands on the string: single-molecule AFM studies on the interaction of antibodies and substrates with the Na^+-glucose co-transporter SGLT1 in living cells. *J. Cell Sci.* **119**: 2960–2967.
- Reid, S. W. *et al.* (2006). The maximum number of torque-generating units in the flagellar motor of *Escherichia coli* is at least 11. *Proc. Natl. Acad. Sci. USA* **103**: 8066–8071.
- Saffman, P. G. and Delbruck, M. (1975). Brownian motion in biological membranes. *Proc. Natl. Acad. Sci. USA* **72**: 3111–3113.
- Saga, Y. *et al.* (2002). Fluorescence emission spectroscopy of single light-harvesting complex from green filamentous photosynthetic bacteria. *J. Phys. Chem. B* **106**: 1430–1433.
- Scheuring, S. *et al.* (2001). High-resolution AFM topographs of *Rubrivivax gelatinosus* light-harvesting complex LH2. *EMBO J.* **20**: 3029–3035.
- Schütz, G. J., Schindler, H. and Schmidt, T. (1997). Single-molecule microscopy on model membranes reveals anomalous diffusion. *Biophys. J.* **73**: 1073–1080.

- Schütz, G. J. *et al.* (2000a). 3D imaging of individual ion channels in live cells at 40 nm resolution. *Single Mol.* **1**: 25–31.
- Schütz, G. J., Gerald, K., Pastuchenko, V. P. and Schindler, H. (2000b). Properties of lipid microdomains in a muscle cell membrane visualized by single molecule microscopy. *EMBO J.* **19**: 892–901.
- Seyfert, K. *et al.* (2011). Subunit rotation in a single FoF1-ATP synthase in a living bacterium monitored by FRET. *Proc. SPIE* **7905**: 79050K1–79050K9.
- Simons, K. and Ikonen, E. (1997). Functional rafts in cell membranes. *Nature* **387**: 569–572.
- Sleep, J., Wilson, D., Simmons, R. and Gratzer W. (1999). Elasticity of the red cell membrane and its relation to hemolytic disorders: an optical tweezers study. *Biophys. J.* **77**: 3085–3095.
- Sonnleitner, A., Mannuzzu, L., Terakawa, S. and Isacoff, E. Y. (2002). Structural rearrangements in single ion channels detected optically in living cells. *Proc. Natl. Acad. Sci. USA* **99**: 12759–12764.
- Sowa, Y. *et al.* (2005). Direct observation of steps in rotation of the bacterial flagellar motor. *Nature* **437**: 916–919.
- Stabley, D. R., Jurchenko, C., Marshall, S. S. and Salaita, K. S. (2011). Visualizing mechanical tension across membrane receptors with a fluorescent sensor. *Nature Methods*. **9**: 64–67.
- Sturgis, J. N. *et al.* (2009). Atomic force microscopy studies of native photosynthetic membranes. *Biochemistry* **48**: 3679–3698.
- Tajima, M., Crane, J. M. and Verkman, A. S. (2010). Aquaporin-4 (AQP4) associations and array dynamics probed by photobleaching and single-molecule analysis of green fluorescent protein-AQP4 chimeras. *J. Biol. Chem.* **285**: 8163–8170.
- Tardin, C., Cognet, L., Bats, C., Lounis, B. and Choquet. D. (2003). Direct imaging of lateral movements of AMPA receptors inside synapses. *EMBO J.* **22**: 4656–4665.
- Ulbrich, M. H. and Isacoff, E. Y. (2007). Subunit counting in membrane-bound proteins. *Nature Methods* **4**: 319–321.
- van Oijen, A. M. *et al.* (1999). Unraveling the electronic structure of individual photosynthetic pigment-protein complexes. *Science* **285**: 400–402.
- Veatch, S. L. *et al.* (2008). Critical fluctuations in plasma membrane vesicles. *Am. Chem. Soc. Chem. Biol.* **3**: 287–293.
- Watanabe, N. and Mitchison, T. J. (2002). Single-molecule speckle analysis of actin filament turnover in lamellipodia. *Science* **295**: 1083–1086.
- Waterman-Storer, C. M., Desai, A., Bulinski, J. C. and Salmon, E. D. (1998). Fluorescent speckle microscopy, a method to visualize the dynamics of protein assemblies in living cells. *Curr. Biol.* **8**: 1227–1230.
- Xu, Y. *et al.* (2011). Dual-mode of insulin action controls GLUT4 vesicle exocytosis. *J. Cell Biol.* **193**: 643–653.
- Yasuda, R., Noji, H., Yoshida, M., Kinosita Jr., K., and Itoh, H. (2001). Resolution of distinct rotational substeps by submillisecond kinetic analysis of F1-ATPase. *Nature* **410**: 898–904.

Questions

FOR THE LIFE SCIENTISTS

Q8.1. Molecular turnover, for example as measured for stator components of the flagellar motor, has been observed in several other molecular machines, and some people now argue that all components in molecular machines turn over, it is just a matter of how fast or slow. But if components in a molecular machine in a living cell turn over in a time period which is relatively short compared to the finite lifetime of that component due to cellular mechanisms for degrading older molecules in the cell, then the cell will need constantly to produce new molecular components otherwise the machine will simply run out of components. This process ultimately requires energy, and one might assume that molecular turnover is therefore a selective disadvantage, namely if a mutant cell were to evolve to have far less rapid turnover then the cell as a whole would use up less energy, require less nutrient input and so eventually become the dominant cell type in a typically highly competitive real physiological environment in which the food supply is scare. Why then do components in many molecular machines exhibit relatively rapid turnover?

Q8.2. In the tethered cell assay of section 8.3.1 the cell body rotated at the same time as the fluorescence observation of the motor was made, indicating that these single-molecule measurements were made on a functional molecular machine. However, in terms of the forces experienced by the stator and rotor units it could be argued that this assay is unphysiological. (a) Why? (b) If the cell was in effect turned upside-down so that the motor under observation was free from the coverslip what effect would this have on the forces experienced? (c) What additional problems might there be with observing the fluorescence from single motors in this geometry? (d) What other examples of single-molecule live-cell assays can you think of in which a molecular machine under study could definitively be said to be functional? (Hint: For more ideas leaf through the next chapter.)

Q8.3. (a) What is the standard free energy change for the hydrolysis of a single mole of ATP to ADP in units of joules? In a cell, the free energy of hydrolysis per molecule of ATP to ADP and inorganic phosphate was measured to be in the range -90 to -100 pN nm. (b) If the inorganic phosphate concentration roughly equals the ATP concentration at 1 mM but the ADP concentration varies, what is the intra-cellular range of this ADP concentration? In the F1 rotation experiment of Noij *et al.* (1997) a ca. 1 μm long F-actin filament of 5 nm diameter was observed to rotate at ~6 revolutions per second. (c) Assuming that the water viscosity is 0.001 N m^{-2} s estimate the frictional torque in pN nm and the energy efficiency of the coupling of ATP hydrolysis to F1 rotation at an ADP concentration of 0.1 mM.

Q8.4. (a) Describe how single molecules of glucose are transported from the gut across the cells that line the human intestine into the blood, and show how ATP is used to energize this process.

Q8.5. (a) What generates osmotic pressure across a cell membrane? (b) In what ways do prokaryotes have a different strategy for dealing with this osmotic pressure compared to eukaryotes, and why?

FOR THE PHYSICAL SCIENTISTS

Q8.6. A lipid bilayer of thickness 5 nm and relative dielectric constant $\varepsilon_r = 2.5$ contains bacteriorhodopsin (BR) molecules and Na^+ channels, and forms a spherical vesicle of radius r = 100 nm. (a) Use Gauss' law to calculate the capacitance of the bilayer neglecting the effect of the proteins. The vesicle is illuminated with the sodium channels closed. (b) How many protons must be pumped across the bilayer by BR to establish a protonmotive force of -170 mV? At zero time the light is removed and a total of 15 sodium channels open instantaneously with a single-channel conductance of ca. 10 pS. (c) Initially the solutions are 100 mM NaCl, both inside and outside the vesicle: how many Na^+ ions are there initially inside the vesicle? (d) Obtain an expression for, and make a labelled sketch of, the membrane voltage as a function of time after the channels open, justifying any assumptions you make. (e) If the external medium were changed to 50 mM NaCl, what would the membrane voltage and the excess osmotic pressure inside the vesicle be, after equilibration?

Q8.7. (a) For one-dimensional diffusion of a single membrane protein complex, assuming that the drag force F is related to the speed v by $F = \gamma v$ where γ is frictional drag coefficient, and that after a time t the mean square displacement is given by $2Dt$ where D is the diffusion coefficient, show that if all the kinetic energy of the diffusing complex is dissipated in moving around the surrounding lipid fluid, then D is given by the Einstein–Stokes relation $D = k_B T/\gamma$. For diffusion of TatA protein complexes (see Leake *et al.*, 2008) two models were considered for the conformation of TatA subunits, either a tightly packed cylinder in the membrane in which all subunits packed together to generate a roughly uniform circular cross-section perpendicular to the lipid membrane itself, or a cylindrical shell model having a greater radius for the same number of subunits, in which the subunits form a ring cross-section leaving a central pore. Using step-wise photobleaching of YFP-tagged TatA molecules in combination with single particle tracking the mean values of D were observed to vary as $\sim 1/N$ where N was the estimated number of TatA molecules per complex. (b) Show with reasoning whether this supports a tightly packed or a cylindrical shell model for the complex (Hint: Use the equipartion theorem and assume that the sole source of the complex's kinetic energy is thermal.)

Q8.8. In Figure 8.1D obtained from Ulbrich and Isacoff (2007), in which each subunit of a tetrameric ion channel was labelled with a single molecule of GFP, there appears to be heterogeneity in the brightnesses of the fluorescent foci across the field of view. (a) Why is this? For a more general case of using step-wise photobleaching with a greater number of subunits n per foci (such as Leake *et al.*, 2006), the mean intensity for a single GFP molecule was first estimated to be ca. 5300 ± 1000 counts on the camera detector ($\pm$standard deviation). (b) If this initial intensity of the multimer is measured using an accurate exponential fit to the fluorescence intensity data obtained by photobleaching all of the GFP molecules in a given foci, and the stoichiometry is calculated by dividing the initial intensity by the brightness of a single GFP, estimate, showing your reasoning, the values of n for which the estimate for counting total number of GFP molecules present ceases to be precise down to the level of a single molecule.

Q8.9. (a) Derive the Michaelis–Menten equation for the rate of F1-mediated hydrolysis of ATP to ADP and inorganic phosphate in the limit of excess ATP, stating any assumptions you make. How does this rate relate to the rate of rotation of the F1 γ subunit rotor shaft? (b) How would applying a load opposing the rotation of γ affect this rate of ATP hydrolysis? (c) By assuming that the ATP-independent rotation step of $\sim$30°–40° can be modelled by one irreversible rapid step due to a conformational change in the F1.ADP complex correlated with hydrolysis of ATP with a rate constant k_4, followed at some time afterwards by another rapid irreversible rotational step due to the release of the ADP with a rate constant k_5, derive an expression for the probability that the conversion from the pre-hydrolysis F1.ADP state to the post-ADP-release F1 state takes a time t, and explain why this results in a peaked distribution to the experimentally measured wait time distribution for the $\sim$30°–40° step. (Hint: Look back to Q5.7.)

Q8.10. Stepping rotation of a flagellar motor powered by a flux of Na^+ ions was observed by monitoring the position of a latex microsphere attached to the flagellar filament using back focal plane detection (see Chapter 4), which is conjugated to the 'rotor' shaft of the motor via a 'hook' component. The hook can be treated as a linear torsional spring exerting a torque $\tau_1 = \kappa \Delta\theta$, where κ is the spring constant and $\Delta\theta$ the angle of twist of the sphere, and the viscous drag torque on the sphere is $\tau_2 = \gamma \mathrm{d}(\Delta\theta)/\mathrm{d}t$, where γ is the viscous drag coefficient. By reducing the external Na^+ concentration to below 1 mM, the motor can be made to rotate at ~1 revolution per second or less. (a) Estimate the maximum number of steps per revolution which might be directly observable if $\kappa = 1 \times 10^{-19}$ N m rad^{-1} and $\gamma = 5 \times 10^{-22}$ N m rad^{-1} s. In the experiment ~26 steps per revolution were actually observed, which could be modelled as 'thermally activated barrier-hopping', namely the local free energy state of each angular position was roughly a parabolic potential but with a small incremental difference between each parabolic potential. (b) If a fraction α of all observed steps are backwards, derive an expression for the 'sodium-motive force', which is the electrochemical voltage difference across the cell membrane due to both charge and concentration of Na^+ ions embodied by the Nernst equation, assuming that each step is coupled to the transit of exactly two Na^+ ions into the cell. (c) If $\alpha = 0.2$ and the internal and external concentrations of the driving ion are 12 mM and 0.7 mM respectively, what is the voltage across the membrane? If a constant external torque of15 pN nm were to be applied to the bead to resist the rotation, what would be the new observed value of α?

Q8.11. (a) Show that the electrochemical potential energy in moving a single proton across a narrow proton-pore in the mitochondrial inner membrane in which there is a ca. −0.15 V potential difference is equivalent to ~(5–6)k_BT. The pH in different membrane-bound mitochondrial compartments was measured directly in single mitochondria using fluorescence imaging of pH-sensitive fluorescent proteins. In the compartment immediately outside the inner membrane the pH was measured at 6.9, whereas in the cytosol compartment on the other side of this membrane the pH was measured at 7.6. (b) What is the ratio of proton concentrations either side of this membrane? What is the free energy of a single proton translocated across this membrane from the outside to the inside via an FoF1 complex? (c) How many such translocated protons are required to synthesize a single molecule of ATP? How does this tally with the model of ATP formation in F1? (Hint: As a reminder of some of the properties of ATP, see Chapter 2.)

Q8.12. Rhodamine-tagged lipid molecules expressed in a planar lipid bilayer at low surface density at a room temperature of 20 °C were excited with a 532 nm laser and imaged using high intensity narrow-field illumination (for example, Schütz *et al.*, 1997) sampling at just 1 ms per image to capture rhodamine fluorescence emissions peaking at a wavelength of ~560 nm, which allowed typically 10 ms of fluorescence emission before each rhodamine molecule on average irreversibly photobleached, which indicated that some molecules exhibited Brownian diffusion at a rate of ~7 μm^2 s^{-1} while others were confined to putative lipid rafts of diameter ~100 nm. (a) Show that for lipid molecules exhibiting Brownian diffusion $p = r/r_0 \exp(-r^2/r_0^2)$ is a solution of the diffusion equation, where p is the probability that the lipid lies in an annulus between r and $r + dr$ after time t. Experiments on bulk samples suggested that by raising the bilayer temperature by a further 20 °C it might be possible to disrupt the corralling effect of a lipid raft. From intensity measurements it was estimated that in some of the rafts two fluorescently tagged lipid molecules were confined, but it was not clear whether these lipid molecules were bound to each other or not as each raft could only be resolved as a single fluorescent foci. An experiment was devised such that one image frame was first recorded, the sample temperature was then increased by 20 °C very rapidly taking less than 1 ms, and fluorescence imaging was continued until the sample was bleached. The experiment was performed 1000 times on separate rafts possessing two tagged lipid molecules. In 202 of these experiments two distinct fluorescent foci could just be resolved near to the site of the original raft at the end of the imaging, while in the rest only a single fluorescent foci could be resolved throughout. (b) Explain with reasoning whether this supports a complete disruption or partial disruption model for the effect of raising temperature on the lipid raft.

FOR THOSE WHO HAVE NOT MADE UP THEIR MIND

Q8.13. A fluorescence imaging experiment was performed at video-rate on a bespoke inverted microscope using rod-like bacteria that had a shape very close to a cylinder of length 2 μm capped by hemispheres of diameter 1 μm, in which a diffuse cell membrane protein was tagged using GFP, with bulk biochemical assays suggesting ca. 200 proteins per cell, using a simple flow-cell that consisting of a glass coverslip in optical contact with a high numerical aperture objective lens via immersion oil, with a bacterial cell stuck to the coverslip surface in a water-based buffer, with the walls of the flow-cell being ~100 μm high stuck on the upper side to a glass microscope slide, above which was then simply air. The cells were first imaged using epifluorescence in which the emergent 488 nm laser excitation beam travelled from below then straight up through the sample, resulting in a halo-like appearance in fluorescence to cells which were stuck with their long axis parallel to the glass coverslip surface (i.e. a bright fluorescence at the perimeter of the cell when setting the focal plane to be at the midpoint of the cylinder). (a) Explain these observations. The angle of incidence of the excitation beam was then increased from zero (epifluorescence) which resulted in the beam emerging from the top of the microscope slide at shallower and shallower angles. Eventually, the emergent beam angle was so shallow that it just dipped below the horizon and could not be seen exiting the microscope slide. At the point the experimentalists concluded that the system was set for TIRF imaging and proceeded to take images of several cells as before. However, they were surprised to still see a halo-like appearance to the fluorescence images of the cells when they had expected to see a brighter region which marked the cell bodies. (b) Explain why they had expected not to see halo-like images, but why they actually did. (Hint: Think about both the microscope slide as well as the coverslip.)

Q8.14. Data from energetics calculations indicates that >1 ion (either Na^+ or H^+), perhaps on average ~1.5 ions, are required to effect a 'power stroke' for a single stator unit in a flagellar motor – how could this actually work in terms of molecular mechanisms?

Q8.15. Some experiments indicate that flagellar motor stator units operate in a cooperative fashion, whereas other experiments indicate that it is possible to observe single rotational step events from cell strains which have been engineered to have only very few, possibly only single, stator units in the motor. How is it possible that both of these observations can be true?

Q8.16. The effective surface area of a single light-harvesting protein complex was estimated using AFM imaging to be ca. 300 nm^2. Sunlight of mean wavelength ~500 nm and intensity equivalent to ca. 3×10^{21} photons $m^{-2}\ s^{-1}$ was directed onto the surface of bacteria containing light-harvesting complexes in their cell membrane, whose energy was coupled to pumping protons across a cell membrane of protonmotive force of ca. −170 mV. To be an efficient energy transfer one might expect that the effective transfer time for transmembrane pumping of a single proton would be faster than the typical rotational diffusion time of a protein complex in the membrane of ca. nanoseconds, otherwise there may be danger of undesirable energy dissipation away from the proton pump. Comment, with reasoning, whether this is an efficient energy transfer.

9 Inside cells

I will arise and go now, for always night and day
I hear lake water lapping with low sounds by the shore;
While I stand on the roadway, or on the pavements gray,
I hear it in the deep heart's core.

(WILLIAM BUTLER YEATS, THE LAKE OF INNISFREE, 1888)

GENERAL IDEA

Here we venture beneath the surface of the cell membrane to explore some of the key biological processes that occur in the core of cells, which have been investigated using single-molecule biophysics techniques either in living samples or in physiologically relevant settings in the test tube.

9.1 Introduction

As we saw in the previous chapter, the cell membrane, with its various associated integrated protein complexes, is a key structure being the first point of contact for the cell with the outside world. However, the meat of the cellular machinery for *metabolizing* nutrients, *manufacturing* new molecular material, *responding* to signals detected at the cell membrane surface and for *storing* its genetic code are all located in the innards of the cell, either occurring in the cytoplasm often associated with a variety of cellular sub-structures or, in the case of eukaryotic cells, in specialized membrane-bound organelles. Previously, we discussed some details of two such organelles in the context of membrane-localized processes, the chloroplasts that perform photosynthesis in plant cells and mitochondria that generate the universal cellular fuel of ATP. Here, we will also extend the discussion to processes occurring in the cell *nucleus*, how the genetic code is *packaged* and ultimately *replicated*, and the means by which this code is converted into molecules of protein. But we will begin the chapter outside the nucleus and discuss the biophysical properties of the cytoplasm, and the mechanisms by which molecular cargo is controllably *trafficked* and *sorted* inside the cell.

9.2 Free, hindered and driven molecular diffusion in the cytoplasm

Several cellular processes depend critically on the mechanical elasticity of the cytoplasm, including the transport of cargo inside the cell. Both the properties of the molecular motors that ultimately drive cargo translocation and the elastic behaviour of the watery cytoplasmic environment can be probed using a variety of physiologically relevant single-molecule techniques.

9.2.1 Using single particle tracking to study free and hindered diffusion in the cytoplasm

The cytoplasm of the cell represents a highly *viscoelastic* environment. In other words, it possesses both viscous and elastic properties in much the same way as treacle, so that there is a drag-related resistance to deforming the cytoplasm which is manifest as *hysteresis* energy losses upon stressing and releasing a local portion of cytoplasm. This viscoelasticity has an effect both on the time dependence of the physical shape of the cell, for example when responding to changes in the cytoskeleton architecture (see Chapter 8) and on the free, hindered and driven diffusion of molecules in the cytoplasm. By using single particle tracking under physiologically relevant conditions, it is possible to explore the flow effects of cytoplasm (for a good review see Wirtz, 2009), namely the *microrheology*, to explore this over the length scale of the cell (micrometres), and of *nanorheology* resulting from the length scale of the constituent molecules (nanometres), and to increase our understanding of the biological processes that require molecules to move through this cytoplasm.

One of the key parameters in live-cell microrheology and nanorheology is viscosity, since this directly affects the drag force experienced by mobile molecules and complexes in the cytoplasm. However, the viscosity of intra-cellular regions in effect depends on the length scale over which one is observing. On the nanometre scale this interstitial viscosity is 20–40% higher than that of water at the same temperature, whereas over the length scale of micrometres the drag effects of fibrous structures in the cytoplasm need to be considered, resulting in a measured apparent viscosity typically 100–1000 times that of the viscosity of water.

The SI units of viscosity are $\mathrm{N\,m^{-2}s}$, with the viscosity of pure water being $\sim 10^{-3}$ $\mathrm{N\,m^{-2}s}$ at room temperature, though many researchers in the field use a unit of *mPa s*, i.e. milliPascal seconds, since 1 Pascal is the unit of pressure equivalent to 1 N of force over an area of $1\ \mathrm{m}^2$. Historically, however, the most popular unit of viscosity is the *centipoise*, unit ‘cP’, such that 1 cP = 1 mPa s, and so the viscosity of pure water at 20 °C is ~1.0020 cP.

An ideal method for measuring live-cell microrheology and nanorheology is single particle tracking. Measuring the diffusion of a single particle in the cytoplasm gives a direct measure of its mechanical properties because of the intimate contact between it and the tracked probe, far superior to the next best approaches which rely on some direct contact between the probe and the cell surface (for example with AFM spectroscopy) which suffer a significant disadvantage of not being able to distinguish independently the source of the measured force on the probe tip between the cytoplasmic membrane and the actual components in the cytoplasm itself. Relatively large probes have been used for tracking, for example endocytosed QDs or fluorescent beads of diameter ~100 nm in eukaryotic cells (Hale *et al.*, 2008) in combination with fluorescence imaging which, since the probe size is relatively large compared to the effective mesh size of the cytoplasm, give a *mesoscale* estimate for the local viscosity.

Smaller probes, such as a single dye molecule, offer better spatial resolution when measuring diffusion but come with potential problems. One difficulty is speed. If the viscosity of the cytoplasm is of the order of 100–1000 times smaller than that of the highly crowded and heterogeneous lipid membrane environment (see Chapters 7 and 8) then the Einstein–Stokes relation (see Equation 8.1) implies a larger effective diffusion coefficient by the same factor. Therefore, a potential issue when attempting single particle tracking in the cytoplasm is that nanoscale particles move so fast that they are too blurry to be detected using conventional video-rate tracking. Simple analysis can be applied to obtain an idea of how fast one must image a particle in order to detect it *unblurred*. A particle moving with random Brownian diffusion with effective diffusion coefficient D in n spatial dimensions will move a mean square displacement $\langle r^2 \rangle$ after a time Δt given by:

$$\langle r^2 \rangle = 2Dn\Delta t \tag{9.1}$$

PHYSICS-EXTRA

In practice, when the experimental mean square displacement is measured using single particle tracking one needs to consider the **errors** due to **localization** and those due to **dynamics** of a tracked particle within the finite time sampling window (see Savin and Doyle, 2005). Each orthogonal coordinate axis in space has an associated localization precision variance, σ^2_i where i is an integer in the range 1–3, for example corresponding to $\sigma_x = \sigma_1, \sigma_y = \sigma_2, \sigma_z = \sigma_3$, but generalizing to other coordinate systems, which will each contribute to the measured mean square displacement. Also, if the sampling exposure time is δt per image, then the observed mean square displacement measured after the sampling is complete at time $t + \delta t$ will be higher than the actual mean square displacement at time t by an average amount of $2D\delta t$. Thus, a more general experimental expression for mean square displacement observed experimentally using single particle tracking after a time Δt would be:

$$\langle r^2 \rangle = 2Dn(\Delta t - \delta t/n) + \sum_{i=1}^{n} \sigma_i^2 \tag{9.2}$$

To see a particle unblurred during a single image exposure of time Δt we can say, as a broad rule of thumb, that the distance it has diffused laterally must be less than its own point spread function width, w (see section 3.9), assuming we are using a far-field imaging system such as a fluorescence microscope, which indicates that:

$$\Delta t < 0.19\lambda^2/Dn(\mathrm{NA})^2 \tag{9.3}$$

This equation applies to a microscope imaging system with numerical aperture NA (typically set by the objective lens, which is in the range 1.2–1.6) and mean emission wavelength λ (500–600 nm for typical fluorescence imaging). In the cell membrane, diffusion is limited to two spatial dimensions, and D for typical membrane proteins and small complexes can be in the range (1–20) $\times$ 10^{-3} μm^2 s^{-1}. This gives a minimum required exposure time of ~100 ms, so in other words video-rate sampling of around a few tens of milliseconds per image is fine. However, for cytoplasmic diffusion, there are three spatial dimensions, with an effective D more typically of 5–10 μm^2 s^{-1}, which indicates a minimum exposure time of just a few milliseconds.

Thus, the first technical challenge is to image at such fast rates. This requires not only a fast camera detector but also a high excitation intensity for the fluorescence imaging employed, as otherwise there are insufficient photons detected from the fluorescence emission signal per dye molecule per image above the level of detector noise. This can be achieved using *narrow-field epifluorescence* and *slimfield* approaches (see section 3.6) for which the laser excitation field is shrunk in size at the sample to create a very high local excitation intensity. However, this can lead to a second difficulty in that at such high excitation intensities there is a potential risk of photodamage to the cell, due primarily to free-radical formation, and of rapid photobleaching of the dye resulting in relatively short duration tracks.

Furthermore, there is a third technical hurdle of image contrast. Since the dye tags are likely to be at least several hundred nanometres away from the cell membrane the popular image enhancement technique of TIRF is in general not effective (although under certain

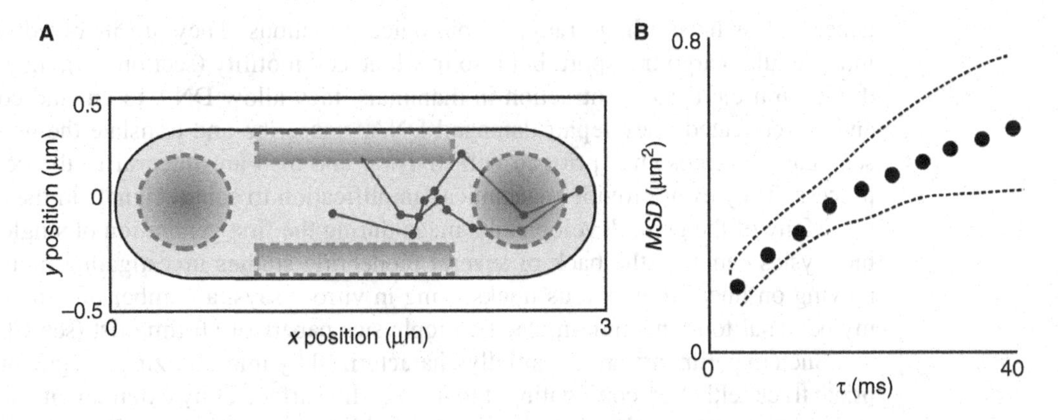

FIGURE 9.1 Cytoplasmic diffusion in bacteria of the stress-response protein RelA. **(A)** Schematic depiction of single tracked RelA-Eos2 particle in a bacterial cell (circle points), with zones of high (dotted rectangles) and low (dotted circles) particle mobility indicated both from experimental data and simulation. **(B)** Schematic of an experimental mean square displacement (MSD) for a single track (circles) indicating good agreement with diffusion coefficient $D = 13\ \mu m^2\ s^{-1}$ with error bounds at $\pm 0.5\ \mu m^2\ s^{-1}$ (dotted lined).

conditions of sparse fluorophore density there might be a small contrast enhancement provided that the fall-off in excitation intensity of the TIRF evanescent field with distance from the sample glass coverslip surface is less rapid than the drop in intensity due to defocusing away from the focal plane, see Q9.5), so other enhancement strategies need to be employed.

One recent study investigated the diffusion of the cytoplasmic protein RelA, found in bacteria (English *et al.*, 2011). RelA appears to function at the heart of bacterial adaptation to *starvation* and *stress*. Under starvation conditions of very low external concentrations of amino acids, RelA remodels the bacterial physiology by manufacturing signalling complexes called *alarmone guanosine phosphates* in a manner which is regulated by RelA binding to ribosomes (see Chapter 2), which in turn globally affect rates of DNA transcription, mRNA translation, and DNA replication in that single cell.

Here, the researchers used the photoactivatable fluorescent protein Eos2 fused to RelA in bacterial cells in combination with narrow-field epifluorescence illumination which resulted in sufficiently sparse active fluorophore labels which could be imaged with an exposure time of ~1 ms per image, allowing them to perform single particle tracking of RelA in the whole of the cytoplasm (Figure 9.1A). At such high excitation intensities the single-molecule dye tag would typically bleach after only 10 consecutive image frames, and so to overcome the limitation of needing to observe mobility effects over time scales longer than ~10 ms, the researchers strobed the illumination (see Chapter 3) to generate a single 1 ms excitation laser pulse every 4 ms.

This allowed observation of the mobility of individual molecules over a time scale of more like ~40 ms before bleaching occurred. The results indicated anomalous diffusion observable over time scales of a few tens of milliseconds which would have been very difficult to spot at shorter time scales (Figure 9.1B), with a sub-population of tracked particles showing relatively hindered diffusion in specific regions of the cells near to the poles, consistent with a ribosome-bound state of the RelA (Figure 9.1A).

9.2.2 Trafficking of molecular cargo in the cytoplasm

Driven or *directed* diffusion in the cytoplasm is due principally to one phenomenon: the mobility of motor proteins running on *cytoskeletal filament tracks* (for a good review of this, and other related aspects of whole cell motility, see Bray, 2000). Motor proteins as a general class have a huge range of biological functions. They are involved not only in intra-cellular cargo transport, but also in whole cell motility (section 8.3), they are key to driving muscle tissue contraction in mammals, they allow DNA to unwind controllably and be replicated, they repair damaged DNA, transcribe and translate the genetic code, segregate chromosomes prior to cell division and also are integral to the cell division process. They even provide mechanical amplification to sound signals in the ear.

Many of the great developments made during the first generation of single-molecule biophysics came on the back of several pioneering studies investigating motor proteins moving on linear filamentous tracks using in vitro assays, a number of which feature in my personal top-ten single-molecule biophysics papers of all time list (see Chapter 6).

Such experiments are essentially characterized by immobilizing a segment of appropriate track, either by conjugating it to a coverslip surface or by extending it using flow or optical tweezers, and then having a relatively high concentration of the appropriate motor protein in the sample solution allowing a sufficient number of observable *stochastic* binding events to the track. The motor protein is labelled with a suitable probe (of which the list of candidates is extensive, including small fluorescent nanoscale beads, quantum dots, gold-coated nanospheres for laser darkfield tracking, microspheres for back focal plane detection and single fluorescent dye molecules). These elegant test tube studies generated significant insight into the mechanisms of motor protein movement on tracks, allowing observations of individual steps and even steps *within* steps at the nanometre and sub-nanometre length scale corresponding to individual stepping motions of the molecular motors (for a comprehensive review, see Veigel and Schmidt, 2011).

Our knowledge of the mechanisms of action of individual molecular motors from elegant single-molecule test tube experiments is advanced, building from several seminal earlier multi-molecule biophysical investigations which were done primarily on the myosin motor protein, which interacts with F-actin tracks in contracting muscle tissue (see Kolomeisky and Fisher, 2007). There are primarily three types of motor protein relevant to vesicle trafficking: myosins (which use actin as a track), kinesins and dyneins (both of which use microtubules as tracks, but are biased to travel in *opposite* directions on it). Extensive single-molecule studies in vitro have been performed on all these classes of motor proteins.

To move our knowledge yet closer towards understanding what really happens in far more physiologically relevant environments has proven technically challenging. For live-cell work in particular, the noise of the signals of detected probes is much higher than in test tube studies, both due to background signals from the rest of the cell and due to independent movement of the molecular track, making it far more difficult to detect nanoscale stepping motions of the motor proteins compared to the earlier test tube studies. However, important developments have been made recently (for a detailed description of the practical developments see Cai *et al.*, 2010).

Linear molecular motors share many common features with those seen previously for the FoF1 rotary motor (see section 8.3). Firstly, their mode of action can be described in terms of *Brownian dynamics*, namely that in addition to any interaction forces between molecules and frictional drag force there is a random force related to the thermal fluctuations of the surrounding water solvent molecules (the *Langevin force*, see sections 1.4 and 5.2). Secondly, the interaction of molecular motors with their associated tracks is *cyclical*. The motor molecules undergo various chemical and conformational changes involving ATP binding, hydrolysis and release, ultimately resulting in the generation of mechanical force and driven motion along the track, but they eventually return back to their initial state from which the cycle can repeat.

The specifics of the motor mechanism obviously depend upon the type of molecular motor involved and its specific corresponding molecular track. Historically, most experimental research to elucidate the stages involved in molecular motor motion have used the myosin class of motors moving along an F-actin track as an experimental system, but there are features of the cycle that are reasonably general to all motor mechanisms.

(i) The motor is bound tightly to a binding site on the track.
(ii) ATP binds to the motor, resulting in a molecular conformational change so that the motor's affinity to the track's binding site is substantially lowered.
(iii) The motor unbinds from the track.
(iv) The bound ATP molecule is hydrolysed, with the released energy *coupled* to generating *strain energy* (with a resultant molecular conformational change) in the motor, increasing the motor's affinity to the binding site again. This conformational change also results in some directional bias with regard to the motor's active *head*.
(v) The motor re-binds to the nearest available binding site, subject to random thermal fluctuations of surrounding water molecules (this means there is a chance it can go *backwards* as well as *forwards* along the track, or even rebind to the original binding site, however the point is that the ATP binding step (iv) above biases the binding to be more likely in the forwards direction).
(vi) Inorganic phosphate from the ATP hydrolysis is released which triggers a *power stroke* force-generation step in the molecular motor.
(vii) ADP from ATP hydrolysis is released from the motor.

The cycle (i)–(vii) can then repeat once more. Overall, each generic cycle above catalyses the hydrolysis of one molecule of ATP and couples this to one power stroke. All transitions are *reversible*, but the free energy of ATP hydrolysis leads in general to net stepping motion biased in one particular direction relative to the track, relating to the conformation change induced in step (iv). This is because in general the track is polar. For example, a common cytoplasmic track for molecular cargo transport is formed from a microtubule, with corresponding molecular motor proteins of the classes dynein and kinesin which bind to it. The active heads of either dynein or kinesin only bind to a microtubule in one orientation relative to the track axis, while ATP binding gives each step its biased direction through a process known as *neck linker zippering*. In terms of thermodynamics, the forward stepping motion has a significantly lower free energy barrier ΔG compared to the reverse stepping motion.

For example, kinesin on a microtubule is estimated to have a forward stepping ΔG of 8–9 times k_BT, which compares with a value of approximately twice that for the reverse process which, if we compare the equivalent Boltzmann factors assuming *detailed balance* (see Chapter 5) of $\exp(-\Delta G/k_BT)$, indicates that the forward stepping process is almost an order of magnitude more likely than the reverse process (for an accessible review of the Brownian stepping behaviour in molecular motors see Yanagida *et al.*, 2007). Basic thermodynamics indicates that $\Delta G = \Delta H - T\Delta S$, where ΔH and ΔS are the enthalpic and entropic changes for the process respectively. The size of ΔH for the well-studied kinesin–microtubule system is roughly the same for the forward and reverse processes at $\sim 18k_BT$, however the value of $T\Delta S$ for the forward step process is $\sim 9k_BT$ compared to only $\sim 4k_BT$ for the reverse process, so in other words forward stepping results in more favourable *disorder* in the kinesin–microtubule system.

Microtubules are composed of alternating tubulin monomers called α and β monomers which polymerize to form repeating $\alpha\beta$ dimer subunits, which then form helical protofilaments of ~ 13 $\alpha\beta$ units each. These bundle laterally involving up to ~ 14 such protofilaments to form the actual microtubule. The directionality of microtubules is defined by reference to the *plus* end which undergoes molecular turnover of α and β monmers and so can grow in length, and the capped *minus* end which does not turn over.

For most cells the minus ends are located in specialized *microtubule organizing centres*, such that the plus ends spread out towards the periphery of the cell.

Most kinesins perform a net movement towards the plus end of a microtubule, thus entailing transporting cargo from the centre of the cell towards the periphery, whereas most cytoplasmic dyneins make a net movement in the opposite direction. Some kinesins are biased towards movement to the minus end of microtubules, and these appear to be more prevalent in organisms that lack dynein, such as the cells of *flowering plants*. There is also another form of dynein called *axonemal dynein*; this has a similar molecular mode of action to cytoplasmic dynein but is expressed in *cilia* resulting in a beating motion.

BIO-EXTRA

A **cilium** is a specialized cellular organelle in eukaryotes which protrudes from the body of the cell. Cilia share some similarities to flagella (see section 8.3) and perform a variety of motile functions ranging from whole cell swimming behaviour of certain micro-organisms, to causing a **beating motion** of surrounding fluid which can move mucus in the lining of **respiratory tracts** and waft **ova** towards the uterus lining the walls of the **fallopian tubes** of mammals. They also perform non-motile chemical, thermal and mechanical sensory functions. They are based on a core **axoneme** which has a characteristic arrangement of nine parallel microtubule filaments arranged radially with two central core microtubule filaments (a so-called **'9+2' arrangement**), which utilize the motor protein **axonemal dynein** in a highly coordinated fashion between the filaments to bring about a periodic bending/beating motion of the whole axoneme structure.

Most motor proteins involved in cargo transport are characterized by having two broadly symmetrical components tightly entwined, each component possessing a long α-helical stalk region a few tens of nanometres in length which separates a *cargo-binding domain* at the tail from a *motor region* at the opposite head end, joined to the *stalk* by a flexible neck whose length depends on the specific motor protein (Figure 9.2A). Both heads in the motor domain can bind to the relevant track and have ATPase activity (i.e. can bind and hydrolyse ATP), making them very similar to the myosin motor protein responsible for muscle contraction on which much of the formative structural and biochemical research was done. In general, the heads appear to operate in such a way that one head remains bound while the other steps over in a *hand-over-hand* mechanism to the next nearest binding site.

For a number of years there was debate as to whether the mechanism was hand-over-hand or more an *inchworm* type model, but this has been largely resolved by a series of elegant single-molecule experiments in vitro, which started with the seminal FIONA paper on a type of myosin with a particularly long neck region (which made it easier to study) called myosin V moving on F-actin (Yildiz *et al*., 2003). Similar studies confirmed this mode of action for other myosins and other in vitro single-molecule techniques demonstrated this for kinesins. Asbury *et al.* (2003) used optical tweezers both for trapping microspheres conjugated to kinesin protein molecules and for monitoring their position to nanometre precision on a microtubule track, and Yildiz *et al.* (2004) used a FIONA strategy on small organic Cy3 dye labels attached to kinesin.

The type of cargo transported in the cell by these motor proteins is extensive. It includes several membrane-based structures most usually in the form of vesicles which contain the cargo, but also organelles such as *mitochondria* and *peroxisomes* (see Chapter 2), mediated through binding to an *adapter receptor protein complex* integrated in the membrane at the tail end of the motor protein (Figure 9.2A). The contents of the vesicles can be highly varied since they often originate from the *Golgi apparatus* where newly manufactured proteins of several different types are packaged prior to secretion, and thus can include a wide range of different proteins for export.

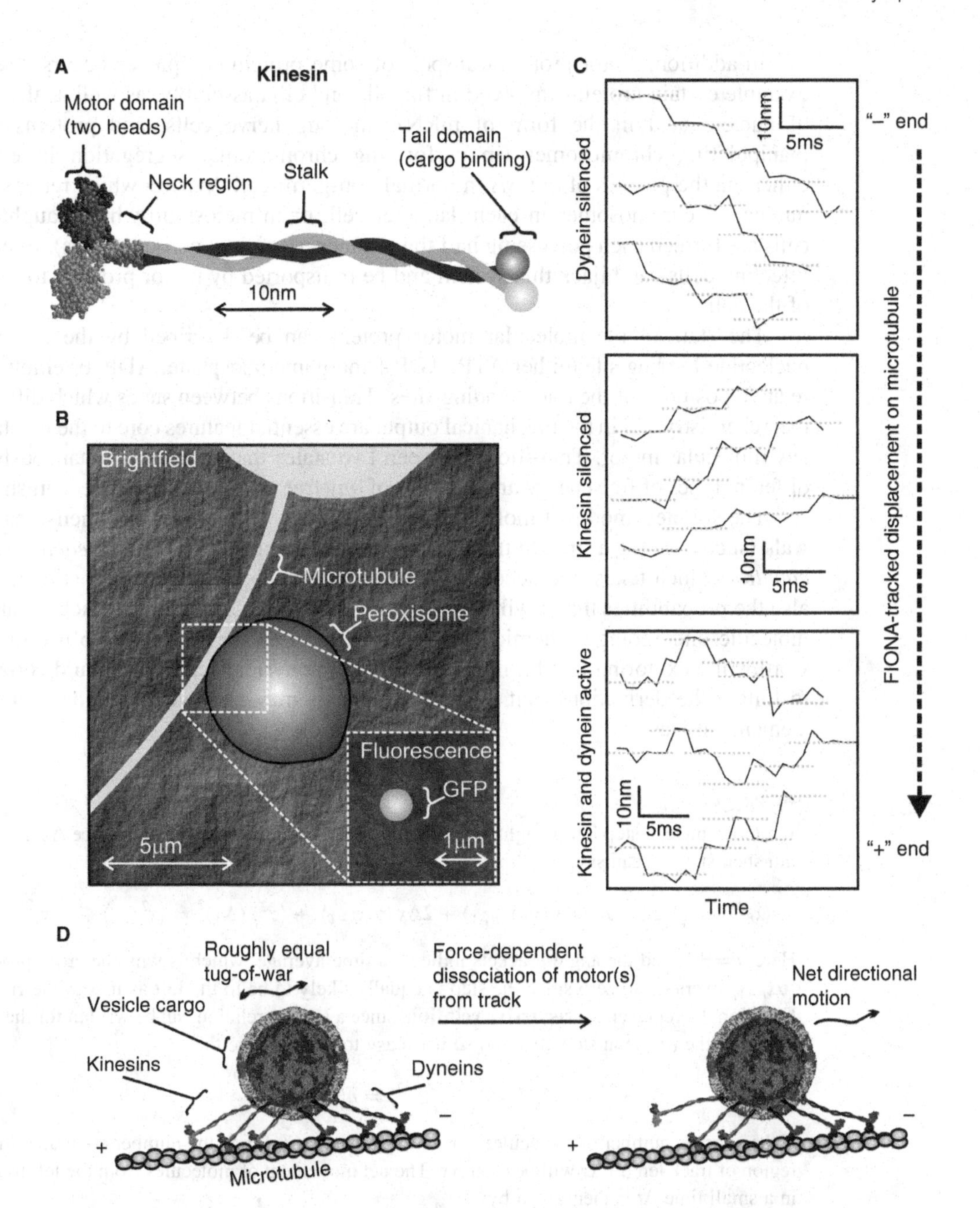

FIGURE 9.2 **(A)** Structure of kinesin motor protein based on known motor domain structure PDB:ID 3KIN, with cartoon for stalk and tail extrapolated. **(B)** Schematic depiction of brightfield and (inset) fluorescence images of microtubule with fluorescence tag on a transported peroxisome, and **(C)** schematic for displacement versus time traces based on iterative FIONA Gaussian fitting of point spread function images of tracked fluorescent dye molecule showing stepping motions of integer multiple of ~8 nm each of kinesin (upper panel), dynein (middle panel) and both kinesin and dynein motors (lower panel), putative step 'levels' indicated by grey dashed lines. **(D)** Schematic suggesting how force-dependent track-dissociation of a motor which is part of several which are engaged in a tug-of-war can result in net movement of a trafficked molecular cargo.

In addition, motor protein transport of some protein complexes occurs directly, for example certain proteins involved in flagella and cilia assembly, as well as the transport of nucleic acid in the form of mRNA in long nerve cells, and proteins used for manipulating chromosomes (in performing chromosomal segregation in eukaryotes either via the process of *mitosis* in normal non-germ-cell division which retains the total number of chromosomes in each daughter cell, or in meiosis in which daughter germ-cells are formed each possessing half the number of original chromosomes). Even viruses infecting cells can hijack this system and be transported by motor proteins to other parts of the cell.

The state of the molecular motor protein can be described by the contents of a nucleotide-binding site (either ATP, ADP + inorganic phosphate, ADP, or empty) and the relative positions of the track binding sites. Transitions between states which differ both in their chemistry and in the mechanical output are essential features core to the mechanism of any molecular motor. Transitions between two states that depend simultaneously on two different types of free energy are the basis of *any* free energy coupling mechanism.

The simplest model of molecular mobility starts with just a one-dimensional random walk (such as along a straight track), but can be generalized to give the *reaction–diffusion equation* which takes into account the diffusion of the molecule, its frictional drag and also the probabilities that it will step forwards and backwards along a track, related to the molecule undergoing a chemical transition in making the step. It is so useful that it is worthwhile exploring in a bit more depth below; physical scientists should certainly aim to follow the derivations, but bold life scientists may also wish to understand the key elements at least.

PHYSICS-EXTRA

If a molecule can step left or right along the track with equal likelihood a distance Δx, then for the nth such step we can say:

$$\langle x_n{}^2\rangle = \langle (x_{n-1} + \alpha\Delta x)^2\rangle = \langle x^2{}_{n-1}\rangle + 2\Delta x\,\langle \alpha.x_{n-1}\rangle + \langle \alpha^2\rangle\,(\Delta x)^2 = \langle x^2{}_{n-1}\rangle + (\Delta x)^2 \quad (9.4)$$

Here $\alpha = \pm 1$, and the angle brackets indicate a time average, which is why the cross-product $\langle \alpha x_{n-1}\rangle$ averages to zero since the step is equally likely to be to the left as it is to the right. Equation 9.4 represents a **recursive relation**, since a similar relation can be written for the $(n-1)$th step, and the $(n-2)$nd step etc., and so it is easy to demonstrate that:

$$\langle x_n{}^2\rangle = n(\Delta x)^2 \quad (9.5)$$

If the average number of molecules per unit distance is $C(x)$ then the number contained in a small region of track length Δx will be $C(x)\,\Delta x$. The net movement of molecules from the left to the right in a small time Δt is then given by:

$$J(x)\Delta t = (C(x) - C(x + \Delta x))\Delta x/2 \quad (9.6)$$

Here $J(x)$ is simply defined as the flux of molecules at a distance x, with the factor of 1/2 emerging because a molecule in a given small region of track with have a probability of one half of stepping either to the left or to the right, with $C(x+\Delta x)-C(x)$ equal to the small change $\Delta C(x)$ in number of molecules per unit length between x and $x+\Delta x$, equal. **Fick's first law of diffusion** is defined as:

$$J(x) = -D\frac{dC(x)}{dx} \quad (9.7)$$

Thus, in the limit of very small Δx and Δt, Equations 9.6 and 9.7 indicate that:

$$D = 2(\Delta x)^2/\Delta t \tag{9.8}$$

Using Equation 9.8 we can then easily arrive at Fick's second law (the **diffusion equation**):

$$\frac{\mathrm{d}C(x)}{\mathrm{d}t} = D\frac{\mathrm{d}^2C(x)}{\mathrm{d}x^2} \tag{9.9}$$

But also, using the result of Equation 9.5 we have that:

$$\langle x_n^2 \rangle = 2D\Delta t \tag{9.10}$$

This gives us the well-known result that the **mean square displacement** for Brownian diffusion scales **linearly** with the **time**. If we then consider that a real molecule will also be experiencing a drag force $\gamma v = \gamma \Delta x/\Delta t$ where v is its speed and γ its frictional drag coefficient, and use the equipartition theorem to equate its kinetic energy with $(1/2)k_BT$, then it is easy to prove the Einstein–Stokes relation as seen previously (see Equation 8.1). This drag force also modifies Fick's first law to give a version of the **continuity equation**:

$$J(x) = -D\frac{\mathrm{d}C(x)}{\mathrm{d}x} + \frac{FC(x)}{\gamma} \tag{9.11}$$

This then modifies the diffusion equation to give the **Fokker–Plank equation** (also known as the **Smoluchowski equation**):

$$\frac{\mathrm{d}P}{\mathrm{d}t} = D\frac{\mathrm{d}^2P}{\mathrm{d}t^2} - \frac{\mathrm{d}}{\mathrm{d}x}\left[\frac{F}{\gamma}P\right] \tag{9.12}$$

Here the factor of $C(x)$ has been normalized by the total number of molecules present, and so P then represents the equivalent probability for finding a molecule at position x. In practice, the mechanism by which individual motor proteins step on their track involves a chemical transition between two different states. If we label these initial and final states as i and f respectively, with associated probabilities for appearing at distance x being P_i and P_f with rate constants for the forward and reverse chemical state transition being k_{if} and k_{fi} respectively, then we arrive at the one-dimensional form of the **reaction–diffusion equation**:

$$\frac{\mathrm{d}P_i}{\mathrm{d}t} = D\frac{\mathrm{d}^2P_i}{\mathrm{d}t^2} - \frac{\mathrm{d}}{\mathrm{d}x}\left[\frac{F}{\gamma}P_i\right] + \sum_f (k_{fi}P_f - k_{if}P_i) \tag{9.13}$$

This reaction–diffusion *continuum* equation (i.e. a purely analytical model which assumes that the variables in the equation can take any 'continuous' value between permitted limits) has proven remarkably powerful in modelling the properties of motor proteins on tracks (see Q9.6). The simplest model assumes three transitions of the molecular motors on the track.

(i) A *binding event* to the track.
(ii) A *power stroke* of the molecular motor, in which the molecule undergoes a conformational change resulting in a force exerted on the track.
(iii) An *unbinding event.*

The three transitions will in general be coupled to chemical processes relating to ATP binding, hydrolysis and product release. In practice, many such equations may need to be solved, requiring significant computational approaches.

However, the reaction–diffusion continuum approach does not really explore the nanoscopic, discrete behaviour of an individual motor, such as generating a prediction for the distribution of dwell times between steps which could be compared against experimental data. To really probe the singular nature of molecular motors requires *discretized simulation*, for example utilizing *Monte Carlo* simulation approaches (see section 2.10).

Monte Carlo simulations have the advantage that individual runs reveal the nanoscopic behaviour of motor proteins such as step sizes, distribution times for intervals between steps and so forth. However, the incremental time step between successive iterations of the simulation must be small compared to the reaction time to ensure that the simulated motor is in one particular unique chemical state during each time step. This can mean that there is a very high computational load for long simulations, with other disadvantages that several such long runs may be needed to determine average properties such as motor speed, dependence upon external load etc. In practice, significant *coarse-graining* is applied to such simulations (for example modelling the motor protein as uniform head shapes with simple linkers and a neck of well-defined stiffness) and so the computational load is not as intensive as for true atomistic level simulations.

A popular recent approach which also minimizes computational load is to *combine* elements of continuum reaction–diffusion modelling with discrete Monte Carlo approaches (interested physical scientists should read Fange *et al.*, 2010). In addition, these may be combined with other more *complex* continuum formulations for molecular motor stepping (for example, the *thermal ratchet*).

The internal organization of eukaryotic cells depends upon trafficking of molecular cargo inside the cell, with motor proteins responsible for pulling a diverse array of attached vesicles and cell organelles. As is evident, much is known about the functioning of individual motors from in vitro single-molecule investigation, but the motors in vivo are far more complex. Firstly, there is the attachment of the cargo itself, which is regulated via several complex mechanisms involving a combination of phosphorylation, calcium ion signalling and controlled protein assembly and degradation (see Akhmanova and Hammer, 2010). Secondly, there are deviations from our observations of the motor mobility in the cell compared to those in vitro.

For example, using a combination of TIRF and confocal imaging (Mashanov *et al.*, 2010), researchers could monitor vesicles of GFP-labelled ion channels being transported inside the cell and ultimately docking with the cell membrane, exhibiting significant deviations from simple directed or Brownian motion with evidence for heterogeneity of mobility not seen in vitro.

Most interestingly, there is significant recent evidence from several studies which indicate that cargos in vivo are carried by *multiple* motors which can involve significant *coordination*. Individual motors are usually observed moving in a net single direction along a filament track in vitro. However, many different cargos in real cells are observed to move *bidirectionally* along the same filament. Using a GFP fusion tag on a signalling protein present in peroxisome organelles (see Chapter 2), researchers could use TIRF to localize the stepping movement of motor-trafficked single peroxisomes near to the cell membrane of live insect cells to within a few nanometres precision (Figure 9.2B), with a sampling time of just $\sim$1 ms per image (Kural *et al.*, 2005). By using *RNA interference* (RNAi), either dynein or kinesin production could be controllably inhibited (see section 2.9).

This indicated clear differences in both step size, direction and speed for the associated dynein and kinesin motors, but most significantly was consistent with a scheme in which a single organelle was moved along a given filament by a process akin to a *tug-of-war* between several competing dynein and kinesin motors pulling on the cargo in opposite directions (Figure 9.2C, D). This process has also been observed for two different types of myosin motors, *myosin Va* and *VI*, which pull in opposite directions on a single F-actin filament, demonstrated by a recent test tube study (Ali *et al.*, 2011).

Furthermore, a recent complex single-molecule in vitro study which monitored the forces and positional displacement of optically trapped beads in a network of both microtubule and F-actin filaments, in which beads were conjugated both to microtubular motors in the form of kinesin and dynein, and to an F-actin based motor in the form of *myosin V*, suggested that at junctions between microtubule filaments it was possible for such a bead literally to change track. Even more interesting, at junctions between a microtubule and an F-actin filament, it was again possible to change track, but here using a completely different type of molecular motor (Schroeder *et al.*, 2010). This, therefore, suggests that having many motors of different types bound to a given molecular cargo can in principle permit extensive trafficked movement in three spatial dimensions, as opposed to just a single dimension as indicated by driven motion along a single filament.

One final point to note with regard to in vivo versus in vitro studies in investigating molecular trafficking is that the motor forces estimated in vivo on trafficked particles are typically 50–100% higher than those measured in equivalent in vitro experiments. This might perhaps be some effect relating to multiple motors in the native cell, but it may also be an indication that there are other additional regulatory factors that are important in the trafficking process which have yet to be identified and which thus are absent from the existing test tube level studies.

Individual microtubule steps have also been observed in live cells using endocytosed QDs (Nan *et al.*, 2005). Here, the superior photostability and brightness of QDs compared to fluorescent proteins permitted a localization precision of ~1 nm with a sampling time of only 0.3 ms per image. This allowed very precise measurement of the nanoscale stepping and directional switching of endosomes on microtubule filaments, due again to the opposing actions of multiple dynein and kinesin motors.

In another recent study, researchers used GFP tagging to monitor the single-molecule dynamics of both kinesin and tubulin on the microtubule track in living mammalian cells (Kner *et al.*, 2009). Here, they utilized TIRF microscopy to image components of the cytoskeleton just beneath the cell membrane, but applied structured illumination as well (see Chapter 3). This allowed super-resolution imaging to be performed at a precision of ~100 nm with a frame rate of ~11 Hz.

Visualizing trafficked cargo in live cells is still not trivial since the standard approaches of attachment of either a relatively large probe or bright organic dye to a motor protein are very difficult to achieve inside cells. Because of this, one popular type of vesicle trafficking studied using single particle tracking involves *melanosomes*. These are vesicles which naturally contain coloured *melanophores* which are exported to the cell membrane via microtubule tracks, again mediated through the opposing action of dynein and kinesin, and since they contain a natural chromophore they have a distinct advantage of being visible using standard brightfield microscopy with no potential hindering dye label or probe (Kural *et al.*, 2006).

Recently, collaborative single-molecule cellular biophysics research has allowed investigators to monitor cargo trafficking in living cells using internalized optically trapped microspheres (work in progress from the research groups of Erika Holzbauer and Yale Goldman, University of Pennsylvania). The cells used are *macrophages* which normally function in the immune response of vertebrate animals to engulf foreign particles in a process called *phagocytosis*, which has elements in common with the more general cellular internalization process of endocytosis (see Chapter 7). Microspheres could be coated with a selection of different molecular motors used in cargo trafficking, prior to internalization by the cell.

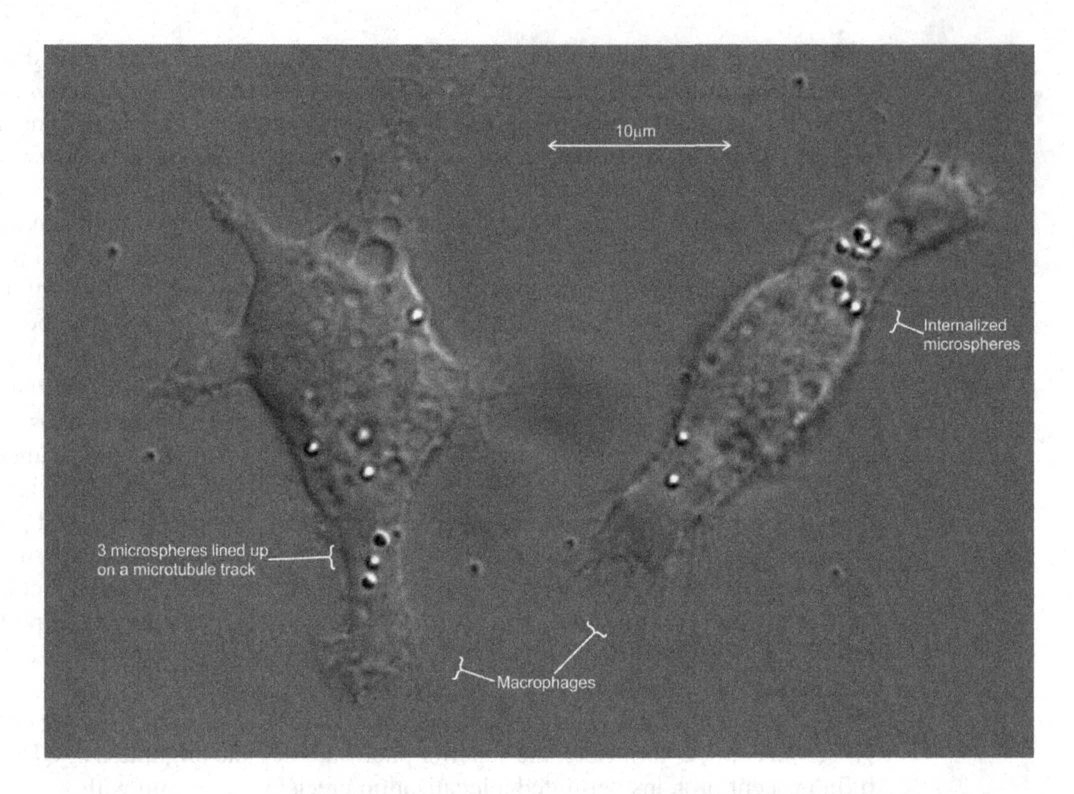

FIGURE 9.3 Live-cell optical trapping to study molecular cargo trafficking. A DIC image of 1 μm diameter latex microspheres that have been phagocytosed into mouse macrophage cells (cell strain J774a.1). Once internalized, the microspheres are encased into a native organelle that recruits the kinesin motor proteins (called kinesin-1 and kinesin-2), and cytoplasmic dynein. These motors then transport the latex microspheres into sub-cellular compartments bidirectionally along individual microtubules. A technique to calibrate the optical trap directly in living cells is used, which takes into account the complex local viscoelastic properties of the cytoplasm. (Original data used with permission of Adam G. Hendricks, Erika L. F. Holzbaur and Yale E. Goldman, Pennsylvania Muscle Institute and Department of Physiology, University of Pennsylvania, Philadelphia, PA.)

Such microspheres could then be observed to move stochastically along internal molecular tracks (Figure 9.3). This has permitted accurate in vivo estimates of the molecular forces involved, and is now opening the door to manipulating cargo trafficking in individual cells.

9.3 Chromosomes and DNA: their architecture and replication

The genetic material of the cell is embodied in its DNA, located inside the cytoplasm. For most of its lifetime this DNA is condensed into units which permit *concerted management* of DNA in coordination with many other activities. Changes in local DNA packing are known to play very critical roles in regulating the ultimate production of peptides and proteins from the genes. Its robust transmission to subsequent generations of cells requires *faithfully replicated* DNA, *efficiently transferred* to daughter cells, involving coordinated interaction between replication and segregation of the DNA, and of cell division. In eukaryotic cells there is a system for manipulating delays between different cell cycle stages, whereas in prokaryotic cells such as bacteria there is no such system so these processes are *highly integrated*: understanding any one system requires knowledge of events elsewhere.

The core features of DNA replication are conserved across many cell types. In eukaryotes the way in which DNA is segregated into the next generation is through complex *mitotic machinery*. In prokaryotes, although there is emerging evidence for potential similarities in the methods employed for pulling apart replicated chromosomes, it is not obvious that any such molecular machinery exists to the level of complexity as is found in eukaryotes. Rather, this prokaryotic DNA segregation process may depend more sensitively on exceptional levels of *positioning control* over individual molecules utilizing coordinated activities of different nanoscale molecular machines.

9.3.1 Probing the architecture of DNA packing using super-resolution imaging

The state of DNA condensation is correlated to specific stages in a cell's cycle. During *interphase*, eukaryotic chromosomes are organized in the cell nucleus into large, relatively low density assemblies of length scale 30–300 nm composed of long, continuous, linear sections of DNA wrapped around protein complexes called *nucleosomes* (see Chapter 2). As the cell cycle proceeds into *metaphase*, the packed DNA becomes denser until eventually it forms distinct, mitotic chromosome structures. Studies from bulk biochemistry and EM suggested the presence of DNA–protein higher order fibrous structures potentially corresponding to intermediate levels of folding before the compact chromosome structure is reached. However, until recently, detailed structural observations under physiologically relevant conditions have not been possible.

In a recent study, investigators were able to label one of the *histone* proteins that comprises the nucleosome with GFP in one of the mitotic chromosomes of the fruit fly *Drosophila*. They found that by exposing the GFP to a reduced form of the vitamin B derivative *riboflavin* they could convert it stochastically to a form that emitted fluorescence in the red as opposed to the green when excited with blue laser light. This enabled super-resolution images of the mitotic chromosome substructures to be reconstructed in live cells, in much the same manner as used for STORM (see Chapter 3).

Using Gaussian fitting of point spread function images of the 'red' GFP in combination with specialized *de-noising algorithms* of the images to enhance the GFP intensity allowed spatial precisions of ~20 nm, which for the first time indicated the clear presence of the intermediate filamentous structures in functional cells. In another study, researchers were able to perform green-red dual-colour super-resolution imaging on two different chromosomal proteins labelled with GFP and red fluorescent protein, RFP, respectively to within a ~20 nm precision (Gunkel *et al.*, 2009). These proteins were for a nucleosome marker, a histone protein called H2A, and an ATPase subunit called Snf2H that is used in certain *chromatin remodelling complexes* (nanoscale machines capable of folding the chromatin into different, tightly packed three-dimensional conformations).

Although this study suffered the disadvantage of being performed on chemically *fixed* (i.e. dead) cells to minimize measurement noise in the optimization of localization precision of the dye tags, it demonstrated great promise in revealing details of the interaction of remodelling machine *hotspots* in the chromosomes, co-localized with the chromatin itself, at an unprecedented effective spatial resolution.

Bacteria, like other prokaryotes, lack a cell nucleus, and the DNA, apart from a few exceptions, is not composed of linear segments in general but rather is present as one or more circular chromosomes which are not packed into the characteristic mitotic chromosomal structures as for eukaryotic cells. They do not possess nucleosomes, however there are a variety of *condensin* proteins which are responsible for significant compaction of the DNA. The compacted bacterial DNA is present in a cytoplasmic *nucleoid* region, analogous to the nucleus of eukaryotes but not delimited by a specific nuclear membrane.

The typical DNA compaction ratio is huge at 10^3–10^4 with at least 20% of the cell volume occupied by nucleoid (microbiologists often refer to this as the *excluded volume*). In practice, it seems likely that the compacted DNA acts as a mesh such that many proteins are excluded from its core, implying that many proteins in the cytoplasm that either bind to the DNA or, at various points in their lifetime, to the cell membrane may be diffusing between polar and mid-cell regions in highly confined cytoplasmic volumes, resulting in potentially dramatically elevated local concentrations, which could have significant implications on the kinetic behaviour of several different biological processes.

Being so compact makes standard live-cell fluorescence imaging techniques challenging, since the length scale of the nucleoid is comparable to the standard optical resolution limit. However, several recent studies have emerged using a variety of super-resolution approaches to monitor the architecture of the nucleoid in live cells. Fluorescent proteins, in addition to undergoing irreversible *photobleaching* to a dark state, also exhibit reversible *photoblinking* under certain conditions of high excitation intensity in which the proteins *alternate* stochastically between light and dark states. This blinking, therefore, can be used to ensure that the mean nearest neighbour distance between light-state fluorescent proteins is greater than the standard optical resolution limit, thus allowing super-resolution Gaussian fitting to each distinct point spread function for fluorescence intensity to be performed (see section 3.9). Using YFP tagged to the non-specific DNA binding protein *hydroxyurea* (HU), researchers were able to reconstruct two-dimensional projections of the nucleoid in live *Caulobacter crescentus* bacterial cells to within ~10 nm precision (Lee *et al.*, 2011), showing indications of local clustered architecture.

9.3.2 Replicating DNA

DNA replication is carried out by a highly complex molecular machine called the *replisome*, which consists of several different protein components which function in a highly coordinated fashion. Significant recent live-cell single-molecule research into DNA replication has been done on model bacterial systems. In *E. coli* there are at least 13 different proteins that are involved in the replication process, the complexity of which reflects the size of the challenge of ensuring highly faithful replication of the original DNA.

Different proteins are required for a wide range of tasks: proteins to locate the start site for replication and to act as a point of assembly for the other proteins in the replisome (a *primase*, *clamp* and *clamp-loader*), a *helicase* to unwind the helical DNA controllably into a fork-like structure consisting of two single-stranded DNA sections called the *leading* and *lagging strands* depending upon whether their orientation is 5′–3′ from the fork centre to the end of the strand, or 3′–5′ (see section 2.6), a *DNA polymerase* to read off the sequence of bases in each strand either continuously in the case of the leading strand or in truncated *Okazaki fragments* of 1000–2000 bases each for the lagging strand (the polymerase is actually composed of three different proteins, and it can only move in the 5′–3′ direction which thus results in the discontinuous replication on the lagging strand), proteins to generate replicated strands via base-pairing to the respective strand template as opposed to allowing single-stranded sections to base-pair spontaneously with other parts of itself (a protein called *single-strand binding protein* or Ssb), a *DNA ligase* to join up sections of replicated DNA, and even a protein called DnaE to *proof-read* the newly replicated sections.

Using YPet fluorescent protein tags, researchers were able to monitor all of the key protein components of the replisome in live *E. coli* cells (Reyes-Lamothe *et al.*, 2010).

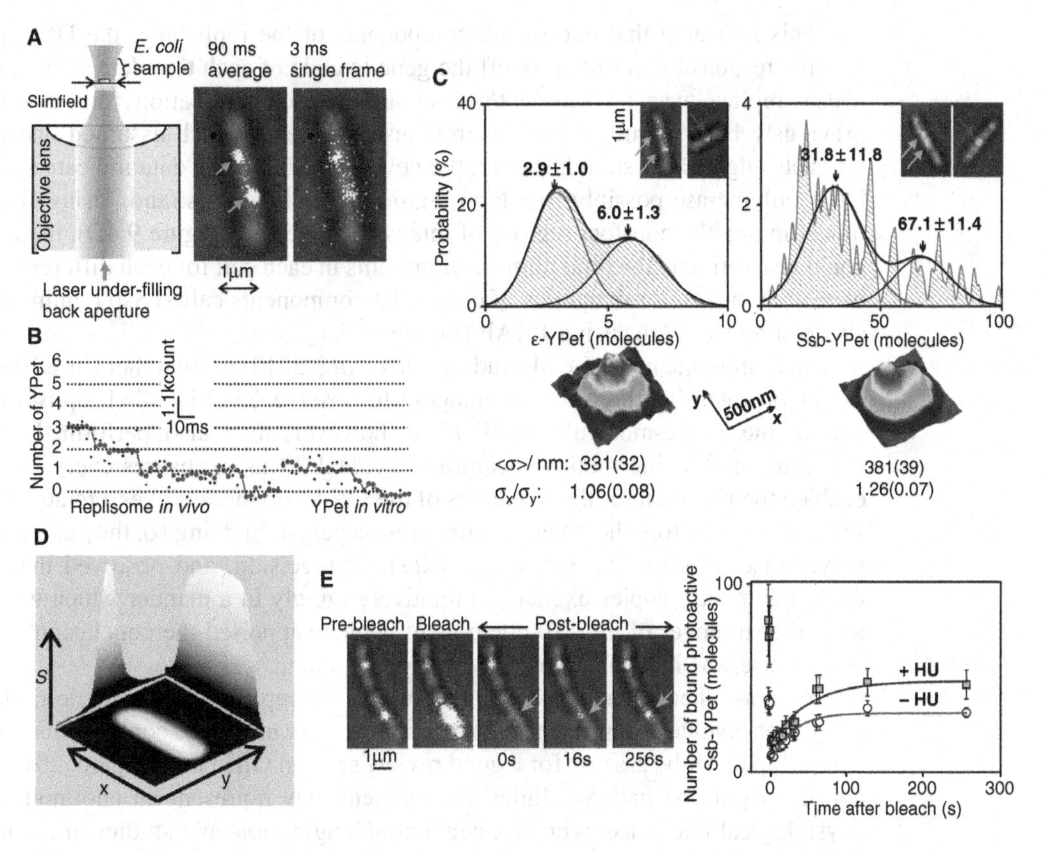

FIGURE 9.4 The single-molecule architecture of DNA replication. **(A)** Slimfield illumination is used to image the individual protein components of the replisome at ~millisecond time resolution, resulting in **(B)** step-wise photobleaching of individual YPet fluorescent proteins which can be used **(C)** to estimate stoichiometry of the different replisome components, shown here with the ε component of the DNA polymerase and the Ssb protein, with dual Gaussian fits shown with peaks corresponding to cells with either two or one foci each (upper panel). Mean spatial distribution of the foci intensity, indicating that core components such as the DNA polymerase are symmetrical and reasonably compact with a ca. 50 nm diameter centred on the replication fork, whereas non-core components such as Ssb extend along the long axis of the cell (lower panel). **(D)** Convolution modelling of the non-foci intensity can be used to estimate the total cellular content of each replisome component. **(E)** FRAP on individual complexes of fluorescently labelled Ssb indicates significant turnover over a time scale of tens of seconds, which is markedly increased in the presence of the replication inhibitor hydroxyurea (HU). (Author's original data, for full details see Reyes-Lamothe *et al.*, 2010.)

They used *slimfield* illumination (see section 3.6) to generate high excitation intensities localized to individual cells which permitted observation of single protein components over a sample time of just 3 ms. Most cells contained two distinct fluorescent foci per cell, consistent with two replication forks moving in opposite directions from a starting origin around the single circular chromosome of DNA, though a minority contained just one focus per cell which was consistent with two replication forks being too close to be observed distinctly (Figure 9.4A). During continuous slimfield excitation, the YPet molecules irreversibly, stochastically photobleached in a step-like manner (Figure 9.4B). This allowed the distribution of multimer stoichiometry to be estimated for all the replisome proteins (see section 8.3), as well as permitting measurement of the spatial distribution of the YPet intensity in each focus (Figure 9.4C).

This indicated that one of the components of the replisome, the DNA polymerase enzyme responsible for reading off the genetic code of each template strand and creating a new replica, was present in *three* copies in each replication fork, not *two* as had previously been surmised for the previous ~40 years which assumed one polymerase for each single DNA strand. Rather, the new single-molecule data indicated an additional DNA polymerase possibly in a loop region on the lagging strand. Using deconvolution modelling on the non-foci regions of intensity in the cell (Figure 9.4D), the investigators could also estimate the total number of proteins in each cell for each different protein, and were able to monitor dynamics of one of the components called Ssb required to stabilize single-stranded DNA, using FRAP (Figure 9.4E).

In a subsequent live-cell study (Lia *et al.*, 2012), researchers used similar high excitation intensity fluorescence imaging to observe YPet-labelled replisome components at the single-molecule level. Here, however, they also performed *stroboscopic illumination* allowing them to acquire individual 20 ms exposures every 200 ms, which enabled them to monitor the dynamics of the DNA polymerase over a relatively long time scale of ~10 s before the YPet fluorophores bleached. In doing so, they could monitor the polymerase *turnover dynamics* with excellent precision, and observed that one of the three polymerase copies exchanged relatively rapidly in a manner which was correlated to similar turnover of the Ssb component, which supported the conclusion that the third polymerase was indeed located on the lagging strand.

Several other investigations have studied the replisome at a single-molecule level using not live-cell approaches but complex, *reconstituted* components in elaborate bottom-up test tube assays (for a good review see van Oijen and Loparo, 2010). Although not directly in the native cellular environment they represent an enormous increase in physiological relevance over first generation single-molecule studies in the number and complexity of the single-molecule components involved.

One such assay (Georgescu *et al.*, 2011) utilized an artificial lipid bilayer to localize reconstituted DNA strands, with multiple replisome components present including the helicase, clamp and polymerase. The DNA was labelled with an *intercalating* fluorescent dye present in the solution which was known not to impair replication significantly and, since the DNA strand was close to the planar lipid bilayer surface, high-contrast TIRF imaging could be applied. By flowing solution over the lipid bilayer the replicating DNA could be stretched out to form a fluorescent curtain of DNA, and the length of the curtain could be monitored with time to indicate the rate of DNA replication.

The investigators were able to generate multi-subunit polymerase complexes that contained either two or three polymerase monomers, and found that those containing three had a faster and more efficient replication rate than those containing two polymerases, and left fewer gaps (presumably unligated DNA sections) between replicated segments of DNA, thus suggesting a clear role for the third polymerase in the native cell.

9.3.3 Segregating the genetic material

The process of *chromosome segregation* is critical to the transfer of replicated genetic material to subsequent generations of cells. In eukaryotes this occurs both during the germ-cell production process of *meiosis* and during non-germ-cell, or *somatic*, reproduction of *mitosis*, during which the parent cell replicates its complement of chromosomes into two identical sets of *chromatids* which are subsequently pulled to separate halves of the cell ultimately to form the set of chromosomes in each new daughter cell prior to the parent cell dividing.

The process of pulling the chromatids apart involves a complex mechanism that, despite exemplary genetics studies and live-cell multi-colour fluorescence imaging investigations using different colour tags to a variety of different protein components implicated in the process, is still not well understood in great detail at the mechanistic level of single molecules. What is apparent is firstly that a *centrosome* substructure is replicated in the vicinity of the middle of the cell, and these two centrosomes are then pushed to opposite ends of the cell in a process that again is not well understood mechanistically but appears to involve transport on a microtubule track via a *mitotic centromere-associated kinesin* that travels anomalously towards the *minus* end of microtubules.

Once there, each centrosome then starts to generate multiple microtubules that polymerize to spread out back towards the middle of the cell to form *mitotic spindles*. Eventually these reach the chromosomes located towards the middle of the cell. Once attached to a chromosome, a combination of microtubule polymerization and depolymerization, in addition to ATP-dependent molecular motor activity at the point of attachment, is believed to cause the chromatid sister pair to be pulled apart.

The movement of pulled chromatids in live cells can be tracked dynamically and automatically in three spatial dimensions with super-resolution precision (Thomann *et al.*, 2002). Here, the investigators labelled chromosomes in budding yeast cells by inserting a special tandem of array of 100–250 *tetracycline* sequences into the DNA, which act as binding sites for a protein called TetR which was fused to GFP, thus constituting a very bright chromosomal marker. Each typical array spans 11,000 nucleotide bases of DNA, equivalent to a potentially very large molecule, ~3.5 μm in its linear 'B' form (see Chapter 2).

However, the DNA in mitotic yeast cells is compacted by a factor of around 200 and so the chromosome tags here occupied an equivalent sphere of only ~25 nm diameter. The tetracycline arrays were inserted ~2000 nucleotide bases away from the centromere region, denoting the physical centre of the chromosome where sister chromatids are linked together, equivalent to 5–10 nm length of compacted DNA. This meant that, using full three-dimensional Gaussian fits to the intensity profiles of the observed point spread function using standard widefield epifluorescence microscopy, the *xyz* centroid coordinates could be extracted for the locus of the microtubule attachment with the chromosome to a very high precision owing to the high brightness of the tracked foci, in the range of 5–20 nm.

The specific point of attachment on microtubules of the individual chromatid is a protein complex called the *kinetochore*, located at the chomatid midpoint. Recently, researchers were able to investigate the substructure of the kinetochore using super-resolution imaging (Ribeiro *et al.*, 2010). Previous data suggested that kinetochores might be formed from chromatin fibres that are compacted around structures similar to nucleosomes found in chromosomes but using imaginatively named CENP, standing for *centromere proteins*, which have a variety of modifications from the related histones normally in nucleosome structures (see Chapter 2).

Here, researchers were able to tag one of the key proteins, CENP-A, with the photoactivatable fluorescent protein Dronpa via genomic fusion, and then extract this chromatin from the cell and label other potential histone modifications using a series of different specific primary antibodies followed by incubation with a secondary antibody labelled with the fluorescent red dye Alexa647 (see Chapter 3). Using repeated cycles of laser activation at 405 nm wavelength and excitation at 488 nm wavelength allowed the researchers to image stochastically single Dronpa molecules and subsequently perform super-resolution Gaussian fits to their respective point spread functions.

By using dual-colour green-red imaging their super-resolution 'green' localization could then be reconstructed in images against the standard 'red' optical resolution intensity distribution of the other histone-like components, in effect a chimera between conventional widefield microscopy and super-resolution PALM. This indicated that the kinetochore is a substantially folded structure with bridging cross-links between the folds which mechanically stabilizes the structure as it is pulled during mitosis.

In prokaryotes, it is still not clear what precise mechanisms are used to segregate newly replicated chromosomes prior to cell division. It was thought until very recently that no equivalent mitotic machinery existed, however in the bacterium *Caulobacter crescentus* there is now evidence for a partitioning (*Par*) system which has many features equivalent to eukaryotic mitotic spindles (Ptacin *et al.*, 2010). Investigators used a combination of in vitro evidence for EM with single-molecule sensitive dual-colour fluorescence imaging both on live cells and on chemically fixed cells to achieve a super-resolution precision. This suggested that one of the Par proteins, ParA, appeared to form filaments which could attach to original and replicated bacterial chromosomes through an adapter protein ParB to a centromere-like region of DNA called *parS*. Depolymerization of the ParA filament could potentially then result in driven retraction of a chomosome pair to opposite ends of the cell in what the researchers termed a *burnt bridge Brownian ratchet* scheme.

It is not clear how many other prokaryotic systems utilize such a method, but studies suggest that even relatively simple bacteria have evolved remarkably intricate methods at the level of single molecules to bring about highly efficient segregation of chromosomes.

Rapidly following chromosomal segregation is the process of cell division. Again, the study of this process in eukaryotes at the single-molecule level is still in its infancy, but recent developments have been made in investigating this process in bacteria. Bacterial cell division is achieved using a protein called FtsZ, which is homologous to the eukaryotic protein tubulin, to form a mid-cell *Z-ring* which is located in the centre of dividing cells to within a very consistent ~2% precision of its total length. With the assistance of in excess of 10 different *divisome* proteins, the Z-ring then constricts to bring about cell division. STED has recently been used to reconstruct super-resolution images of the Z-ring in different stages of assembly using fluorescent markers to FtzZ on chemically fixed cells, which identified key helical structures that may be used in intermediate stages of the Z-ring formation (Jennings *et al.*, 2010). PALM has also been used on live bacterial cells to generate super-resolution images of the intact Z-ring, suggesting tightly packed *protofilaments* arranged in multiple layers in each ring (Fu *et al.*, 2010).

The final stages of cell division in bacteria require more than just FtsZ alone since curvature measurements of Z-ring protofilaments suggest that the smallest Z-ring possible has a diameter of ~20 nm which is still large enough to allow chromosomal material to pass through, which could lead to incorrect DNA segregation. It seems that a remarkable motor protein called FtsK may be implicated here (it is one of the *strongest* known motor proteins capable of kicking off all other proteins bound to DNA that are in its path) and it is likely that single-molecule experiments on this system in the near future may provide significant insight.

9.4 Translating, transcribing and splicing the genetic code

The central dogma of molecular biology states in essence that the genetic code embodied in a cell's DNA is first transcribed into mRNA which in turn is translated into polypeptides and proteins (see section 2.9). Over recent years this has been investigated in great detail using the techniques of single-molecule cellular biophysics (for a good, modern review see Li and Xie, 2011). This includes monitoring the nature of *gene expression bursts* from the DNA, investigating the single-molecule localization and dynamics of *transcription factors* that regulate this gene expression, imaging the synthesis and distribution of mRNA directly, studying the makeup of the *nuclear envelope* and *pores* of the nucleus through which mRNA molecules must translocate following translation, and probing various mechanisms by which the transcribed genetic code is modified before it is finally translated into peptide chains.

9.4.1 Gene expression bursts and transcription factors

In a seminal study using live *E. coli* bacterial cells, gene expression was probed one molecule at a time using single-molecule fluorescence imaging (Yu *et al.*, 2006). The researchers used the model gene expression system of the *lac* operon, well known to biochemistry undergraduate students as being an exemplary system by which gene expression is suppressed via the influence of a DNA binding protein called a transcription factor, such that translation of the genetic code of that gene is only allowed to occur once the transcription factor experiences a reduced binding affinity to the DNA, due in general to the binding of a specific ligand to the protein (Jacob and Monod, 1961).

BIO-EXTRA

An **operon** is the genetics term for a cluster of genes on the same section of the chromosome which are all under control of the same **promoter**, all of which are transcribed and translated in the same continuous run. The promoter is a short nucleotide base sequence on the DNA which acts as an initial binding site for **RNA polymerase** (see section 2.9) and thus determines where transcription of an mRNA sequence translated from the DNA begins. The classic description of the operon comes from studies of the bacterial *lac* operon. Here there is an **operator** region, a section of DNA between the promoter and the operon genes, which binds a regulator protein which acts as repressor to prevent the movement of the RNA polymerase along the DNA. In the *lac* operon, there are three genes which express proteins that form enzymes which are involved either in digesting the disaccharide **lactose** into the monosaccharides **glucose** and **galactose**, or in transporting lactose across the cell membrane. However, decreases in the intra-cellular concentration of lactose results in reduced affinity of the repressor protein to the *lacI* gene which in turn is responsible for generating the LacI protein repressor molecule which prevents expression of the operon genes and is by default normally switched 'on' thus preventing operon gene expression. Thus, in a roundabout way a decrease in cellular lactose concentration results in an increase in the genes required to transport lactose from outside the cell and digest it inside the cell. This system is also regulated in the opposite direction by a repressing protein called **CAP** whose binding in the operator region is inversely proportional to intra-cellular glucose concentration. In other words, there is negative feedback between the rate constants of gene expression and the products of gene expression. The non-native chemical **isopropyl-β-D-thio-galactoside** (IPTG) is known to bind to the LacI repressor and inactivate it, thus causing the operon genes to be expressed. This phenomenon is widely utilized in genetic studies involving controllable expression of genes for protein research using model *E. coli* bacterial cell systems. Typically, a gene which is desired to be expressed is fused upstream of the *lac* promoter region in the *lac* operon and genetically spliced into a short circular sequence of DNA called a plasmid. These plasmids can then be amplified in number and **transfected** into suitably **competent** bacterial cells, so that when these bacteria grow and divide they will also replicate the plasmids which will be passed on to subsequent generations. A common approach is to insert another gene into the plasmid not under control of the *lac* promoter which is always **constitutively active** which makes continuously a suitable enzyme known to confer a specific **antibiotic resistance** to the bacterial cell. Thus, if the bacteria are grown in a solution which contains that type of antibiotic in the growth medium (or streaked out on agar plates if first selecting for individual bacterial colonies) then only the cells which truly contain the plasmids will survive. After a few generations of cell division almost all of the cells in the growth culture will contain the plasmid. Then, if IPTG is added to the growth medium it will be ingested by the cells and the repressing effects of LacI will be inactivated, thus

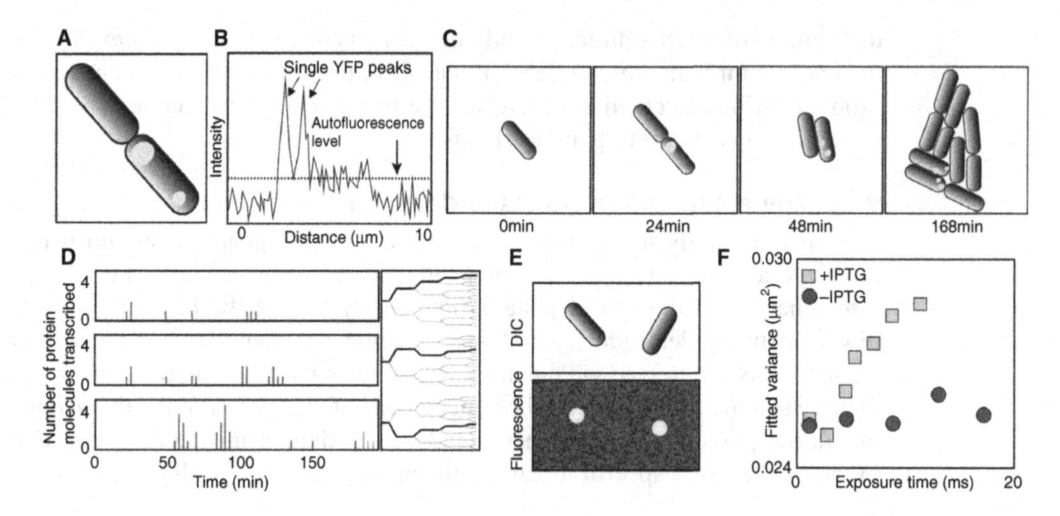

FIGURE 9.5 **(A)** Schematic depictions of brightfield (DIC) and fluorescence images overlaid for Tsr-YFP with **(B)** schematic for fluorescence intensity indicating single molecules above the level of cellular autofluorescence which can be monitored **(C)** over multiple cellular generations to yield **(D)** kinetic statistics from clusters of gene burst activity (schematic). **(E)** Schematic of brightfield (upper panel) and fluorescent foci (lower panel) for LacI-YFP, showing **(F)** differences in measured fluctuation behaviour in the presence or absence of the chemical IPTC which binds to LacI and reduces its binding affinity to DNA, thereby allowing gene expression to proceed for the *lac* operon genes (schematic).

the protein of interest will start to be made by the cells, often at levels far above normal **wild-type** levels since it is difficult to prevent a large number of plasmids from being present in each cell. Since the IPTG does not have an infinite binding affinity to the LacI there is still some degree of suppression of protein production, but similarly the LacI repressor is not permanently bound to the operator region and so even in the absence of IPTG a small amount of protein can be produced (colloquially described by biochemists as being due to a **leaky plasmid**). In principle it is possible to tailor the concentration of IPTG used to a desired cellular concentration output of protein, however the response curve for changes in IPTG concentration is typically steeply sigmoidal meaning that in practice there is more of an all or nothing response to IPTG concentration. Another operon system used for genetics research in *E. coli* is the **arabinose** operon which uses arabinose itself as the equivalent repressor-binder, and here the steepness of the sigmoidal response is in general less than that of the IPTG operon system making it feasible to control the protein output by varying the external concentration of arabinose.

The investigators were able to encode genetically a membrane protein Tsr fused to YFP in the genome but under control of the *lac* promoter to act as a reporter for gene expression from that operon. This particular Tsr protein was selected firstly since the normal unfused Tsr with no fluorescent tag attached is expressed as part of a membrane receptor complex at levels of around thousands per cell, and so the small expression bursts from the *lac* operon do not significantly alter the native physiology.

Secondly, as newly expressed Tsr is ultimately transported to and localized into the cell membrane its mobility is sufficiently low that conventional epifluorescence using a relatively slow sampling time of 100 ms per image could be applied to image single unblurred Tsr–YFP molecules (Figure 9.5A). The researchers recorded live-cell fluorescence images up to eight cell generations (Figure 9.5B), allowing a comprehensive picture of gene burst activity to be assembled (Figure 9.5C). Using the intensity of observed fluorescent foci due to YFP allowed them to estimate the number of Tsr molecules that were made in each burst of activity (Figure 9.5D) and to correlate this to the number of mRNA molecules in each burst. They concluded that a single molecule of mRNA was associated with each gene expression burst, but that the number of Tsr molecules translated from this varied with a mean of roughly 4–5 for each single mRNA molecule.

This indicated that translation of a single mRNA molecule by one or more *RNA polymerase* enzyme complexes results in the generation of multiple copies of a protein molecule before the mRNA is degraded. To date, the direct microscopic visualization and counting of individual protein molecules for measurement of gene expression measurements has only been demonstrated in prokaryotic cells. In eukaryotes, the lifetime of a typical mRNA molecule is longer before it is controllably degraded by enzymatic action. As a result, this will present a challenge to monitoring gene burst activity since the number of molecules present in each burst will consequently be significantly higher, making it more difficult to achieve single-molecule counting precision.

In a similar study to that utilizing the Tsr–YFP as a gene expression reporter, researchers used YFP fused directly to a *lac* repressor protein to monitor the dynamics and kinetics of this individual transcription factor at the single-molecule, single cell level (Elf *et al.*, 2007). Here, the investigators again used the *lac* operon in the model *E. coli* system, but fused the repressor protein LacI to YFP. By using high laser-excitation intensity they could image single LacI–YFP molecules in the cytoplasm with a laser open-shutter time down to just 1 ms per image, sufficient to allow unblurred images to be recorded (Figure 9.5E), and by applying stroboscopic illumination they could make observations over tens of milliseconds before the YFP photobleached.

By monitoring the spatial fluctuations and displacements of the LacI repressor under different external concentrations of the chemical IPTG (Figure 9.5F), which is known to bind to the repressor and thus permit *lac* operon gene expression to proceed, the investigators could monitor the kinetics of binding and dissociation of the LacI transcription factor at a single-molecule level. This also indicated that for ~90% of the time when the LacI is conjugated to the DNA it is actually bound non-specifically in a process of facilitated passive diffusion along the DNA track until it finally finds the operator binding site. This suggested a generic method which might be utilized by many such transcription factors in other cell types.

A subsequent study (Li *et al.*, 2009) has further demonstrated that crowding of the DNA in the bacterial nucleoid can actually facilitate *hopping* between relatively separate looped sections of DNA, thereby permitting transcription factors to find their desired binding site more efficiently than by simply diffusing along long stretches of DNA alone.

The data from single cell, single-molecule experiments can be used to estimate the total content of specific proteins within the cell, but on a cell-by-cell basis. For example, by applying image deconvolution to the single cell fluorescence images described previously in this chapter, and using knowledge of the fluorescence intensity associated with a single fluorescent protein, typically using the single photobleach step molecular signature of a fluorescent protein, one can estimate the number of proteins in a cell typically to better than ~30% precision (if typically the size of the photobleach step is comparable to the measured noise on that single-molecule signal then we might expect the proportional error on measuring the intensity from N such fluorescent proteins to vary as $\sim 1/\sqrt{N}$ from Poisson sampling statistics, see section 8.3, so for total cellular copy numbers of proteins in the range 10–1000 this would indicate typical errors of a few per cent).

This compares with arguably the next best technique of western blotting (see Chapter 1) which probes the cellular extract from a bulk sample of cells using gel electrophoresis combined with specific labelled antibodies on the protein of interest, which has an associated error of more like 200–300%. Thus, using a single cell, single-molecule approach the distribution of protein content across a population of cells can be estimated very precisely, and the results can be fed back into models for gene expression (see Cai *et al.*, 2006 for a good description of this). In this study, the investigators demonstrated elegantly how high-throughput single-molecule data can be acquired using bespoke automated microfluidic systems – one drawback with much modern single-molecule cellular biophysics research is that it has traditionally been very slow to acquire large datasets corresponding to significant populations of cells, so use of automated flow-cell systems which can be combined with on-the-fly image analysis to identify candidate cells from brightfield images, and home in on them with subsequent fluorescence imaging, is a significant step forward.

The researchers further showed how many aspects of these data can be combined successfully with data from bulk *proteomics* studies (a variety of *multi-molecule* biochemical and biophysical methods which can be used to estimate protein content from a population of cells) from construct robust models for gene expression. Classical gene expression studies use just the isolated mRNAs or polypeptides/proteins from cell populations to determine expression profiles. However, these approaches throw away the vital spatial and temporal information that is present in living cells and most importantly only present a static picture (proteins, as we have now seen from multiple single-molecule cellular biophysics studies, are often localized to discrete regions of the cell and their behaviour is often dependent upon the timing of specific events triggered by complex feedback between several different cellular process).

The utility of this model is significant and merits some greater explanation. The essence of the formulation assumes that, at *steady-state*, protein production, from the process of gene transcription and subsequent mRNA translation, is balanced by a dilution effect from cell growth and division. The production component of the model is the gene expression, which refers to the sum of processes that result in a particular level of a specified mRNA and protein in the cell. This can be depicted in its simplest most general form as a three-stage model and can be represented by the *random telegraph model* (interested physical scientists might wish to read Peccoud and Yeard, 1995). The promoter of the gene (or operon cluster of genes) of interest can undergo transition between two discrete 'on' and 'off' states, and transcription only occurs if the promoter is active in the on state; both translation and transcription (along with competing degradation processes for mRNAs and proteins) can be modelled as first-order chemical reactions.

The end result is a typical *Gamma*-associated probability distribution $p(x)$ for the steady-state probability distribution of number of molecules x of a specific type of protein in each cell (see Q9.8), determined by the average number of gene expression bursts in cell cycle, a, and the average number of protein molecules that are produced in each burst, b:

$$p(x) = \frac{x^{a-1} e^{-x/b}}{b^a \Gamma(a)} \tag{9.14}$$

Here, $\Gamma(a)$ is the *Gamma function*. The parameters a and b can be determined separately from the two *moments* of the Gamma distribution by its mean value m and standard deviation σ, such that:

$$a = m^2/\sigma^2 \tag{9.15}$$

$$b = \sigma^2/m \tag{9.16}$$

Using a discrete, *geometric distribution* to describe burst size ultimately gives rise to a *negative binomial distribution* for the equivalent steady-state discrete values of $p(x)$, which converges to a Gamma distribution as the size of the discrete bins becomes infinitesimally small.

Single cell studies have become an essential tool in understanding the rules that govern gene expression, especially as a result of advances in single-molecule fluorescence imaging techniques, which now provide firm quantification of the full probabilistic distribution of numbers of protein molecules per cell as opposed to just a mean and a standard deviation. This is coupled with additional advantages in the ability to monitor these distributions both dynamically and in real time. There is a significant parallel advance in mathematical modelling being fed by such high-quality data and this marriage between imaging and theory has rooted single cell/single-molecule methods at the heart of modern systems biology (for a good review of this general field see Larson *et al.*, 2009).

Regulation of gene expression is clearly highly complex, and involves more than the binding kinetics of transcription factors. In eukaryotes one further issue is the presence of nucleosomes which provide a structure around which DNA can condense but similarly represent a free energy barrier towards the RNA polymerase enzyme reading off the genetic code in the process of transcription. Also, although prokaryotes lack nucleosomes they still possess DNA binding proteins and complexes that appear to function similarly in condensing the DNA.

Evidence has emerged from elegant physiologically relevant in vitro experiments using dual optical tweezers to tether a single DNA molecule in which the RNA polymerase was actively transcribing the DNA and a nucleosome was either bound to the DNA or absent (Hodges *et al.*, 2009). This allowed the researchers to identify the separate sub-nanometre displacement fluctuations due to the DNA or the nucleosomes. The results suggested that instead of attempting to detach the bound DNA from the histone proteins of the nucleosomes, the RNA polymerase actually acts more as a *ratchet* to rectify the thermal fluctuations of individual nucleosomes, with the effect of reducing the associated free energy barrier for transcription. The full story of regulation of gene expression is still far from well understood, but what is clear is that the whole complex milieu in which the DNA finds itself is highly critical to the regulation processes, hence the need to push forward with single-molecule studies that are suitably complex in their physiological relevance.

9.4.2 mRNA, nuclear pores and post-translational modification

Several methods have been employed to permit microscopic visualization of mRNA molecules (for a review see Weil *et al.*, 2010). A popular approach has involved in situ FISH (see Chapter 3) to target key mRNA target sequences specifically with fluorescent probes. This has been used to good effect to quantify levels of specific mRNA content in mammalian cells (Raj *et al.*, 2006) and yeast cells (Zenklusen *et al.*, 2008) at the cell-by-cell level. Knowledge of the dynamic cellular localization of mRNA is also important in understanding its mode of action. Although mRNA is manufactured in the transcription process of the DNA genetic code at the location of the DNA, the actual mRNA molecules need to interact with ribosomsal complexes before the next step leading to protein biosynthesis can occur.

In eukaryotic cells this involves having to translocate across the nuclear membrane out of the nucleus. The *nuclear pore complex* spans the double-membrane of the *nuclear envelope* and regulates trafficking of multiple different types of molecules between the cytoplasm and the nucleus in a *two-way direction*. The types of molecules translocated are diverse. Those translocated into the nucleus include transcription factors, polymerase enzymes for both DNA and RNA, histone proteins for the assembly of nucleosomes, and ribosome and RNA processing proteins, whereas components translocatedout of the nucleus include, most importantly, mRNA. Small molecules less than ~30 kDa appear to transit pores through passive diffusion, however larger substrates can be translocated using carrier proteins, and these substrates can include enormous molecules as large as ribosomal subunits, making these pores essentially unique among transporter complexes.

Recent studies using the techniques of single-molecule cellular biophysics have substantially increased our understanding of the architecture of the nuclear envelope and of the pores through which the mRNA must pass before it can emerge into the rest of the cytoplasm. Structured illumination (SI) has been used to generate super-resolution three-dimensional images of the periphery of the nucleus (Schermelleh *et al.*, 2008). In this study, the investigators were able use multi-colour fluorescence imaging in combination with SI to reconstruct simultaneous images of chromatin, nuclear pore complexes and the dense *nuclear lamina* network inside the nucleus composed of intermediate filaments of several tens of nanometres in thickness (Figure 9.6A).

This indicated several features that could not be detected using conventional optical microscopy, including evidence that individual nuclear pore complexes co-localize with channels in the lamina fibrillar network inside the nucleus and with chromatin localized towards the edge of the nucleus. Also, the investigators were able to image double-layered invaginations of the nuclear envelope that had previously only been detected using EM.

The pore-translocation process itself has been the target of several recent studies. Using permeabilized cells, researchers were able to apply single-molecule fluorescence imaging to track molecules of the pore-translocation of molecules composed of a fusion of the peptide NLS with GFP (Yang *et al.*, 2004). NLS is a *nuclear localization signal* sequence composed of 10–20 amino acid residues which functions to tag a protein which is destined for import into the nucleus, and thus which is recognized in some way by the nuclear pore complex which can utilize this in a *gating/filtering* mechanism.

Several models have been proposed for this selectivity. There is test tube evidence that certain pore localized proteins, *FG proteins*, may aggregate to form a *selective phase hydrogel* through which substrate molecules possessing the correct sequence tag can selectively diffuse. It has also been proposed that FG proteins may act as an *entropic brush* forming a transient barrier that can be opened on binding of the correct tagging sequence, but these ideas are still in the realm of speculation. The results indicated that the substrate molecule typically spent most of its ~10 ms pore transit time randomly moving in the pore channel, such that the rate-limiting step in the translocation process was escape from this central pore.

Also, the maximum accumulation rate of NLS at each pore was measured to be ~1000 molecules per second, thus indicating that each pore must be capable of translocating ~10 substrate molecules simultaneously – in other words, if the substrate molecules were queuing to get into the popular night-club of the nucleus then it is not a case of one-in-one-out but rather if one molecule enters then so ultimately do its nine other friends.

Translocation through nuclear pore complexes has also been imaged at the single-molecule level in vivo using *pair correlation functions* (Cardarelli and Gratton, 2010). This method estimates the time a molecule takes to move from one location to another within the cell in the presence of several molecules of the same type to within approximately millisecond time resolution. The correlation in both space and time data between any two locations in the cell acts as a local map of molecular mobility and can indicate the local presence of diffusional barriers. Here, again, an NLS–GFP fusion was used as the substrate reporter molecule, using a fast laser-scanning confocal microscope to

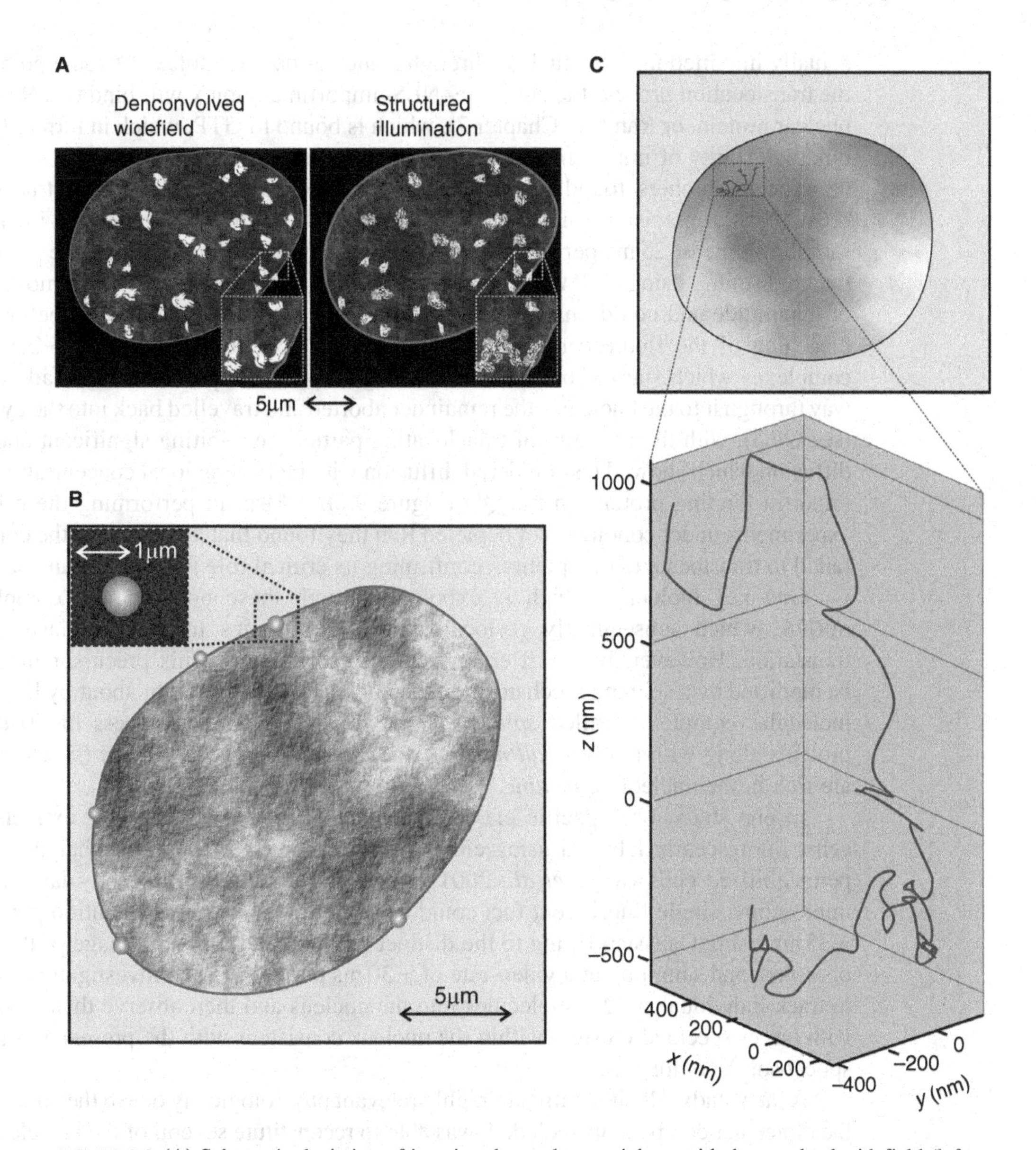

FIGURE 9.6 **(A)** Schematic depiction of imaging the nuclear periphery with deconvolved widefield (left panel) and clearly greater resolution using structured illumination (right panel). **(B)** Schematic of QD-importin translocation through nuclear pore complexes in brightfield with positions of QDs marked, with inset showing schematic of an expanded fluorescence image of a single QD. **(C)** Schematic of tracking of single molecules of GFP-labelled mRNA in a live cell (upper panel) using the double-helix point spread function method to generate three-dimensional tracks (lower panel).

acquire images. The results indicated significant heterogeneity to diffusion both within compartments in the nucleus and in crossing through the pore complex itself.

The actual selectivity mechanisms of the nuclear pore complex have also been investigated using single-molecule techniques (Lowe *et al.*, 2010). In this elegant study researchers fused target substrate molecules using QDs that were functionalized with a protein known to act as a binding molecule for a transport protein called *importin*. Importin is expressed at the same time as NLS, and it is known that NLS-tagged protein will bind strongly to importin, and it is this substrate–NLS–importin complex that is actually imported into the nucleus through a nuclear pore complex. At some point during the translocation process the substrate–NLS–importin complex will bind to a Ras-related nuclear protein, or Ran (see Chapter 7), which is bound to GTP, which in turn reduces the binding affinity of importin causing it to break free.

The researchers found that single QD-labelled complexes could be tracked with ~6 nm spatial precision using Gaussian fitting to the point spread function images, at a sampling time of 25 ms per image, for several seconds continuously, unlike earlier NLS tracking studies using GFP which would have a worse spatial precision by almost an order of magnitude and could only be tracked for ~10 consecutive data points before photobleaching of the fluorescent protein. The investigators found that only ~20% of the complexes which started to translocate through a pore complex actually made it all the way through into the nucleus – the remainder aborted and travelled back into the cytoplasm (see Q9.3), with the mobility of translocating particles exhibiting significant anomalous diffusion which showed less hindered diffusion with increasing local concentrations of the importin binding protein on the QD (Figure 9.6B). Also, in performing these live-cell experiments under conditions of depleted Ran they found that almost all of the complexes failed to translocate to completion, confirming its critical role in nuclear translocation.

One key molecule which is exported through these nuclear pore complexes is mRNA, which subsequently co-localizes with ribosomes in the cytoplasm prior to translation. However, before it emerges into the cytoplasm this precursor mRNA can be modified by a splicing mechanism in the nucleus. This is brought about by large multimolecular complexes called *spliceosomes*, which comprise in excess of 70 different proteins along with a few *small nuclear ribonuclear protein molecules* (snRNPs) which are rich in the nucleotide *uridine*.

In one study, one specific class of snRNPs, the U1 snRNPs, were extracted from cells, fluorescently labelled using either Alexa488 or Cy5, and re-incubated back with permeabilized cells (Kues *et al.*, 2001). Using high excitation intensity laser confocal microscopy, single fluorescent foci could be tracked with a super-resolution precision of ~35 nm using Gaussian fitting to the distinct point spread function image of the dye tag observed, and sampling at a video-rate of ~30 ms per image. The investigators were able to track individual snRNP molecules into the nucleus and then observe them co-localize with larger speckled clusters within the nucleus consistent with the presence of multiple spliceosome complexes.

A later study, albeit in vitro but highly relevant physiologically due to the complexity of the molecular components included, was able to reconstitute several of the key elements of the spliceosome complex. Using step-wise photobleaching the investigators could estimate the number of separate molecules present in each spliceosome of a specific protein called *polypyrimidine tract-binding protein* (PTB), known to regulate tissue-specific splicing (Cherny *et al.*, 2010). This indicated that 5–6 individual PTB molecules are required to bring about splicing regulation. It suggests that PTB molecules occupy mainly multiple overlapping sites on the spliceosome and that potentially the regulation process involves multiple, coordinated steps as opposed to being due to a simple single protein binding event, which potentially gives scope for fine-tuning of the splicing regulation.

Clearly, the precise localization of mRNA after it has emerged from the nucleus, and its associated protein complexes, is a very important factor in the ultimate regulation of gene expression. Using super-resolution imaging, researchers have been able to track single molecules of mRNA particles in living cells (Thompson *et al.*, 2010). The investigators fused GFP to mRNA, which resulted in a variable number of GFP molecules per mRNA molecule, averaging ca. 30–40, which thus provided a bright signal even at a reasonably fast sampling exposure time 15 ms per image. Using the *double-helix point spread function* approach (see Chapter 3), the investigators were able to track individual mRNA molecules with a super-resolution precision of ~25 nm in x and y dimensions and ~50 nm in the z dimension (Figure 9.6C).

This indicated that the mobility of mRNA in the cytoplasm was mostly consistent with Brownian diffusion, however with a significant minority of tracks exhibiting signs of either confined diffusion (potentially due at least in part to binding to ribosomes) or directed diffusion. The confined diffusion occurs mostly in domains of effective radius 100–300 nm, though sometimes these are as large as ~500 nm. This relatively large confinement length scale may indicate potential interaction with sub-cellular features such as organelles, parts of the cytoskeleton or some other intra-cytoplasmic compartment. The directed diffusion component was largely eradicated with the treatment of chemicals known to disrupt the cytoskeleton, and this was consistent with mRNA being ferried via a cargo transport mechanism (see section 9.2).

THE GIST

- By increasing the local fluorescence excitation intensity, single tagged molecules in the cytoplasm can be imaged with approximately millisecond time resolution.
- Strobing of laser excitation allows unblurred imaging of cytplasmic molecules over much longer time scales than the exposure time.
- Single-molecule mobility in the cytoplasm is in general highly heterogeneous.
- Molecular trafficking can be monitored in vivo down to the nanometre scale to reveal a tug-of-war between dynein and kinesin molecular motors on microtubules, as well as different types of myosin on actin filaments.
- The architecture and stoichiometry of several different proteins that bind to chromosomes, including the replication machinery, can be measured in live cells to a single-molecule precision.
- Gene regulation of transcription can be monitored one molecule at a time in real time in model bacterial systems.

TAKE-HOME MESSAGE The challenges of monitoring single-molecule processes that occur deep in the inside of cells is far greater than the challenges of investigating processes which occur in, or in the vicinity of, the cell membrane. However, by using a combination of very fast, super-resolution fluorescence imaging it is possible to monitor cytoplasmic processes in model cellular systems.

References

GENERAL

- Bray, D. (2000). *Cell Movements from Molecules to Motility*, 2nd edition. Garland Science.
- Cai, D. *et al.* (2010). Recording single motor proteins in the cytoplasm of mammalian cells. *Methods Enzymol.* **475**: 81–107.
- Jacob, F. and Monod, J. (1961). On the regulation of gene activity. *Cold Spring Harbor Symp. Quant. Biol.* **26**: 193–211.
- Kolomeisky, A. B. and Fisher, M. E. (2007). Molecular motors: a theorist's perspective. *Annu. Rev. Phys. Chem.* **58**: 675–695.

- Larson, D. R., Singer, R. H. and Zenklusen, D. (2009). A single molecule view of gene expression. *Trends Cell Biol.* **19**: 630–637.
- Li, G. W. and Xie, X. S. (2011). Central dogma at the single-molecule level in living cells. *Nature* **475**: 308–315.
- van Oijen, A. M. and Loparo, J. J. (2010). Single-molecule studies of the replisome. *Annu. Rev. Biophys.* **39**: 429–448.
- Veigel, C. and Schmidt, C. F. (2011). Moving into the cell: single-molecule studies of molecular motors in complex environments. *Nature Rev. Mol. Cell Biol.* **12**: 163–176.
- Weil, T. T., Parton, R. M. and Davis, I. (2010). Making the message clear: visualizing mRNA localization. *Trends Cell Biol.* **20**: 380–390.
- Wirtz, D. (2009). Particle-tracking microrheology of living cells: principles and applications. *Annu. Rev. Biophys.* **38**: 301–326.
- Yanagida, T. *et al.* (2007). Brownian motion, fluctuation and life. *Biosystems* **88**: 228–242.

ADVANCED

- Akhmanova, A. and Hammer, J. A. III (2010). Linking molecular motors to membrane cargo. *Curr. Opin. Cell Biol.* **22**: 479–487.
- Ali, M. Y. *et al.* (2011). Myosin Va and myosin VI coordinate their steps while engaged in an in vitro tug of war during cargo transport. *Proc. Natl. Acad. Sci. USA* **108**: E535–E341.
- Asbury, C. L., Fehr, A. N. and Block, S. M. (2003). Kinesin moves by an asymmetric hand-over-hand mechanism. *Science* **302**: 2130–2134.
- Cai, L., Friedman, N. and Xie, X. S. (2006). Stochastic protein expression in individual cells at the single molecule level. *Nature* **440**: 358–362.
- Cardarelli, F. and Gratton, E. (2010). In vivo imaging of single-molecule translocation through nuclear pore complexes by pair correlation functions. *PLoS One* **5**: e10475.
- Cherny, D. *et al.* (2010). Stoichiometry of a regulatory splicing complex revealed by single-molecule analyses. *EMBO J.* **29**: 2161–2172.
- Elf, J., Li, G. W. and Xie, X, S. (2007). Probing transcription factor dynamics at the single-molecule level in a living cell. *Science* **316**: 1191–1194.
- English, B. P. *et al.* (2011). Single-molecule investigations of the stringent response machinery in living bacterial cells. *Proc. Natl. Acad. Sci. USA* **108**: E365–E373.
- Fange, D., Berg, O. G., Sjöberg P. and Elf, J. (2010). Stochastic reaction-diffusion kinetics in the microscopic limit. *Proc. Natl. Acad. Sci. USA* **107**: 19820–19825.
- Fu, G. *et al.* (2010). In vivo structure of the *E. coli* FtsZ-ring revealed by photoactivated localization microscopy (PALM). *PLoS One* **5**: e12682.
- Georgescu, R. E., Kurth, I. and O'Donnell, M. E. (2011). Single-molecule studies reveal the function of a third polymerase in the replisome. *Nature Struct. Mol. Biol.* **19**: 113–116.
- Gunkel, M. *et al.* (2009). Dual color localization microscopy of cellular nanostructures. *Biotechnol. J.* **4**: 927–938.
- Hale, C. M. *et al.* (2008). Dysfunctional connections between the nucleus and the actin and microtubule networks in laminopathic models. *Biophys. J.* **95**: 5462–5475.
- Hodges, C. *et al.* (2009). Nucleosomal fluctuations govern the transcription dynamics of RNA polymerase II. *Science* **325**: 626–628.
- Jennings, P. C., Cox, G. C., Monahan, L. G. and Harry, E. J. (2010). Super-resolution imaging of the bacterial cytokinetic protein FtsZ. *Micron* **42**: 336–341.
- Kner, P. *et al.* (2009). Super-resolution video microscopy of live cells by structured illumination. *Nature Methods* **6**: 339–342.
- Kues, T. *et al.* (2001). High intranuclear mobility and dynamic clustering of the splicing factor U1 snRNP observed by single particle tracking. *Proc. Natl. Acad. Sci. USA* **98**: 12021–12026.
- Kural, C. *et al.* (2005). Kinesin and dynein move a peroxisome in vivo: a tug-of-war or coordinated movement? *Science* **308**: 1469–1472.
- Kural, C. *et al.* (2006). Tracking melanosomes inside a cell to study molecular motors and their interaction. *Biophys. J.* **90**: 318–327.

- Lee, S. F. *et al.* (2011). Super-resolution imaging of the nucleoid-associated protein HU in *Caulobacter crescentus. Biophys. J.* **100**: L31–L33.
- Li, G. -W., Berg, O. G. and Elf, J. (2009). Effects of macromolecular crowding and DNA looping on gene regulation kinetics. *Nature Phys.* **5**: 294–297.
- Lia, G., Michel, B. and Allemand, J. F. (2012). Polymerase exchange during Okazaki fragment synthesis observed in living cells. *Science* **335**: 328–331.
- Lowe, A. R. *et al.* (2010). Selectivity mechanism of the nuclear pore complex characterized by single cargo tracking. *Nature* **467**: 600–603.
- Mashanov, G. I. *et al.* (2010). Direct observation of individual KCNQ1 potassium channels reveals their distinctive diffusive behavior. *J. Biol. Chem.* **285**: 3664–3675.
- Nan, X., Sims, P. A., Chen, P. and Xie, X. S. (2005). Observation of individual microtubule motor steps in living cells with endocytosed quantum dots. *J. Phys. Chem. B* **109**: 24220–24224.
- Peccoud, J. and Ycard, B. (1995). Markovian modelling of gene-product synthesis. *Theor. Popul. Biol.* **48**: 222–234.
- Ptacin, J. L. *et al.* (2010). A spindle-like apparatus guides bacterial chromosome segregation. *Nature Cell Biol.* **12**: 791–798.
- Raj, A. *et al.* (2006). Stochastic mRNA synthesis in mammalian cells. *PLoS Biol.* **4**: e309.
- Reyes-Lamothe, R., Sherratt, D. J. and Leake, M. C. (2010). Stoichiometry and architecture of active DNA replication machinery in Escherichia coli. *Science* **328**: 498–501.
- Ribeiro, S. A. *et al.* (2010). A super-resolution map of the vertebrate kinetochore. *Proc. Natl. Acad. Sci. USA* **107**: 10484–10489.
- Savin, T. and Doyle, P. S. (2005). Static and dynamic errors in particle tracking microrheology. *Biophys. J.* **88**: 623–638.
- Schermelleh, L. *et al.* (2008). Subdiffraction multicolor imaging of the nuclear periphery with 3D structured illumination microscopy. *Science* **320**: 1332–1336.
- Schroeder, H. W. III *et al.* (2010). Motor number controls cargo switching at actin-microtubule intersections in vitro. *Curr. Biol.* **20**: 687–696.
- Thomann, D., Rines, D. R., Sorger, P. K. and Danuser, G. (2002). Automatic fluorescent tag detection in 3D with super-resolution: application to the analysis of chromosome movement. *J. Microsc.* **208**: 49–64.
- Thompson, M. A. *et al.* (2010). Three-dimensional tracking of single mRNA particles in *Saccharomyces cerevisiae* using a double-helix point spread function. *Proc. Natl. Acad. Sci. USA* **107**: 17864–17871.
- Yang, W., Gelles, J. and Musser, S. M. (2004). Imaging of single-molecule translocation through nuclear pore complexes. *Proc. Natl. Acad. Sci. USA* **101**: 12887–12892.
- Yildiz, A. *et al.* (2003). Myosin V walks hand-over-hand: single fluorophore imaging with 1.5-nm localization. *Science* **300**: 2061–2065.
- Yildiz, A., Tomishige, M., Vale, R. D. and Selvin, P. R. (2004). Kinesin walks hand-over-hand. *Science* **303**: 676–678.
- Yu, J. *et al.* (2006). Probing gene expression in live cells, one protein molecule at a time. *Science* **311**: 1600–1603.
- Zenklusen, D., Larson, D. R. and Singer, R. H. (2008). Single-RNA counting reveals alternative modes of gene expression in yeast. *Nature Struct. Mol. Biol.* **15**: 1263–1271.

Questions

FOR THE LIFE SCIENTISTS

Q9.1. Ficoll is an artificial polymer of sucrose that is often used to change osmotic pressure and/or solution viscosity in single-molecule biology experiments. A version of Ficoll with molecular weight ~40 kDa is known to have viscosities relative to water at room temperature of [1, 5, 20, 60, 180, 600] corresponding to (w/v)% in water of [0, 10, 20, 30, 40, 50] respectively. If a fluorescein-labelled IgG antibody has an effective Stokes' radius of 8 nm, and the viscosity of water at room temperature is 0.001 N s m^{-2}, estimate the molarity of the Ficoll needed to be present to observe free labelled antibodies in vitro unblurred in solution using a widefield epifluorescence microscope of 1.45NA capable of imaging at 40 ms per frame.

Q9.2. A single mRNA molecule has a molecular weight of ca. 98 kDa. (a) How many nucleotide bases does it contain? (Hint: See Chapter 2.) (b) Assuming that the mRNA molecule can be modelled as a Gaussian chain with a segment length equal to ca. 4 nm, estimate the Stokes radius of the whole molecule. (Hint: See Chapter 5.) The longest cells in a human body are nerve cells with axons running from the spinal cord to the feet (see Q9.4), roughly shaped as a tube of circular cross-section with a tube length of ca. 1 m and diameter ca. 1 μm. (c) Estimate the diffusion coefficient of a single mRNA molecule as above in the cytoplasm of the axon in $\mu m^2\ s^{-1}$, assuming the viscosity of cytoplasm is ~30% higher than that of water. How long will the mRNA molecule on average take to diffuse across the width of the axon, and how long would it take to diffuse from one end to the other? (d) With reference to your answers explain why certain cells do not rely on free diffusion alone to transport mRNA, and discuss the other strategies they employ instead.

Q9.3. For a particle exhibiting one-dimensional Brownian diffusion we might expect the probability p for being at a distance x after a time t to satisfy $p = t^{½}\exp\{-x^2/(2\langle x^2\rangle)\}/(4\pi D)^{1/2}$ where $\langle x^2\rangle$ is the mean square displacement, equal to $2Dt$. In a live-cell single particle tracking experiment on QD-labelled importin molecules being imported into the cell nucleus (for example see Lowe *et al.*, 2010), the QD–importin complex was estimated as having a Stokes radius of ~30 nm. (a) What is its diffusion coefficient D in the cytoplasm? (b) After a time t the mean square displacement through an unrestricted channel might be expected to be $2Dt$; why is this not $4Dt$ or $6Dt$? (Hint: See the result of Equation 9.1 and/or the analysis of Equations 9.4–9.10.) It was found that out of ~850 trajectories that showed evidence of initiating translocation into the nucleus, only ~180 translocated all the way through the pore complex of channel length ~55 nm into the nucleus, a process which took on average ~1 s. (c) Explain with quantitative reasoning why the translocation does not appear to be consistent with Brownian diffusion. (d) What could account for this non-Brownian behaviour?

FOR THE PHYSICAL SCIENTISTS

Q9.4. (a) Write down the Brownian diffusion equation for numbers of molecules n per unit length in x that spread out by diffusion after a time t. (b) Show that a solution exists of the form $n(x,t) = \alpha t^{½}\exp\{-x^2/(4Dt)\}$ and determine the normalization constant, α, assuming there are n_{tot} molecules in total present. (c) Show with reasoning what the mean expected values are for $\langle x\rangle$ and $\langle x^2\rangle$, and sketch the solution of the latter for several values of $t > 0$. (d) What does this imply for the location of the particles at zero time? The longest cells in a human body are neurons running from the spinal cord to the feet (see Q9.2), roughly shaped as a tube of circular cross-section with a tube-length of ca. 1 m and diameter ca. 1 μm with a small localized cell body at one end which contains the nucleus. Neurotransmitter molecules are synthesized in this cell body but required at the far end of the neuron. When neurotransmitter molecules reach the far end of the neuron they are subsequently removed by a continuous cellular process at a rate which maintains the mean concentration in the cell body at 1 mM. (e) If the effective diffusion coefficient is $\sim 10^3\ \mu m^2\ s^{-1}$ estimate how many neurotransmitter molecules reach the end of the cell in each second due to diffusion alone. In practice, neurotransmitters are packaged into lipid-based synaptic vesicles (see Chapter 7) which each containing ca. 10^4 neurotransmitter molecules. (f) If ca. 300 of such molecules are required to send a signal from the far end of the neuron to a muscle, what signal rate could be supported by diffusion alone? (g) What problem does this present, and how does the cell overcome this? (Hint: Consider the frequency of everyday activities which involving moving your feet, such as walking.)

Q9.5. A bacterial cell in a water-based medium is expressing a fluorescently labelled cytoplasmic protein at a low burst rate level resulting in a mean concentration in the cytoplasm of molarity N is immobilized to a glass coverslip that assembles into 1 or 2 distinct cytoplasmic complexes in the cell of mean stoichiometry S. (a) If complexes are ca. half a cell's width of 0.5 μm from the glass coverslip surface and the depth of field of the objective lens used to image the fluorescence and generate a TIRF evanescent excitation field is d μm, generate an approximate expression for the signal-to-noise ratio of the foci using such TIRF excitation. (b) Showing reasoning, explain under what conditions might it be useful to use TIRF to monitor foci in the cytoplasm.

Q9.6. A dynein molecular motor was labelled with a QD and its driven motion along a microtubule track was monitored using fluorescence imaging in combination with single particle tracking and iterative Gaussian fitting of the point spread function image of the fluorescent QD. It was found that the intrinsic diffusive motions of the dynein were negligible compared to the motion due to two state transitions along the track, a transition from a bound state (probability P_b) to an unbound state (probability $P_u = 1 - P_b$) which proceeded at a rate k_1 at distance x along the track such that $x_0 < x < x_0 + \Delta x$, and the reverse transition which goes from the unbound to the bound state which proceeds at a rate k_2 such that $x > 0$. (a) Show that the governing equation of motion can be represented here by

$$\frac{dP_b}{dt} = k_1(1 - P_b) - k_2 P_b - v\frac{dP_b}{dx}$$

(b) It was observed that the mean speed v of the dynein along the track was roughly constant. If you assume a steady-state solution (why?) then show that this equation can be solved in three regions of track: (i) a 'binding' zone around $x = x_0$ (assume initial conditions of no binding, so that $P_b(x_0) = 0$, and that $k_2 = 0$), (ii) a 'power stroke' zone $x_0 < x < 0$ (assume $P_b = \text{constant}$), and (iii) a unbinding 'drag-stroke' zone of $x > 0$ (assume $k_I = 0$). Comment on the description of each zone in the light of the mode of action of a molecular motor on a linear track, and sketch solutions of P_b versus x for a 'fast' dynein molecule (high v) and a 'slow' dynein molecule (low v). (c) If the stiffness of each chemical bond made between a bound dynein molecule and a microtubule is κ and the mean distance between neighbouring dynein binding sites on the microtubule is d, so that the average force $\langle F \rangle$ exerted per bound dynein molecule on the microtubule is:

$$\langle F \rangle = \frac{\kappa}{d}\left(\frac{x_0^2}{2} - \frac{v^2}{k_2^2}\right)\exp\left(1 - \frac{k_1 \Delta x}{v}\right)$$

(Hint: This is an easier problem than it first appears; for the force exerted by a single dynein on the microtubule assume a simple Hookean spring, integrate this force with respect to distance and the probability that a dynein molecule is bound to get the total work done, and equate this to the average force multiplied by the average distance between binding sites.)

FOR THOSE WHO HAVE NOT MADE UP THEIR MIND

Q9.7. Bacterial chemotaxis (see section 7.2) can be modelled as a biased random walk in three spatial dimensions. (a) Explain what this means. In a uniform chemo-attractant concentration gradient it was found that *E. coli* bacteria swam in a straight line at a mean speed of ~10 μm s^{-1} in between tumbling, in which the cell stopped swimming for ~0.5 s before swimming off in another random direction, with a typical tumbling frequency of ~0.5 Hz. An individual swimming cell in a microscope's large sample container could be tracked automatically in three spatial dimensions by rapidly feeding back the position of each cell to a nano-piezo stage to recentre the stage so that the cell always appears at the centre of the field of view in the microscope. (b) If a cell is first tracked when the stage is at its zero position, and the stage can move ±50 μm in x, y and z before reaches its deflection limit, how long can a typical cell be tracked for?

Q9.8. Deconvolution analysis was performed on the non-foci cellular regions of slimfield images of ~120 different cells in which a component of the DNA polymerase of the replisome, ε, was fused to the fluorescent protein YPet. Step-wise photobleaching of the fluorescent foci observed indicated ~3 molecules per replication fork. In this cell strain the native gene expressing ε was deleted and replaced with ε-YPet. It was found there was a ca. 25% likelihood for a given cell to contain either ~77 non-foci molecules per cell or ~412 non-foci molecules per cell. (a) What is the mean number of ε-YPet molecules per cell, and the relative error as a percentage of the mean of this estimate? In another experiment a different type of cell strain was used in which the native gene was not deleted but the ε-YPet fusion genes were placed on a plasmid under control of the lac operon. When no IPTG was added the mean estimated number of ε-YPet molecules per cell was ca. 50, and using step-wise photobleaching of the fluorescent replisome foci this suggested only ca. one molecule per focus. When excess IPTG was added the step-wise photobleaching indicated ca. three molecules per focus and the mean number of non-foci molecules per cell was ~900. (b) Explain these observations.

Q9.9. The prediction from the random telegraph model for the distribution of the total number of a given type of protein molecule present in each bacterial cell is a Gamma-related distribution (essentially a distribution which has a unimodal peak with a long tail). However, if you measure the stoichiometry of certain individual molecular complexes then the shape of that distribution can be adequately fitted using a Gaussian distribution. How can this be explained?

Q9.10. Many single-molecule experiments now indicate that cargo trafficking is a tug-of-war process between multiple competing molecular motors pulling in opposite directions. If each associated molecular power stroke requires energization then this can hardly be an energetically efficient process, compared to simply using single molecular motors that do not compete with each other. If there were no ultimate gain to this process occurring to the organism in question then one might expect such behaviour to have been selected out during the process of millions of years of evolution. So why is molecular cargo trafficked in this way?

10 Single-molecule biophysics beyond single cells and beyond the single molecule

Now this is not the end. It is not even the beginning of the end. But it is, perhaps, the end of the beginning. (WINSTON CHURCHILL, 1942)

GENERAL IDEA

Here we take stock of the remarkable developments and innovations in biophysics that have allowed us to address very challenging and fundamental questions about the key cellular processes, and speculate where this might lead in the near future.

10.1 Introduction

The emergence of single-molecule cellular biophysics represents a *coming-of-age* of single-molecule bioscience. The first generation of single-molecule experiments resulted in some exceptionally pioneering developments in terms of the physics of techniques and novel analytical methods and in terms of significantly increasing our understanding of the functioning of isolated biological molecules. But now, as we have seen from myriad investigations discussed in this book, the next generation of single-molecule bioscience has opened up outstanding opportunities to study biological processes under physiologically highly appropriate conditions – in other words to gain enormous insight into how single molecules really function in the context of their native environment of the living cell. In this final chapter we survey the developments that have led us to this point, and ask the question 'what next?' As we will see, there is great potential to apply these novel technologies in areas that may have a large future impact on society, namely those of *bionanotechnology* and *synthetic* biology, *fuel production* for commerical use and *single-molecule biomedicine*.

10.2 Single-molecule biophysics in complex organisms

A schism divides modern biology: on the one hand, *reductionist* approaches yield precise information about the particular; and on the other hand, *integrationist* systems-level biology struggles to grapple with complex networks of interactions at a level of generality that often precludes experimental testing. Biochemistry is traditionally a highly reductionist discipline: its *modus operandi* is the dissection of biological phenomena to yield hypotheses that can be tested. It has furnished us with very precise understanding about what appear to be the smallest 'units' in biology – molecules. In some cases, for example in the Krebs/TCA cycle (see Chapter 2), the painstaking assembly of sequences of molecular interactions has been possible. However, biological molecules function within the context of living cells, representing large systems of interacting molecules. So, what is the realistic scope for applying single-molecule methods to far more complex systems which consist not just of one cell, but of multiple cells, such as a complex *tissue*, *organ* or whole *organism*?

10.2.1 Single-molecule systems biology

Even a single cellular process may involve a very large number of interactions. For example, the signalling pathway for tissue cell attachment in eukaryotic cells involves some ~200 molecules. A single mechanism such as this stretches the reductionist approach to breaking point. Biology has become a coarse-grained discipline, with loose coupling between levels, but very limited capacity for integrative thinking vertically across the length scales. And yet the very nature of biological science makes it unsuited to a coarse-grained way of thinking: events at the molecular level can influence the behaviour of organisms or even ecosystems and thus the development of a *holistic* approach to biological science requires the capacity to integrate molecular-level understanding into systems-level models.

Many cells, of differing types, integrate to form an organ; organs and tissues are integrated in organisms; ecosystems consist of many interacting organisms. At this level biological systems exhibit breath-taking complexity. Indeed, it can be argued that in many important senses biological systems may be *irreducibly complex*. The recognition of this challenge has led to the birth of *systems biology*, in which computational and mathematical approaches are utilized to attempt to model large networks of interacting elements. However, traditional systems approaches remain disconnected from the molecular-scale understanding furnished by biochemistry, the most precise biological knowledge at our disposal. What has now started to emerge from the development of single-molecule cellular biophysics is a combination of single-molecule level precise experimentation with the physiological context beyond just an isolated cell, to higher length scales of multi-cellularity.

It is now feasible to monitor single molecules within one given living cell and also to place that cell in an environment in which it is sufficiently close to neighbouring cells to permit inter-cellular communication to occur. For example, bacteria are known to move in response to external chemical signals via chemotaxis (see Chapter 7), however there is a high-level length scale response which modifies the behaviour of any given cell, which is *quorum sensing* (for a good review see Diggle *et al.*, 2007).

This discovery that bacteria can in effect talk to each other has altered our view of many single, simple organisms. Bacteria are able to measure the concentration of certain signal molecules within a cell population, and to do so in a way which allows them to sense the local concentration of bacterial cells in their environment (cell density). This enables bacteria to *coordinate* behaviour at a multi-cellular level, and as such is potentially a tractable model for understanding multi-cellular behaviour in more settings such as tissues and complex organisms. In addition, new data are resulting in a merging of the lines between 'simple' bacterial cells and 'complex' eukaryotic cells which can form multi-cellular tissues. For example, a bacterium that moves principally by gilding called *Myxococcus xanthus* contains signalling and cell motility mechanisms which show more similarity to those of eukaryotic cells than to those of bacteria (Mauriello, 2010).

Revealing future levels of inquiry might include further development of *high-throughput* imaging approaches to allow several cells in a population to be monitored at the single-molecule or single cell level simultaneously. This could provide significant insight into highly complex systems-level behaviour while still retaining the definitive information available from single-molecule measurements.

10.2.2 Probing even smaller length scales – sub-cellular and quantum biology

As has been suggested earlier in this book, the principal unit of functionality for a biological process is, arguably, that of the single molecule. Single molecules represent a highly appropriate *mesoscale* length scale that reaches out into greater levels of complexity at higher length scales involving multiple molecules in biological machines, ultimately to the level of the cell and beyond. However, single biological molecules are held together by the forces relevant to *quantum mechanics*. For many cellular processes such quantum effects appear to be largely irrelevant in that we can sensibly model the observed behaviour essentially by recourse only to single-molecule biochemistry and/or soft condensed matter physics, which is a world of deformable nanoscale building blocks with interacting binding sites. However, there are some important examples that require an understanding of physics at smaller length scales than the single molecule (for an accessible description of this emerging field see Ball, 2011). These include the breaking and making of hydrogen bonds, for example in forced unfolding and subsequent refolding of regions of a biological molecule whose function is related to its mechanical properties.

Really understanding the process of bond breaking/making requires a knowledge of the specific quantum mechanics of these bonding systems. This is essentially not a tractable analytical problem but must be tackled through *numerical simulation* techniques. This presents a significant challenge at the level of computation, since to simulate the atomic and sub-atomic environment of even relatively simple hydrogen binding systems results in time scales of at best 10^{-6} s worth of simulation data, but more typically two or three orders of magnitude smaller. This is an issue, since the time scale for the physiological occurrence of unfolding and refolding is in general significantly larger than this, and so an undesirable solution at present is to use less than physiological conditions in the simulations (for example, extreme values of external forces, and/or much higher temperatures than would occur naturally).

Other important biological quantum effects include the processes of light conversion to electrochemical energy, such as in photosynthesis, as well as the vital *quantum-tunnelling* effects that occur during electron transfer in oxidative phosphorylation (see Chapter 8). Future important developments in these areas of *quantum biology* are likely to involve improvements to *parallel super-computational efficiency*, and the use of high-tech computational resources such as *cloud-computing* which potentially could spread the onus of the challenging computational workload across a distributed network of powerful servers.

However, without recourse to quantum biology effects it is still feasible to push towards a smaller length scale while still retaining holistic physiological information. One obvious route is to monitor single-molecule effects at the sub-cellular level of single organelles. This may at first seem like a lateral move in terms of optimizing the physiological relevance of experiments, however in complex eukaryotic cells in particular, biological functionality is often physically highly *compartmentalized*, and thus understanding the workings of functional sub-cellular organelles, such as chloroplasts and mitochondria, at the level of single molecules is bound to improve significantly our understanding of how the organelles link in with the rest of the cellular environment and potentially of how they act in a coordinated fashion.

10.2.3 Single-molecule biomedicine

One exciting application of single-molecule cellular biophysics is the development of single-molecule methods to be used specifically in biomedicine, which will ultimately have some future positive impact on *improving human health.* Although one criticism laid in the path of much current live-cell single-molecule research is that the majority of the studies are performed on relatively simple cells, such as bacteria, bacterial systems are arguably one of the most important areas of modern biomedical research at present. Statistics from the World Health Organization indicate that almost half a million new cases of multidrug-resistant tuberculosis emerge annually, causing ~150,000 deaths reported in over 60 different countries to date. In addition, many achievements of modern medicine are put at risk by *antimicrobial drug resistance*, including infections following *surgery*, *chemotherapy* and *tissue transplantation.*

Increasing our understanding of the specific mechanisms that allow bacteria to thrive from generation to generation will have a substantial impact in developing new methods to combat this insidious danger to human health. Single-molecule biophysics studies on live-cell bacterial systems potentially impact human health in at least two ways.

(i) The bacterial molecular machines often studied are prime candidates for future antibiotic developments to combat the next imminent generation of highly resistant microbes.

(ii) These machines in model bacterial systems have translatable paradigm elements which can significantly increase our understanding of normal and pathological human cellular processes, for example concerned with DNA translocation and molecular positioning.

In addition to pushing forward with single-molecule research on model bacterial systems there are likely to be significant future developments on eukaryotic cellular systems of direct relevance to human disease. In fact, the first definitive single-molecule live-cell study involved TIRF imaging of EGF receptors (see Chapter 7, Sako *et al.*, 2000). Mutation of the EGF cellular pathway commonly results in the formation of cancer cells, and so single-molecule studies of these types of cells using ever greater experimental complexity in, for example, monitoring multiple components of such important signalling biochemical pathways, are likely to prove very important in the near future.

There is also reasonable evidence to indicate correlated changes in the mechanical state of cells at the molecular level during cancer cell formation or *metastasis.* Understanding the role of these molecular mechanical forces in live cancer cells is likely to be an important future area of research (for a good review of the physics of cancer see Wirtz *et al.*, 2011). A point to note is that, by definition, cancer is a disease which involves multiple cells, as opposed to just a single, isolated cell. Therefore, high-throughput single-molecule imaging techniques which permit the monitoring of multiple metastatic cells either simultaneously or in very rapid succession are likely to improve the biomedical relevance of future studies.

Currently, single-molecule cellular studies utilize cultured cells. For bacterial research these cultures can be made very close to the native physiological environment, though clearly the majority of eukaryotic cells in complex organisms are expressed not in isolation but as part of a surrounding tissue structure composed of both extra-cellular biological material and other closely packed cells. Future developments in this area are likely to be twofold. Firstly, to generate more robust multi-cellular *mimetic* systems in culture, that is in effect to grow *artificial tissues* and perform experiments on these; developments in *stem cell* research are likely to be significantly beneficial here, since stem cells can be programmed using a cascade of chemical cues to differentiate into a variety of different cell types associated with specific different tissues in the human body. Secondly, to develop aspects of *native-tissue* single-molecule imaging methods.

The biggest challenge with native-tissue single-molecule experimentation is the obvious one of inaccessibility. The usual candidate technique, as should now be apparent to readers of the earlier chapters of this book, will most probably involve some level of optical microscopy, ideally fluorescence imaging. In exceptional cases, such as research on cells that grow on the outside of the body, for example those on the skin or the outer layer of the eye or even on the outer surface layer inside the gut viewed using suitable fibre optic probes, it may be possible to use existing single-molecule imaging methods with relatively little modification to the actual microscopy hardware, but in general this is not the case.

For cells buried greater than a few cell layers deep, there is the problem of efficiently probing fluorescent dye molecules, which tag intra-cellular molecules of interest, with excitation light, of limiting the excitation light to the specific cells of interest, and of collecting the emitted fluorescence due to scattering from the multiple cellular layers. The first issue of localized, efficient excitation may benefit from future improvements in multi-photon excitation methods (see Chapter 3), for example cells absorb and scatter significantly less infrared light than they do light of visible wavelengths. The point spread function of a focused laser beam is limited to a depth of typically a few micrometres parallel to the optic axis of the objective lens, but using Bessel beams can potentially circumvent this (see Chapter 3). Thus, Bessel beam multi-photon fluorescence imaging using near infrared lasers may prove a highly productive area of tissue-level single-molecule research in the near future.

Another potential development for *deep-tissue* single-molecule fluorescence imaging is likely to involve *light sheet microscopy* (see Chapter 3). Light sheet microscopy utilizes a thin light sheet to illuminate orthogonal to the detection pathway, and thus only a narrow zone near the sample focal plane is excited, resulting in a substantial increase in imaging contrast. Single-molecule tracking has already been established in imaging single fluorescently labelled mRNA molecules in the nuclei of salivary gland cells at a depth of $\sim$200 μm from the surface (Ritter *et al.*, 2010), so it is highly likely that other biomolecules of interest will be studied in such a manner soon.

Improvements in deep-tissue fluorescence imaging of single molecules are also likely to be made by using new fluorescent proteins which fluorescence at *longer* wavelengths in the infrared then the first generation of developed fluorescent proteins, thereby minimizing the scatter of the fluorescence emission through deep tissue. Another problem associated with deep-tissue imaging is one of *degradation* of the image due to heterogeneity in the sample layers between the focal plane at the detector. A highly active area of current research involves the development of *adaptive optics* techniques to circumvent such problems (see Chapter 3).

Other potential future directions for research include *Raman spectroscopy* and associated SERS imaging (see Chapter 4). Raman approaches have a distinct advantage in that cells are relatively transparent to the emission signal and so in principle Raman techniques can be performed deep in tissues with significantly less drop-off in sensitivity with distance compared to optical imaging. This has already been applied to imaging small biological features in deep tissues such as collections of cells in tumour formation (Keren *et al.*, 2008), and is only likely to improve in the future in the direction of sub-cellular/molecular features.

10.3 Bionanotechnology and 'synthetic' biology

Over the past decade, research into the development of synthetic devices at the nanometre length scale which have been inspired by natural biological machines and processes has exponentiated. Life, to the best estimate, has existed on this planet for ~4 billion years, in which time the process of genetic mutation combined with selective environmental pressures have resulted in highly optimized evolutionary *solutions* to biological *problems* at the molecular level. These are found in many examples of *established bionanotechnology* – instead of attempting to design miniaturized devices *de novo* from the bottom up it makes much more sense to try to learn lessons from nature, both of the *physical architecture* of the actual molecular machines and cellular sub-structure that have evolved to perform certain highly optimized functions, but also of the general *processing structure* of cellular functions at the molecular scale, namely how molecular machines interact with other processes in the cell.

10.3.1 DNA nanotechnology

The highly specific nucleotide base-pairing of the DNA double helix has offered the potential to design artificial DNA nucleotide sequences with the specific intention of creating novel synthetic nanoscale structures that may have useful applications (for a brief, accessible discussion see Turberfield, 2011). This *DNA origami* has developed recently into a strong area of active research both in applying single-molecule biophysics methods in characterizing such DNA nanostructures, and also in utilizing the structures for further single-molecule biophysics research. A significant advantage with such structures is that in general they self-assemble spontaneously with high efficiency from solutions of the different linking sequences, provided they are present in the correct stoichiometric ratios and the sequences have been specifically designed to minimize undesirable base-pairing interactions.

One particularly useful structure already developed includes two-dimensional DNA arrays which can be used as templates for the attachment of proteins (Selmi *et al.*, 2011). This has great potential for generating atomic-level structures of membrane proteins and complexes in particular. Many of these are not suitable for study using x-ray crystallography, owing to the technical challenges of generating stable lipid–protein interactions in a large crystal structure (see Chapter 2). The standard alternative technique of NMR (see Chapter 1) has associated disadvantages in that it is inherently insensitive, requiring purified samples of greater than 95% purity in concentration of several mg per ml. These samples are typically prepared from recombinant protein produced by time-consuming genetic modification of a bacterium such as *E. coli*. The protein needs to be isotopically labelled, which in itself is non-trivial, and NMR will not work with any great precision for either moderate to large proteins or small proteins whose molecular weight lies outside the range 50–100 kDa.

Single particle cryo-electron microscopy allows direct imaging of biological molecules from a rapidly frozen solution supported on an electron-transparent carbon film, and circumvents many of the problems associated with NMR and x-ray crystallography. However, high electron currents can damage samples and there are also potential protein aggregation effects at the high concentrations used, and the random orientation of particles means that analysis is limited to small groups at a time with limited potential for high-throughput analysis. Attaching the target protein to specifically engineered binding sites on a self-assembled two-dimensional DNA template minimizes many of these issues, and opens the possibility for actual two-dimensional crystallography if proteins can be bound to the template in consistent orientations, perhaps using multiple binding sites.

Other uses of two-dimensional synthetic DNA structures include designing tracks for *artificial molecular motors* and *miniaturized logic circuits*. Such developments currently show signs of promise, but artificial motors to date are slow and inefficient and DNA logic circuits are not as reliable as conventional electronic circuits, but in the future they may have significant potential. Other useful DNA nanostructures include three-dimensional shapes. Several have already been developed, but arguably the most elegant for its engineering simplicity is the *nano-tetrahedron*, which has an edge length of just ~7 nm with the potential for acting as a nanoscale brick for more extensive synthetic three-dimensional structures (Goodman *et al.*, 2005).

10.3.2 Nanomedicine

The use of nanotechnology in medicine is already established at the level of *targeted drug delivery*. For example, in the destruction of diseased cells, including those of cancers, nanoparticles coated with specific antibody probes are used. This method is also used indirectly to visualize diseased tissue, for example using antibody-tagged QDs which will specifically bind to tumour tissue and assist in the discrimination between healthy and non-healthy cellular material. In the field of *bio-mimetic engineering*, artificial replacement tissue is being developed for bone, skin, connective tissue and blood vessels and even blood, which all need to have very similar structural and chemical properties to the original material at the molecular level. Such developments have benefited enormously from the formative studies performed using single-molecule cellular biophysics.

Much active research is being done in the area of *bio-sensing*; an ultimate aim is to develop *lab-on-a-chip* systems in which diagnosis can be made by detection and analysis of microlitre quantities of a patient's blood fed through a miniaturized bio-molecular detection device. Such devices potentially involve complex elements of nanofluidics and optical engineering for detection. Many systems under development utilize some of the contrast-enhancement techniques of fluorescence imaging discussed previously in the context of single-molecule cellular biophysics (see Chapter 3).

A *Holy Grail* of the bio-sensing field is the ability to *sequence* single molecules of DNA efficiently. At present, no single technique can perform this robustly on long real DNA sequences, however the method of ion conductance measurement through engineered nanopores, either solid-state or manufactured from protein adapters such as α-haemolysin (see Chapter 4) look the most likely to deliver this capability in the near future. In fact in 2012 an Oxford-based bionanotech company released a disposable sequencing device using such technology that could interface with a PC via a USB port and which cost less than $1000. The device has a stated accuracy of ~4%, which makes it not sufficiently precise for some applications, but even so with the capability of sequencing DNA segments up to ~1 million base-pairs in length this represents a significant step forward.

Bionano-assisted drug delivery is another area with great future potential. Developments already exist which allow the specific delivery of certain drug compounds into specific cell types *piggy-backing* on the normal process of endocytosis (see Chapter 7). Other areas of active research for delivery involve the use of synthetic nanostructures, for example made from DNA, to act as *molecular cages* to permit the efficient delivery of a variety of drugs deep into a cell while protecting it from normal cellular degradation processes before it is released to exert its pharmacological effect.

Other more speculative nanomedicine developments include the use of bionano-electric devices to interface with nerve and muscle tissue, for example in the brain and the heart, and also in the design of cellular repair nanoscale robots, or *nanobots*. There has been much hype in the media about such devices, and the terrifying consequences were they to go wrong. This is the so-called *grey goo* catastrophe concerning an imagined apocalyptic scenario in which uncontrolled self-replicating nanobots consume all matter on Earth in order to replicate themselves (for those interested, self-replicating machines were originally speculated by the mathematician John von Neumann, but for the science fiction see Drexler, 1987). However, the truth is that currently it is technically very challenging to get even natural nanoscale machines to work unmodified outside of their original enviroinment, let alone to generate *de novo* molecular machines that can be programmed by the user to perform cellular tasks. Perhaps the most realistic development in this arena will come from the application of stem cells combined with some aspects of relatively well-characterized nanotechnology such as the use of nanostructures of some form to act as a growth template to permit stem cells to assemble in highly specific regions of space to facilitate generation of new tissues.

KEY POINT

There is a lot of outlandish media coverage concerning nanotechnology in general, however behind the hype significant steady advances are being made in areas of **nanomedicine** in particular which have utilized the developments of single-molecule cellular biophysics and may pave the way to significant future **health benefits**.

10.3.3 'Biofuel' production

It has been said that advanced biofuels may end up being crucial to building a clean energy economy. With depletion of planetary supplies of fossil fuels and the decommissioning and debated safety and environmental issues of nuclear power, the development of natural biofuels offers a not insignificant attraction. Currently, the state of biofuel development which utilizes molecular-level technology is at a relatively early stage. For example, certain nanoparticles are known to increase the efficiency of the biofuel production process which uses enzyme catalysis to convert the polysaccharide *cellulose* from plant matter into smaller sugars from which chemical energy can ultimately be extracted. Also, there have been promising developments in designing *bio-batteries*, or miniaturized electrical charge storage devices that utilize biological material in some way. One such development was already mentioned previously in the context of *nanodroplet arrays* of lipid vesicles (see Chapter 4), however these are currently low power and inefficient.

More practicable designs involve the use of *nanowire* microelectrodes which can run off a fuel of relatively few molecules of natural *redox* enzymes (Pan *et al.* 2008). The biggest challenge with miniaturizing electrical wires is that conventionally the electrical resistance varies inversely with the cross-sectional area, so that a material which obeys *Ohm's law* of electrical resistance in this way could have a significantly high electrical resistance for wires whose width is of the nanometre length scale, resulting in highly inefficient generation of heat.

Active methods for tackling this resistance problem currently include non-biological condensed matter research into the use of *superconductive* materials whose electrical resistance can be made exceptionally low. Currently these are limited to use at relatively low temperatures (for example, that of liquid nitrogen), however it is likely that viable room-temperature superconductive material will be available in the near future.

Biological-inspired research into this area of varying success has included developing nanowires from single DNA molecule templates periodically labelled with alternating electron-donor and electron-acceptor probes, thereby allowing electrical conduction through a process of serial quantum tunnelling events rather than by the conventional means of electron drift. Related approaches have also been investigated using *molecular photonics wires* that utilize typically a DNA molecule containing multiple FRET acceptor–donor pairs along its length which permit the transfer of optical information through space via FRET, thus potentially acting as a controllable *optical switch* at the nanometre length scale (Heilemann *et al.*, 2004).

In 2010, the estimated total power consumption of the human world was ~16 TW (1 TW = 1 *terawatt* = 10^{12} W), equivalent to the output of around 8000 *Hoover Dams*. The total power production estimated from *global photosynthesis* is more like 2000 TW, which may indicate a sensible route forward to develop methods to harness the single-molecule energy transduction properties of natural photosynthetic systems in a more controllable artificial context. That being said, the total power available to the Earth from the Sun is ~200 PW (1 PW = 1 *petawatt* = 10^{15} W), and so in this sense natural photosynthesis is arguably relatively inefficient at only extracting ~1% of the available solar energy (much of the remaining solar energy which is not reflected/scattered away from plant leaves is actually transformed into heat as opposed to being locked in high-energy chemical bonds in sugar molecules).

It may be that future single-molecule experiments on photosynthetic systems may result in significant improvements in this efficiency. One route being developed is in the modification of naturally photosynthetic *cyanobacteria*, potentially to develop synthetic chloroplasts by introducing them into other more complex organisms. This has currently been tested on mammalian macrophage cells and zebrafish embryos, to increase their photosynthetic yield (Agapakis *et al.*, 2011).

Modification of bacteria has also been applied in the development of cell strains which can directly generate alcohol-type fuels, such as *butanol* (Bond-Watts *et al.*, 2011), by manipulation of the normal *fermentation process* in modifying the key acetyl-CoA complex which forms an essential *hub* between many metabolic pathways, most importantly in the Krebs/TCA cycle (see Chapter 2). The prime difficulty here is that the ultimate alcohol product is toxic to the cell and so achieving commercially feasible concentration yields is technically challenging. However, an understanding of this process from single-molecule cellular biophysics, namely of how *molecular homeostasis* of the acetyl-CoA is achieved and potentially can be manipulated, may be key to future progress here.

10.4 The outlook for single-molecule cellular biophysics

As a scientific discipline, single-molecule cellular biophysics is undergoing an enormous expansion and is likely to be a key field in revealing the real mechanistic features of biological processes in living cells. This will have implications for the shape of biological, biophysical and biomedical research in the future and, as we have speculated here, may have significant implications for other related areas of research. The commercial incentive to miniaturize synthetic bio-inspired devices is already starting to feed back into academic research laboratories in catalysing a general *down-sizing* approach for measurement apparatus.

For example, much of the first generation of single-molecule biophysics was characterized by bespoke microscopy using cumbersome vibration isolation tables and custom-built equipment requiring a highly specialized knowledge of practical physics to operate. However, there is now a new trend to consolidate the physical footprint of such devices and to design them from scratch with the explicit intention of ease of use so that they have the potential for being relatively *portable*, *transferable* onto other similar machines and *user-friendly*, such that researchers with no formal training in the physical sciences can operate them. In addition, analysis code is being developed which does not require an intensive knowledge of the use of efficient computer algorithms. However, the directions of movement in the field will, as with others, be dictated largely by what happens at its *edges*, namely, how it *interfaces* with the other *related fields* around it.

There is a compelling need to push this area of physiologically relevant single-molecule bioscience forward significantly, and this can only be truly facilitated by future generations of life and physical scientists *talking to each other*. In a world of such rapid technological advances it is sometimes difficult to take a step back to see the beautiful, fundamental science behind them and really appreciate that, as a general rule, science cannot be compartmentalized into strict areas of academic discipline but should be allowed to explore its own interfaces to find the best solutions to challenging, unresolved questions. Physical and life scientists in particular traditionally blend like oil and water, such immiscibility often stemming from the unfortunately early nature of choice of academic study which schoolchildren are asked to make.

However, what is needed now is an appreciation that some of the most fundamental scientific concepts in each discipline can be shared by both the physical and life sciences camps, once elements of unwieldy terminology and mathematics have been put aside. In doing so, physical scientists may start to appreciate the transferable nature of the academic skills they have learnt into uncharted biological waters, and life scientists may perhaps learn that taking a broader view of the processes of the cell which encompass other less traditionally life-biased sciences may help to address some highly fundamental questions about how the cell really works.

The truth is that the outlook for single-molecule cellular biophysics is highly promising, but it is fundamentally driven by the *enthusiasm* of the talented researchers willing to *take a punt* and *cross bridges* into areas of science unknown.

It is the author's strong hope that perhaps the reader may be one such researcher.

THE GIST

- Bionanotechnology and synthetic biology both feed off progress in single-molecule cellular biophysics.
- DNA nanotechnology, biofuel production and bio-sensing are key areas of future exploitation for single-molecule bioscience.
- Nanomedicine techniques are already in practical use and future advances are likely to move towards single-molecule biomedicine.
- Cells in complex organisms do not function in isolation but as part of multi-cellular systems, and there are likely to be developments in applying single-molecule biophysics techniques to investigating these in the near future.
- Physical and life scientists should talk to each other.

TAKE-HOME MESSAGE The techniques and results of single-molecule cellular biophysics will extend significantly into other scientific and technological fields in the near future, allowing ever more complex life processes to be understood, provided physical and life scientists work together.

References

GENERAL

- Ball, P. (2011). Physics of life: the dawn of quantum biology. *Nature* **474**: 272–274.
- Diggle, S. P., Griffin, A. S., Campbell, G. S. and West, S. A. (2007). Cooperation and conflict in quorum sensing bacterial populations. *Nature* **450**: 411–414.
- Drexler, K. E. (1987). *Engines of Creation: The Coming Era of Nanotechnology*. Anchor Books.
- Turberfield, A. J. (2011). DNA nanotechnology: geometrical self-assembly. *Nature Chem.* **3**: 580–581.
- Wirtz, D., Konstantopoulos, K. and Searson, P. C. (2011). The physics of cancer: the role of physical interactions and mechanical forces in metastasis. *Nature Rev. Cancer* **11**: 512–522.

ADVANCED

- Agapakis, C. M. *et al.* (2011). Towards a synthetic chloroplast. *PLoS One* **6**: e18877.
- Bond-Watts, B. B., Bellerose, R. J. and Chang, M. C. Y. (2011). Enzyme mechanism and a kinetic control element for designing synthetic biofuel pathways. *Nature Chem. Biol.* **7**: 222–227.
- Goodman, R. P. *et al.* (2005). Rapid chiral assembly of rigid DNA building blocks for molecular nanofabrication. *Science* **310**: 1661–1665.
- Heilemann, M. *et al.* (2004). Multistep energy transfer in single molecular photonic wires. *J. Am. Chem. Soc.* **126**: 6514–6515.
- Keren, S. *et al.* (2008). Noninvasive molecular imaging of small living subjects using Raman spectroscopy. *Proc. Natl. Acad. Sci. USA* **105**: 5844–5849.
- Mauriello, E. M. F. (2010). Cell polarity/motility in bacteria: closer to eukaryotes than expected. *EMBO J.* **29**: 2256–2259.
- Pan, C. *et al.* (2008). Nanowire-based high-performance 'micro fuel cells': one nanowire, one fuel cell. *Adv. Mater.* **20**: 1644–1648.
- Ritter, J. G. *et al.* (2010). Light sheet microscopy for single molecule tracking in living tissue. *PLoS One* **5**: e11639.
- Sako, Y., Minoghchi, S. and Yanagida, T. (2000). Single-molecule imaging of EGFR signalling on the surface of living cells. *Nature Cell Biol.* **2**: 168–172.
- Selmi, D. N. *et al.* (2011). DNA-templated protein arrays for single-molecule imaging. *Nano Lett.* **11**: 657–660.

Questions

FOR THE LIFE SCIENTISTS

Q10.1. Find a physical scientist and explain to them what you mean by 'life' without mentioning the specific names of any biological molecules.

Q10.2. What are the biological challenges to deliver and release a drug specifically to a given sub-cellular organelle which is caged inside a three-dimensional DNA nanostructure?

FOR THE PHYSICAL SCIENTISTS

Q10.3. Find a life scientist and explain to them why the mean square displacement of a particle exhibiting Brownian diffusion scales linearly with time, without using any written equations.

Q10.4. How is Schrödinger's quantum mechanical wave equation governing the likelihood for finding an electron in a given region of space and time related to the probability distribution function of a single diffusing biological molecule? Comment on the significance of this between the life and physical sciences.

Q10.5. An optical based bionanotechnology data transmission molecule was designed using a 'molecular photonics wire' in which a series of five dye molecules of increasing peak wavelength of excitation from blue, green, yellow, orange through to red were conjugated in sequence to a single molecule of DNA attached to a glass coverslip, with each dye molecule spaced apart by a single DNA helix pitch. The mean Förster radius between adjacent FRET pairs was known to be ~7 nm, all with similar absorption cross-sectional areas of $\sim 10^{-16}$ cm^2. (a) When a stoichiometrically similar mix of these five dyes was placed in bulk solution the ratio of the measured FRET changes between the blue dye and the red dye was ~15%. Comment on how this compares with what you might expect. Doing similar measurements on the single DNA–dye molecule then suggested a blue-red FRET efficiency of more like 90%. (b) Why is there such a difference compared to the bulk measurements? Blue excitation light of wavelength 488 nm was shone on the sample in a square wave of intensity 3 kW cm^{-2}, oscillating between on and off states to act as a data clock pulse signature. (c) If the thermal fluctuation noise of this last dye molecule is roughly $\sim k_B T$ estimate the maximum frequency of this clock pulse that can be successfully transmitted through the DNA–dye molecule. You can assume that the emission from the donor dye at the other end of the DNA molecule was captured by an objective lens of NA 1.49 of transmission efficiency 75%, split by a dichoic mirror to remove low wavelength components which captured 55% of the total fluorescence, filtered by an emission filter of 88% transmission efficiency and finally imaged using a variety of mirrors and lenses of very low photon loss (<0.1%) onto an electron-multiplying CCD detector of 92% efficiency. (d) How might your answer be different if a light-harvesting complex (see Chapter 8) could be coupled to the blue dye end of the photonic wire? (Hint: See Heilemann *et al.*, 2004.)

FOR THOSE WHO HAVE NOT MADE UP THEIR MIND

Q10.6. Find a life scientist and ask them to explain the biochemical process of DNA repair and recombination. Find a physical scientist and explain this process to them with reference primarily to changes in free energy.

Index

3T3. *See* immortal cell
4Pi microscopy, 90

Abbe limit, 81–82, 90
ABEL trap, 142, 147, 206
acetone, 107
acetylcholine, 49
acousto-optic deflector, 131
acronyms, annoying over-use of in fluorescence microscopy, 90
actin, 18, 48, 117, 128, 133, 152, 157–158, 162, 170, 175, 180, 193, 200–201, 208–209, 214–216, 224–226, 231, 247–249
adaptive optics, 78, 257
Adenovirus, 176, 180
A-DNA, 43
Aequorea victoria, 71
AFM (atomic force microscopy), 58, 80, 91, 97, 102–105, 116–120, 134, 136–140, 143–144, 147–148, 152, 156, 172–173, 178, 185, 204, 206–208, 212, 214, 219, 221
AGT (O6-alkylguanine-DNA alkyltransferase), 73
Airy disc, 81
alarmone, 223
aldehyde, 38, 126
ALEX (alternating laser excitation), 94
allophycocyanin, 206
alpha helix, 5, 35, 41, 43, 51, 56, 58–59, 68, 86, 96, 135, 152, 157, 187, 245, 247, 249, 258, 264
alpha-haemolysin, 112–113, 259
amino acid, 6, 8, 34, 36–37, 50–52, 56, 59, 67–69, 72, 113, 137, 147, 160, 188, 244
amino group, 34
AMPAR, 186
ampipathic, 46
angular resolution limit, 60
anisotropy, 7, 95, 97
anomalous diffusion, 79, 176, 188, 214, 223, 246
antibiotic, 189, 239, 256
antibody, 8, 11, 18, 37–38, 66–70, 100, 126, 187, 237, 249, 259
anti-codon site, 51
antigen, 18, 37–38, 67
antiporter, 184
anti-Stokes scattering, 66
apoptosis, 186
aquaporin, 189, 213
arabinose, 86, 240
A-RNA, 43
Arrhenius equation, 139
ASKA library, 74, 98
ATP (adenosine triphosphate), 20, 33, 35, 40, 47–49, 55, 88, 97, 111, 132, 152, 163, 175, 190, 193, 199–203, 205–206, 213, 215–218, 220, 224–226, 228–229, 237
axoneme, 226

back focal plane detection, 115–116, 119, 125, 127, 196–197, 200, 217, 224
bacterial flagellar motor, 119, 142, 193, 215
BaLM (blinking assisted localization microscopy), 89, 209
Baltimore classification system, of viruses, 174
base-pair, 40, 42, 52, 58, 143, 234
B-DNA, 42–43, 59, 145, 156
Bessel beam, 77, 91, 129–130, 144, 214, 257
beta-barrel, 137
BiFC (bifunctional fluorescence complementation), 82, 87, 209
binomial distribution, 243
bio-battery, 112–113
biofuel, 260
biological function, 23
bioluminescence, 71
biomass, 205
bio-mimetic engineering, 259
bionanotechnology, 258, 262
bio-sensing, 91, 113, 115–116, 259, 262
biotin, 67–69, 126
Bjerrum length, 18–19
blinking. *See* photoblinking
blue-green cyanobacteria, 206
Boltzmann distribution, 70
Boltzmann factor, 47, 139, 201
BR (bacteriorhodopsin), 15, 112–113, 206, 216
Bragg's law, 106, 131

brightfield, 10
Brownian diffusion, 79, 142, 176, 181, 211, 218, 221, 229, 247, 250, 263
burnt bridge Brownian ratchet, 238

calorimetry, 3
cAMP, 164–165
cancer, 70, 144, 159–160, 163, 186, 256, 263
cantilever, 104–105, 119
capping, 186
capsid, 32, 173–174, 177, 181
Carboxyl group, 34, 44, 126, 141
cargo sorting, 168
catenation, of carbon compounds, 31
Caulobacter crescentus, 98, 161, 204, 209, 213–214, 234, 238, 249
CCD (charge-coupled device), 12–13, 62–63, 264
CD (circular dichroism), 5
CdSe (cadmium selenide), 69
cell death. *See* apoptosis
cell division, 58, 128, 160–162, 186, 224, 228, 232, 238–239
cell lysis, 178
cell membrane, 33, 47, 73, 79, 84–86, 89, 94–95, 106, 108, 110, 115, 129, 138, 152, 154–155, 158–164, 166, 168–172, 174–178, 181, 183–190, 193–196, 198–199, 202, 204, 206, 208–212, 215–216, 218–220, 222, 230–231, 234, 239–240, 247
cell motility, 38, 164, 168, 183, 193, 197, 223, 254
cell wall, 33, 181, 192, 194, 204, 207, 210, 213
cellulose, 31, 39–40, 260
CENP-A, 237
central dogma, of molecular biology, 44, 50–51, 56, 238
centromere, 237–238
centrosome, 33, 237
CFP (cyan fluorescent protein), 84, 93, 100
CFTR, 190
cGMP, 62
chaperone protein, 190
Che proteins, in bacterial chemotaxis, 166
chemical equilibrium, 17, 23, 26
chemiosmosis, 47, 56
chemotaxis, 164, 180
chemotherapy, 256
chiral, 36
chitin, 40
chlorophyll, 47, 206
chloroplast, 47, 184, 205, 263
CHO. *See* immortal cell
cholesterol, 14, 45–46, 49, 210–211
choline, 45–46
chromatic aberration, 76
chromatid, 237
chromatin, 44, 233, 237, 244
chromophore, 67, 71, 231
chromosome, 18, 40, 43–44, 57, 69, 72, 233, 235–237, 239, 249
Chung–Kennedy filter, 151, 153
cilia, 226, 228
cis–trans isomerization, 62
citric acid cycle, 40
CJD, 50
clamp, in replisome, 234
clathrin, 169–171, 177, 179
Clausius–Mossotti factor, 141–142
CLIP-tag. *See* SNAP-Tag
cloud-computing, 255
collagen, 161
condensation reaction, 36, 39, 44, 46
condensin, 233
confined diffusion, 175, 211, 247
confocal microscopy, 62–63, 74, 77, 79–80, 99, 178, 246
connective tissue, 161, 163, 186, 259
contact mode, 104–105, 118
continuity equation, 229
cornea, 61
corner frequency, 125
co-transporter. *See* symporter
covalent bond, 20–21, 31, 73
crescentin, 209
cristae, 47
critical angle, 87
crossing-over, 43
cryo-electron microscopy, 258
cryo-fixation, 107
crystallography, 9, 14–15, 17
cut-and-paste, using AFM, 136, 138
cyclodextrin, 113
CypHer5, 177
cytoplasm, 33, 51, 99–100, 152, 158, 160, 162, 167, 170, 175, 185, 199, 220–221, 223, 232, 234, 241, 243, 246–247, 250–251
cytoskeleton, 33, 100, 162, 170, 175, 180, 190, 207–209, 211, 213, 221, 231, 247

dark noise, 61
Dawes limit, 82
de Broglie relation, 106, 122
Debye–Hückel length, 19
deep-tissue imaging, 91, 257
deformable mirror, 79
degeneracy, 52
denaturation, 130–131
de-noising algorithms, 233
deoxyribose, 41–42, 57
depolarization, of membrane, 111, 171
depth of field, 251
detailed balance, 22, 70, 139, 147, 152, 198, 225
diabetes, 172, 185
dialysis, 3
DIC (differential interference contrast), 10, 64
dichroic mirror, 63, 100–101, 115, 133
dictyostelium, 165
dielectrophoresis, 141–142
diffraction, 144
diffusion-to-capture approach, 177
DIG (digoxigenin), 69
dihedral angles. *See* torsion angles
dipole orientation factor, 93
directed diffusion, 175, 223, 247
disaccharide, 39–40, 239
dispersion-steric repulsion forces. *See* Van der Waals forces
dissociation constant, 37
disulphide bond, 37, 68
divisome, 238
DNA. *See* nucleic acid
DNA condensation, 233
DNA curtains, 133
DNA origami, 258
DNA polymerase, 234–236
DNA replication, 38, 50–51, 76, 131, 135, 223, 233–236, 249
DNA sequencing, 113, 115–117, 259
double-helical point spread function method, 86
dronpa, 78, 89, 237
Drosophila, 78, 233
dumbbell assay, 133
dynein, 162, 171, 225–227, 230–231, 247–248, 251

E. coli. See Escherichia coli
e-cadherin, 161–162, 164
ECM (extra-cellular matrix), 31, 161
edge-preserving filtration, 151
EDL (electrical double layer), 54
EGF, 160
EGFR, 27, 98, 156, 160–161, 166, 179, 263
Einstein–Stokes relation, 126, 192, 217, 221, 229
elastic scattering, 13–14
electrode potential, 47, 58
electron diffraction, 13, 15, 107
electron microscopy, 5, 24, 100, 102, 106–109, 111, 116
electron transport, 47, 202–203
electrorotation, 140
electrostatic forces, 18, 53
emission filter, 63, 101, 264
endocytosis, 115, 169–171, 174–175, 177, 179, 183, 231, 259
endoplasmic reticulum, 32–33, 190
enthalpy, 3, 16, 22, 58
entropic brush, 244
entropy, 1, 3, 16, 22–23, 28, 49, 58, 129, 198
enzyme, 13, 35, 48, 51, 62, 73, 129, 131–132, 152, 174, 201, 205, 236, 239, 241, 243, 260
Eos, 89
epifluorescence, 74–76, 87, 99, 101, 170, 175, 177, 191–192, 203–204, 209, 219, 222–223, 237, 240, 250
EPR (electron paramagnetic resonance). *See* ESR
equipartion theorem, 217
ergodic processes, 27
Escherichia coli, 44, 74, 97–98, 153, 166, 214, 249
ESR (electron spin resonance), 9
eukaryote, 33
evanescent field, 10, 82, 88, 129–130, 168, 178, 194, 223
exciton, 69
excluded volume, 114, 234
exocytosis, 74, 168, 171–172, 174, 179, 182, 185, 190, 215
exons, 52
exonuclease, 51
Eyring theory, 139

Fab region, 37
facilitated diffusion, 184

Fallopian tube, 226
Faraday's constant, 48
far-field, 81
fat, 33, 45, 185
fatty acid, 30–31, 45–46, 56
Fc region, 37–38, 67
FCCS (fluorescence cross-correlation spectroscopy), 80
FCS (fluorescence correlation spectroscopy), 79
fermentation, 40, 202, 261
fibroblasts, 161, 186
fibronectin, 137
Fick's laws of diffusion, 119, 211, 228
filopodia, 209
FIONA (fluorescence imaging with one nanometre accuracy), 83
FISH (fluorescence in situ hybridization), 68
fixation, 89, 107, 205, 210
flagellar motor. *See* bacterial flagellar motor
FlAsH (fluorescein arsenical helix binder), 66
FLIM (fluorescence lifetime imaging), 81
FLIP (fluorescence loss in photobleaching), 168, 196
flip-flop mechanism, 185
fluctuation theorem, 22, 198, 213
fluid mosaic model, 210
fluorescein, 24, 66–68, 175, 249
fluorescence microscopy, 11, 25, 48, 65, 70, 74, 76, 84, 87, 90, 97–98, 133, 135, 138, 167, 169, 214
fluorophore, 68–70, 77, 81, 84–85, 87, 89–90, 92–93, 99, 156, 177, 187, 203, 208, 223, 249
focal adhesion complex, 161
FoF1-ATP synthase, 193, 199, 201–202, 213, 215, 218, 224
Fokker–Plank equation, 229
force spectroscopy, 58, 136–140, 147–148, 152, 172–173, 178, 185, 204, 206, 208, 212, 221
force-feedback, 138
Förster radius, 92, 99, 264
Fourier transform, 8, 12, 153
FP (fluorescent protein), 71
FPALM. *See* PALM
FRAP (fluorescence recovery after photobleaching), 168
free energy, 1, 3, 13, 22, 27, 46–49, 54–55, 58, 126, 139–140, 144, 163, 169, 184, 189, 193, 198–199, 201, 205, 216, 218, 225, 228, 243, 264
freely jointed chain, 114, 127–129, 146
free-radical, 33, 69, 94–95, 124, 195, 206, 222
freeze-etching, 108
FRET (Föerster resonance energy transfer), 92
FRET efficiency, 93, 95, 99–100, 138, 201, 264
FSM (fluorescent speckle microscopy), 87, 99, 208, 215
FtsK, 134, 146, 238
FtsZ, 238, 248

Gaia hypothesis, 23
galactose, 239
Gamma distribution, 242, 252
ganglion, 62
Gaussian chain model, 128–129
GDP, 163
GFP (green fluorescent protein), 71
Gibbs free energy. *See* free energy
gigaseal, 109
glucose, 35, 38–40, 56, 184–185, 205, 214, 216, 239
GLUT proteins, 185
glutaraldehyde, 107
glycerol, 44–46, 56, 189
glycerolipid, 44
glycogen, 39–40
glycolysis, 40, 202
glycoprotein, 33
glycosidic bond, 39
Goldman equation, 110
Golgi apparatus, 33, 190, 226
Gouy–Chapman layer. *See* EDL
G-protein, 163
gradient force, 121–123, 129–132, 142–143, 146
gramicidin, 188
gram-negative/positive, 194
graphene, 108, 112, 117, 145
Grb2, 163
green sulphur bacteria, 206
ground-state depletion microscopy. *See* STED
GTP, 163
gyrase, 135

halorhodopsin, 207
HaloTag. *See* SNAP-tag
hand-over-hand mechanism, 226
haptan, 69
Hayflick limit, 186
HEK. *See* immortal cell
HeLa. *See* immortal cell
Helfrich forces, 55

helicase, 135
heteroduplex DNA, 43
high-throughput, 15, 131, 242, 254, 256
HILO (highly inclined and laminated optical sheet illumination), 75
histone, 233, 237, 243–244
HIV (human immunodeficiency virus), 177
Holliday junction, 42–43
holographic optical trap, 132
homeostasis, 261
homologous recombination, 43
hopping diffusion, 241
hormone, 49
host cell, 32, 50, 174, 176
HPLC (high-performance liquid chromatography), 3
HS-AFM (high-speed AFM), 105
HU (hydroxyurea), 234
hub, 30, 126, 261
hydration shell, 47
hydrodynamic radius, 128
hydrogen bonding, 30, 35–37, 41, 49, 56, 131
hydropathy, 36
hydrophobic force, 30, 113, 137
hyperpolarization, 62
hysteresis, 140, 221

IgG, 37–38, 67, 69, 172, 179, 249
IgM, 37
immortal cell, 186
immune response, 37, 46, 115, 129, 164, 172, 186, 211, 231
immunofluorescence, 67
immunoglobulin, 37
importin, 245
inchworm mechanism, 226
inelastic scattering, 14, 109
infection, 37, 173–181
influenza, 176–177, 179
insulin, 36, 57, 172, 185, 215
intercalating dye, 236
intermediate filament, 209
internal energy, 22
interphase, 233
intersystem crossing, 65
introns, 52
ion channel, 58, 108–110, 118–119, 186–189, 212–213, 217
ion exchange pump, 111
IPTG (isopropyl-β-D-thio-galactoside), 74, 86, 239–241, 252
ITC (isothermal titration calorimetry), 3

Jablonksi diagram, 63, 65
Jarzynski's equality, 140
Jurkat. *See* immortal cell

KDE (kernel density estimation), 153–154
Kelvin probe microscopy, 105
ketone, 39
kinesin, 128, 133, 145, 152–153, 156, 162, 171, 225–227, 230–231, 237, 247
kinetochore, 237
Kramers theory, 140
Krebs citric acid cycle. *See* citric acid cycle

lab-on-a-chip, 259
Lac operon, 239–241, 252
lactose, 239
lagging strand, 234
Laguerre–Gaussian beams, 133
leading strand, 234
ligand, 3, 8, 10, 110, 152, 158–161, 163–165, 178, 182, 186, 190, 211, 239
ligase, 234
light microscope, 60, 64
light-harvesting complex, 205
lipid, 44
lipid microdomain. *See* lipid raft
lipid raft, 162, 210–213, 218–219
lipopolysaccharide, 194
liposome, 58, 115, 199, 202
low-density lipoprotein, 169
lymphocyte, 186, 211

macrophage, 232, 261
macrostate, 129
magnetic tweezers, 134
magnification, 63
major groove, of DNA, 43
malate, 48
maleimide, 69
maltose, 40

man-made cell, 33
MAP (microtubule-associated protein), 209
Markovian process, 153
mass spectrometry, 7
maturation, of fluorescent proteins, 72
maximum information methods, 150
mCherry, 82, 93
MCP (methyl-accepting chemotaxis protein), 166
MDS (molecular dynamics simulation), 53
mean square displacement, 79, 123, 125, 175, 180, 188, 204, 211, 217, 222, 229, 250, 263
mechanical force detection, 94
mechanosensitive channel. *See* stretch-activated channel
meiosis, 43, 228, 236
melanophore, 231
melanosome, 231
mercury-arc lamp, 74, 101
metabolic map, 34
metaphase, 233
metastasis, 256, 263
Michaelis–Menten kinetics, 201, 217
micro-lens array, 79
microscopic reversibility, 22, 139
microsphere, 116, 119, 122, 128–129, 132–133, 141, 145–146, 193, 198, 200, 217
microtome, 107
microtubule, 152, 176–177, 209, 225–227, 231, 237, 248–249, 251
Mie regime, 122
minor groove, of DNA, 43
mitochondria, 32–33, 44, 47, 58, 184, 199, 202–203, 214, 218, 220, 226, 255
mitosis, 43, 209, 228, 236, 238
mitotic spindle, 237
Moiré fringes, 91
molality, 21
molarity, 21
mole, 21
molecular cage, 259
molecular cargo, 38, 129, 220, 223, 225, 227, 230–231, 252
molecular heterogeneity, 16
molecular motor, 43, 116, 132, 144, 146, 152–153, 162, 169, 171, 175, 192, 220, 224–225, 228–231, 237, 247–248, 251–252, 259
molecular reporter, 64–65, 68–69, 210, 240–241, 244
molecular signature, 113, 136–138, 147, 150, 207, 241
monosaccharide, 38
Monte Carlo simulation, 53, 230
morphogenesis, of tissues, 160
motor proteins. *See* molecular motor
MreB, 162, 204, 209, 213–214
MRI (magnetic resonance imaging), 9
mRNA (messenger RNA), 50
multi-dimensional imaging, 96
multi-photon excitation, 257
multi-plane imaging, 85
muscle, 1, 4, 6, 14, 18, 31, 36, 39–40, 67, 108, 126–128, 133, 137, 145, 152, 162, 184–185, 190, 211, 215, 224, 226, 251, 260
mutation, 24, 72, 186, 256, 258
Mycoplamsa, 33
myofibril, 67
myosin, 18, 88, 97, 99, 105, 108, 117, 119, 128, 133, 144, 146, 152, 156–157, 162, 171, 224–226, 231, 247–248
Myxococcus xanthus, 254

NA (numerical aperture), 62
NALMS (nanometer-localized multiple single-molecule fluorescence microscopy), 84
nanobot, 260
nanodroplet network, 112–113
nano-eye, 112–113
nanofabrication, 111, 263
nanogold probe, 200
nanomedicine, 259
nanometre, 18
nanopore, 108–114, 116–119, 134, 145, 177, 183, 190–192, 259
nanoreactor, 205
nano-tetrahedron, 259
nanothermometer, 70
nanowire, 260
napthalocyanine, 105
narrow-field, 76, 99, 186–187, 210, 218, 222–223
near distance, 60–61
near-field, 87, 91, 97
Nernst equation, 110, 118, 198, 218
neurotransmitter, 49, 62, 171–172, 186, 190, 250
neutron diffraction, 14
neutrophil, 21
NHS (N-hydroxysuccinimide), 69
nicotinamide adenine dinucleotide, 40
Nipkow disc, 76

nitrospirobenzopyran, 78
NLS (nuclear localization signal), 244
NMR (nuclear magnetic resonance), 9
non-contact mode, 105
non-covalent interactions, 54
nonsense codon, 69
notch filter, 75
nuclear envelope, 44, 51, 239, 243–244
nuclear lamina, 244
nuclear pore, 76, 99, 243–246, 248–249
nucleic acid, 40
nucleoid, 33, 43, 98, 234, 241, 249
nucleoside, 40–41, 190, 202
nucleosome, 32, 233, 237, 243
nucleotide, 11, 40, 42, 56, 58, 68–69, 111–112, 115, 129, 131, 163, 214, 228, 237, 239, 246, 250, 258
nucleus, 9, 15, 29, 32–34, 44, 51, 59, 76, 101, 176, 220, 233, 239, 243–246, 248, 250
Nyquist–Shannon information theory, 83

objective lens, 62
oblique epifluorescence, 74
oestrogen, 49
Ohm's law, 260
Okazaki fragment, 234
OLID (optical lock-in detection), 77
oligomer, 4, 162
oligosaccharide, 40
one-photon excitation, 70
operator, 239
operon, 239
opsin, 62
optical axis, 74
optical contrast, 64
optical fibre, 91
optical isomer. *See* chiral
optical microcavities, 91
optical microscope. *See* light microscope
optical resolution limit, 60, 66, 81, 83–84, 87–88, 90, 93, 96, 164, 172, 181, 234
optical spanners, 132
optical stretching, 130
optical traps. *See* Optical tweezers
optical tweezers, 102, 111, 116–117, 121–123, 126–129, 132–135, 137, 139–140, 143–145, 152, 157, 193, 208, 215, 224, 226, 243
ORF (open reading frame), 74
organ, 172, 254
organelle, 33–34, 44, 170, 202, 205, 226, 230, 232, 263
organic dyes, 66
organism, 1, 14, 23–25, 29, 31–32, 49, 59, 71, 164, 173, 186, 197, 252, 254
osmium tetroxide, 107
osmosis, 189
over-expression, 72
oxaloacetate, 48
OXPHOS (oxidative phosphorylation), 40–41, 47, 202–205, 212, 214

paddle motif, 190
PALM (photoactivated localization microscopy), 88
ParA proteins, 238
partition function, 22
patch clamp, 110
patch clamping, 6, 105, 109, 189
patterned illumination, 91
Pauli exclusion principle, 54
PBsome (phycobilisome), 206
PCR (polymerase chain reaction), 131
PDB (Protein Data Bank), 56
Pelagia noctiluca, 71
peptide, 34
peptide bond, 36
peptidoglycan, 194
permittivity, 18, 46, 141–142
peroxisome, 33
pH, 47
phagocytosis, 231
phase-contrast, 10
phase-transition boundary, 163
pHlourin, 74, 170–172, 185
phospholipid, 14, 23, 45–46, 58, 171–172, 181, 184–185, 189–190, 202–203, 207
phosphorescence, 65
photobleaching, 69
photoblinking, 89, 95, 209, 234
photodiode, 12, 104, 115–116, 119, 125, 127
photon bunching, 65
photonic waveguide, 91
photosynthesis, 14, 33, 47, 202, 205–207, 212, 220, 255, 261
photosynthetic reaction centre, 205
phototransduction, 62
phycoerythrin, 24

piconewton, 18
piconewton nanometre, 19
plasmid, 73–74, 86, 239–240, 252
PleC, 161
PMF (protonmotive force), 112, 190–191, 193, 198, 202, 206, 216, 219
PMT (photomultiplier tube), 12
podosome, 209
polarization microscopy, 95
poliovirus, 177
poly(A) tail, 51
polyadenylation, 51
polydisperse diffusion, 80
polynucleotide, 40
polypeptide, 36
polyribosome, 52
polysaccharide, 40
polysome. *See* polyribosome
positive-hole, 69
post-transcriptional modification, 51
power spectrum, 153, 191
power-stroke, 146, 157, 219, 225, 229
primary antibody, 37
primary structure, 5, 37
primase, 234
prion, 50
processive enzyme, 85
prokaryote, 33
promoter, 239
proteasome, 33
protofilaments, 162, 225, 238
proton gradient, 48, 199, 202
pseudo-TIRF, 75
PSF (point spread function), 60, 81–84, 86, 90, 96, 160, 167, 186–187, 203–204, 209, 222, 227, 233–234, 237, 245–247, 249, 251, 257
PspA, 192
PTB, 246
purine, 41
purple bacteria, 206
pyrimidine, 41

QD (quantum dot), 66
quantum mechanical oscillator, 139
quantum tunnelling, 104, 202, 261
quasi-TIRF, 75
quaternary structure, 54, 131
quorum sensing, 254, 263

radioactivity, 11
Ramachandran diagram, 37, 146
Raman scattering, 66
Raman spectroscopy, 66
random coil, 5, 118
random telegraph model, 242, 252
random walk, 129, 166, 228, 252
Ras, 163
ratiometric dye, 74, 171
ray optics, 122
Rayleigh limit, 82
Rayleigh regime, 122
Rayleigh scattering, 66
reaction-diffusion equation, 228–229
ReAsH, 68
RecA, 142
red algae, 206
reductionist versus integrationist, 253
refractive index, 10, 62, 64, 76, 81, 87–88, 93, 121–122, 129–131
RelA, 223
reovirus, 174
replisome, 234
resorufin, 68
respiratory tract, 176
respirazone, 204
retina, 60–61, 99, 207
retinal, 62
retrovirus, 50
reverse transcription, 50
Reynolds number, 124, 126
RFP, 233
rheology, 221
rhodamine, 25, 66–67, 70, 152, 163, 169, 175, 203, 210, 218
rhodopsin, 62, 100
riboflavin, 233
ribose, 38, 40, 42
ribosome, 50–52, 223, 244
RNA, 40
RNA duplex, 43
RNA hairpin, 42, 126
RNA replication, 50

RNAi (RNA interference), 230
RNAP (RNA polymerase), 50–51, 143, 239, 241, 243, 248
rotary motor, 127, 129, 193–194, 198, 224
rotational time scale, 93
rRNA (ribosomal RNA), 51
RuBisCO (ribulose-1,5-bisphosphate carboxylase oxygenase), 205

Saffman–Delbrück equations, 192
Salmonella enterica, 166, 180
sarcoma, 163
sarcomere, 108, 137
saw-tooth pattern, 137
SAXS (small-angle x-ray scattering), 14
scanning confocal microscopy. *See* confocal microscopy
SCIM (scanning ion conductance microscopy), 105
screening effect, 19, 47
Sec pathway, 190
Second law of thermodynamics, 22, 198
second messenger, 46
secondary antibody, 37
secondary structure, 5, 37, 43, 55
secretory vesicle, 168
selectin, 162
selective phase hydrogel, 244
self-assembly, 112, 173, 192, 263
selfish gene hypothesis, 24
SEM (scanning electron microscopy), 109
semipermeable barrier, 110, 181
SERS (surface enhanced Raman spectroscopy), 114
SFM (surface force microscopy), 103
SHREC (single-molecule high-resolution co-localization), 84
SHRImP (single-molecule high-resolution imaging with photobleaching), 84
SI (structured illumination), 91, 213, 244
signal transduction, 159–161, 163–164, 166, 181, 211–212
signal-to-noise ratio, 12, 62, 65, 68, 76, 137, 251
single-beam gradient force trap. *See* Optical tweezers
single-molecule logic switch, 105
siRNA (small interfering RNA), 40
slimfield, 74–76, 98, 222, 235, 252
SLM (spatial light modulator), 132
smFRET (single-molecule FRET), 92, 94, 152, 189, 201
Smoluchowski equation, 229
snap-freeze. *See* cryo-fixation
SNAP-Tag, 73, 171
SNOM/NSOM (scanning near-field optical microscopy), 91
snRNP (small nuclear ribonuclear protein molecule), 246
sodium-motive force, 198, 218
solid phase, of lipids, 163
somatic reproduction, 236
Sparrow limit, 82
spectral overlap integral, 93
spectrin, 208
SPEM (saturated pattern excitation microscopy), 92
spherical aberration, 61, 88, 108–109
sphingolipid, 210
sphingosine, 46
SPIM (selective plane illumination microscopy), 78, 97
spinning disc. *See* Nipkow disc
spliceosome, 51, 246
splicing, 51, 238, 246, 248
SPM (scanning probe microscopy), 102
SPR (surface plasmon resonance), 10, 115
SPRAIPAINT, 89
SPT (single particle tracking), 84, 161–162, 165, 170, 172, 174–176, 178, 181, 190–192, 203, 209–212, 217, 221–223, 231, 248, 250–251
Ssb (single-strand binding protein), 234
SSIM (saturated structured-illumination microscopy), 90
stacking interaction, 42
starch, 39–40, 56
start codon, 52
stator unit, 193, 195–197, 219
STED (stimulated-emission depletion microscopy), 90
stem cell, 31, 256, 260
step-wise photobleaching, 153, 168, 187, 189–192, 195, 204, 212, 217, 235, 246, 252
steroid, 46, 69
sterol, 49
STM (scanning tunnelling microscopy), 102
Stokes law, 99, 125
Stokes radius. *See* hydrodynamic radius
Stokes scattering, 66, 114
Stokes shift, 63, 65
stop codon, 52
STORM (stochastic optical reconstruction microscopy). *See* PALM

streptavidin, 69
streptomyces, 69
stretch-activated channel, 189
stroboscopic illumination, 236, 241
sub-diffusion. *See* anomalous diffusion
sub-stoichiometric labelling, 86
sugar-carrier proteins. *See* GLUT proteins
sulphydryl group, 69
superconductivity, 260
super-resolution, 81–82, 98, 209, 212–213, 248–249
surgery, 256
switch complex, in bacterial chemotaxis, 167
symporter, 184
synapse, 49, 62, 170–172, 186–187
synthetic biology, xii, 192, 253, 262
synthetic gene, 33
systems biology, xiii, 243, 254

Talin, 209
TALM (single particle tracking localization microscopy), 84
tapping mode. *See* non-contact mode
Taq polymerase, 131
targeted drug delivery, 259
Tat, 190
TCA cycle. *See* citric acid cycle
teleological argument, 23
telomere, 186
TEM (transmission electron microscopy), 107
TEM00, 122
tensegrity structure, of cell, 208
termination codon. *See* nonsense codon
tertiary structure, 68, 190
tetracycline, 237
Texas Red, 177
TF (transcription factor), 238
thermal equilibrium, 1, 22–23, 26, 139
thermal ratchet, 230
thermionic emission, 106
thermus aquaticus, 131
thylakoid membrane, 206
TIRF (total internal reflection fluorescence), 75, 82, 85, 87–88, 91, 96, 100–101, 127, 133, 152, 157, 159–160, 164–165, 167–168, 172, 178, 181, 185, 187–189, 193–194, 196, 200, 203–204, 206, 219, 222–223, 230–231, 236, 251, 256
tissue, 1, 6, 10, 18, 27, 31, 36, 68, 71, 78–79, 91, 98, 107–108, 137, 159–161, 164, 171, 181, 224, 246, 254, 256–257, 259–260, 263
titin, 4, 11, 26, 36, 67, 98, 117, 123, 126, 137, 144–145, 152, 156
TLC (thin-layer chromatography), 3
Tom proteins, 203
Topoisomerase, 135
torsion angles, 36
trafficking, of molecules, 129, 175–178, 208–209, 224, 230–232, 243, 247, 252
transcription, 44
transducin, 62
transfection, 239
transferrin, 169–171
translation, 44
transplantation, 256
treadmilling, 210
triplet codon, 51
triplet state, 65
tRNA (transfer RNA), 41
tubulin, 100, 162, 177, 208–209, 225, 231, 238
tug-of-war, 227, 230, 247–248, 252
tumour, 177, 186, 257, 259
tungsten–halogen lamp, 63, 66
turnover, of molecules, 169
two-photon excitation, 77, 97

ultrafiltration. *See* dialysis
unicellular, 29, 33, 59
uniporter, 184
unit of life, 1, 32, 59, 154
uranyl acetate, 5, 107

vacuole, 33
Van der Waals forces, 104
Van't Hoff equation, 208
variable-angle epifluorescence, 75
vimentin, 162
vinculin, 209
virus, xi, 32, 103, 155, 173–182
viscoelasticity, 221, 232
viscosity, 99, 124–126, 129, 147, 158, 200, 216, 221, 249–250
viscous-drag radius. *See* hydrodynamic radius
visual acuity, 60–61

vitamin, 233
voltage clamp, 188
voltage-gating, 109

water channel. *See* aquaporin
Watson–Crick base-pairing. *See* base-pair
wavenumber, 66
western blot, 8
WGM (whispering gallery mode), 91
white blood cell, 21
widefield, 74
width of energy potential, 140
work, 140
worm-like chain, 127–129, 145, 147

xanthenes, 66
X-ray diffraction, 15, 210

YFP (yellow fluorescent protein), 74, 84, 89, 93, 100–101, 161, 163–164, 191, 209, 217, 234, 240–241
YPet, 167–169, 234–236, 252

Z-disc, 137
Z-DNA, 43
zinc sulphide, 70
Z-ring, 238
zwitterion, 34

Printed in the United States
by Baker & Taylor Publisher Services